# Russian Spacesuits

Springer
*London*
*Berlin*
*Heidelberg*
*New York*
*Hong Kong*
*Milan*
*Paris*
*Tokyo*

Isaak P. Abramov and Å. Ingemar Skoog

# Russian Spacesuits

co-authored by

Mikhail N. Doodnik
Guy I. Severin
Anatoly Yu. Stoklitsky
Vitaly I. Svertshek

Dr Isaak P. Abramov
RD&PE Zvezda JSC
Tomilino
Moscow Region
Russia

Dr Å. Ingemar Skoog
Astrium GmbH
Friedrichshafen
Germany

SPRINGER-PRAXIS BOOKS IN ASTRONOMY AND SPACE SCIENCES
SUBJECT *ADVISORY EDITOR*: John Mason M.Sc., B.Sc., Ph.D.

ISBN 1-85233-732-X Springer-Verlag Berlin Heidelberg New York

British Library Cataloguing in Publication Data
Abramov, Issak
Russian spacesuits.—(Springer-Praxis books in astronomy and space sciences)
1. Space suits 2. Life support systems (Space environment)—Russia (Federation)
I. Title II. Skoog, Å. Ingemar
629.4′772′0947

ISBN 1-85233-732-X

Library of Congress Cataloging-in-Publication Data
Abramov, I. P. (Isaak Pavlovich)
Russian spacesuits / Isaak Abramov and Ingemar Skoog.
p. cm.—(Springer-Praxis books in astronomy and space sciences)
Includes bibliographical references and index.
ISBN 1-85233-732-X (alk. paper)
1. Spacesuits—Russia (Federation) I. Skoog, Å. Ingemar. II. Title. III. Series
TL1550_A26 2003
629.47′72′0947—dc21 2003045585


Printed in Germany



Cover design: Jim Wilkie
Project Management: Originator Publishing Services, Gt Yarmouth, Norfolk, UK

Printed on acid-free paper

# Contents

# List of contributors

**Dr Isaak P. Abramov** is Deputy General Designer and Candidate of Technical Sciences at RD&PE Zvezda JSC, Moscow, Russia

**Mikhail N. Doodnik** is Chief Expert at RD&PE Zvezda JSC, Moscow, Russia

**Professor Guy I. Severin** is General Director and General Designer at RD&PE Zvezda JSC, Moscow, Russia and an Academician of the Russian Academy of Sciences

**Dr Å. Ingemar Skoog** is Programme Manager at Astrium GmbH in Friedrichshafen, Germany

**Dr Anatoly Yu. Stoklitsky** is Chief Expert at RD&PE Zvezda JSC, Moscow, Russia and a Candidate of Technical Sciences

**Dr Vitaly I. Svertshek** is First Deputy General Director and Deputy General Designer at RD&PE Zvezda JSC, Moscow, Russia and a Candidate of Technical Sciences

# Preface

By the time this book is published, it will be more than 40 years since the first manned space flight took place. From that first day in 1961 Astronautics, or Cosmonautics as it is known in the Soviet Union, now the Russian Federation, has come a long way, from Gagarin's short single-orbit mission to the multi-year operation of manned orbiting stations. For all these activities spacesuits are indispensable to the safety of humans in space missions. They are the main means of protecting the crew in emergency situations inside the spacecraft and practically the only protective means for activities outside the spacecraft—in free space.

This book describes the history of the development of Soviet/Russian spacesuits—a history in which the activities of the Research Development and Production Enterprise "Zvezda"—RD&PE Zvezda—(initially called "Plant No. 918") and its staff play an integral part.

The Russian authors were among the first group of Zvezda spacesuit engineers who started suit development for Yuri Gagarin in 1959. They still work within Zvezda in leading positions. Thus, they are eminently qualified to represent the total knowledge of the political, technical and personal experience of this history, and furthermore they represent the only group of spacesuit specialists with experience ranging from the very first manned flights up to today's International Space Station (*ISS*) activities.

This is a history with many famous "firsts" in the space race, like Gagarin's flight in 1961, the first "space walk" or extravehicular activity (EVA) in 1965 and the transfer from one spacecraft to another in open space in 1969, not to mention all the famous EVAs from the *Mir* station in the 1980s and 1990s. Another "potential" first, a manned lunar mission, was—as far as the spacesuit programme was concerned—ready in time, but was not realized due to the termination of the Soviet Moon programme. The Russian authors also give insight into their personal memories and the many unpublished events in the Soviet/Russian manned space flight programme and, particularly, in spacesuit history, all documented by photos and detailed descriptions.

The book is intended as a documentary history of the decisions taken in the Soviet/Russian space programme that have influenced spacesuit development, the technical work involved in developing, testing and flying the spacesuits and recollection of the personal memoirs of the participants in these 40 years of intensive work. The Russian authors and the files of Zvezda are sources of information that in many cases have never been published before. It is the intention to give as complete an overview as possible of the organizational, technical and human aspects of this very essential part of the Soviet/Russian space programme. Another important aspect concerns why Soviet/Russian spacesuit development took a path different from that of the USA.

The products described in the book were developed by a large group of Zvezda employees and those of some subcontractors. It should be separately mentioned that Professor Guy I. Severin, General Director and General Designer, who has been at the head of Zvezda for more than 38 years, made an important contribution to both company development and product creation.

As for myself, I am employed by Astrium GmbH, former Dornier System, then Deutsche Aerospace and later DaimlerChrysler Aerospace (DASA), and was the programme manager for European spacesuit development from 1986 to 1994. During this period we enjoyed close cooperation with Russia and Zvezda, planning to develop a West European/Russian spacesuit for the European space plane *Hermes* and the Russian space station *Mir 2*, later the *ISS*. This professional working contact has since then turned into a long-lasting personal friendship. With our mutual interest in the history of space activities the group of authors got together to prepare this documentation of Soviet/Russian spacesuit history.

My very special thanks go to Olga I. Khromova and George A. Rykov, our interpreters and translators, and the team supporting the authors at Zvezda. They have made a fantastic job of translating the manuscript and in remaining patient with the ever-changing inputs from the authors. Olga managed to keep all the co-authors under control at all times, despite the huge amount of work she and George had to cope with, double that of all the authors put together.

*Å. Ingemar Skoog*
Member of the International Academy of Astronautics

# Acknowledgements

The Russian authors kindly thank our colleague and friend Dr Å. Ingemar Skoog for assistance and participation in book-writing. In many respects this book has come about because of his insistence and efforts. He also undertook the hard task of editing the manuscript in the English language.

The authors are grateful to all the people at Zvezda and those working for other organizations who assisted in the selection of the material and gave valuable advice.

The authors would like to join Å. Ingemar Skoog in giving thanks to the translators of the text and express their gratitude to Elena Soldatova, Nadezhda Rozhkova and Vera Komarova for great assistance in preparing the text and illustrations.

This book in our opinion will be both interesting and useful to those involved in manned mission activities and to anybody with a keen interest in the history of Cosmonautics.

*Guy I. Severin*
Member of the Russian Academy of Sciences

*Front cover:*
EVA from the *Mir* orbiting station on 23 July 1999. Cosmonauts V.M. Afanasyev and S.V. Avdeyev in ORLAN-M spacesuits are taking part in the Russian/Georgian experiment "Reflector", designed to evaluate antenna deployment. When unfolded the structure measures 7 m in diameter. In the background the antenna transport cover is being pushed away. Photo taken by French cosmonaut Haigneré.
From Archive Zvezda.

*Back cover top:*
A.A. Leonov during the first ever EVA on 18 March 1965.
Painting from Archive Zvezda.

*Back cover bottom:*
KRECHET-94 fit checks and interface evaluations are held with the use of the Moon lander mock-up at OKB-1.

# Editorial notes

Preparing a manuscript on a Soviet/Russian high-technology subject in a non-Russian language causes some difficulties for editorial handling of the typescript. For example: How should Cyrillic letters be transcribed? What are the proper definitions of technical terms when translating into the target language? How should Soviet/Russian official documents be cited?

Transliteration of Cyrillic names can cause differences in Latin spelling, as each Germanic or Romance language has its own way of transcribing Cyrillic letters. Most of the time the names are identifiable, but may cause some concern to readers not aware of the problems. Thus this text has tried to follow Germanic (in this case English) standards for transcribing the Cyrillic alphabet. The Cyrillic form has only been retained in some acronyms.

In the aerospace community today it has become routine to use acronyms in capital letters for many products, components, operations and even company and institutional names (e.g., NASA, RKA, *ISS*, STS, EVA, etc.). This is also true for the Soviet/Russian aerospace community: in particular for the sometimes very long company and institute names. Based on the fact that some Cyrillic letters are optically identical to Latin ones, but with a different transcription in reality (e.g., "CCCP" in Cyrillic is "SSSR" in Latin), confusion could arise if both types were used simultaneously in the text. Therefore, this text uses Latin-letter acronyms only, but with the original Cyrillic spelling in square brackets at the first time of use (e.g., SI-3 [СИ-3]). A separate list of acronyms is also included on pp. xix–xxiv.

The names of the spacesuits and their life support systems (LSSs) are in many cases the same as those used for aviation suits, aircraft or other space equipment. As the proper description of the development history requires that these non-spacesuit elements are also mentioned, all spacesuits are given in capital letters to distinguish them from non-spacesuit equipment.

Most of the technical terms are self-explanatory or an explanation is given in the text. However, some standard aerospace terms might differ between their use in the

Soviet Union/Russia and in the USA and Europe. A major effort has been taken to avoid ambiguities by giving explanations when regarded as necessary. When it comes to spacesuits, the very subject of this book, the following basic definitions of spacesuit terminology have been used:

- *Spacesuit* is the overall term for equipment that provides cosmonaut/astronaut environmental protection and life support during *any* part of a space mission and consists of the suit enclosure proper and the LSS components arranged on it. Spacesuits are subdivided into:
  - *Rescue suits*, which are used to protect cosmonauts inside the spacecraft's pressure cabin in case of depressurization or on-board LSS failure. These suits are designed for the safety and rescue of cosmonauts and are additional to the protection provided by the spacecraft's cockpit systems. They are mostly dependent on support from the mother vehicle systems for consumables, and are designed for a very low mass. Rescue suits permit the crew to carry out their normal activities (to operate the spacecraft, etc.) both during a nominal flight and in an emergency. Suits of this type include the *Vostok*, *Soyuz* (SOKOL) and *Buran* (STRIZH) suits.
  - *EVA (extravehicular) suits*, which are dedicated to long, operational work shifts in 0 g outside a capsule, station or transport vehicle. These suits are fully or almost fully self-contained for their operation and designed for high mobility when pressurized, yet giving full protection in open space. Suits of this type are the Russian BERKUT (*Voskhod-2*), YASTREB (*Soyuz*) and ORLAN. These suits can be used for activities in the airlock and inside other depressurized modules, if necessary. The BERKUT suit (*Voskhod-2*) was actually universally used as a rescue suit as well.
  - *Lunar suits*, which are autonomous (EVA) suits designed for a lunar environment of $\frac{1}{6}$ g. Suits of this type include the KRECHET and ORIOL.
  - *IVA (intravehicular) suits*, which are designed for work inside a capsule or station module under vacuum. In Russia, IVA suits are considered as synonymous with the above-defined rescue suits. They do not hamper crew members' activities in the spacecraft's pressurized cabin and make it possible to operate the spacecraft (in case of its depressurization) while seated in their workplaces. Thus, the term "IVA suit" is hardly used in this text. After the accidents on board *Mir* and the necessity to work in a depressurized module of the station, the decision was made to perform these activities using the already available ORLAN-DMA EVA suits. In this function the EVA suit was practically used as an IVA suit.

This book covers a period of 50 years of spacesuit history, most of it under the Soviet Union era. In particular, in the 1950s, 1960s and 1970s—during the Cold War's severe political climate and the ongoing space race—a lot of information concerning space activities was classified. Most documents were issued in very small numbers and were classified as secret. Today, some of them no longer exist and special permission must be applied for to gain access to those documents that remain for use in publications. Unfortunately, these circumstances do not make it easy to verify all information and statements given. This has also in the past led to many "stories" about space events, which hopefully to some extent can be set right by this book. Regarding government decisions at various levels, they were neither published in total, nor were the official titles with official numbers made publicly accessible. Government directives and resolutions are mainly known in the industry by their title and date of issue and not by number. Thus the authors of this book have agreed to use titles and dates only for official governmental documents, and numbers are only used if they can be verified by the authors themselves through available files or by reference to sources published in the Soviet Union or the Russian Federation.

In this book a government resolution means, as a rule, a joint resolution by the Council of Ministers and the Central Committee of the Soviet Union Communist Party, while a government directive means a directive of the Military–Industrial Committee at the USSR Council of Ministers.

Some terms have slightly different meanings in Russia and the West and need to be defined more precisely:

- In Russia "initial data" are understood to be the preliminary specifications issued by the customer to the contractor for the beginning of activities.
- Performance specifications are understood to be the set of technical requirements issued by the customer, which the final product must meet in the long run.
- Technical requirements are understood as requirements to be met by a product.
- Technical characteristics are understood as properties inherent in a product.
- "Preliminary design" and "technical project" are phases of a product development programme used in the Russian Federation (the requirements in respective phases are determined by the performance specifications).
- The functions of these phases approximately correspond to the following phases adopted in the West:
    - preliminary design is synonymous with preliminary design and ends with a preliminary design review;
    - technical project is synonymous with detailed design and development and ends with a critical design review;
    - mock-up approval board is a breadboard review.
- After approval of the technical project the working documentation for the product is issued.

It should also be mentioned that the first publication in the Soviet Union on the design and use of Soviet spacesuits did not take place until 1970. The actual names of the suits' designers first appeared in articles and papers in the late 1980s.

## ILLUSTRATIONS

Most of the illustrations and photographic material presented in the book are taken from the archives of Zvezda or provided by the authors. The photographic material for illustrations, whose ownership is not identified, came from earlier published photo-chronicle materials of *TASS* and other public organizations, as well as from films made by cosmonauts during space missions.

# Acronyms

| | |
|---|---|
| AKhSG | *Buran* PLSS cooling/drying unit |
| AL | Airlock |
| APM | *Columbus* Attached Pressurized Module |
| APS | Auxiliary power system |
| ASO | Aerostatic support test facility, for testing of UPMK |
| B-1M | On-board interface unit between spacesuit LSS and lunar MOS |
| 1B-2M | On-board interface unit between spacesuit LSS and lunar MS |
| B-3 | On-board spacesuit interface units for *Almaz* |
| BG-1M | Cooling unit for SOKOL-KV suit in *Almaz* programme |
| BO-1M (BO-2M) | *Buran* PLSS contamination control assembly |
| BP-1 | *Buran* PLSS oxygen supply module |
| BPG-1 | Gas supply unit for SOKOL-KV suit in *Almaz* programme |
| BPS | Backpack subsystem for ESSS (earlier called ELSM) |
| BR-1 | On-board distribution unit for SOKOL-KV-2 suit in *Soyuz* programme |
| BRS-1M | Main on-board unit of *Buran* PLSS |
| BRTA | Removable suit assembly with the battery, radio and telemetry equipment for the ORLAN suit |
| BSS-1 | On-board spacesuit interface unit for *Salyut-6* |
| BSS-2 | On-board spacesuit interface unit for *Salyut-7* and *Mir* |
| BSS-2M | On-board spacesuit interface unit for *Mir* |
| BSS-4 | On-board spacesuit interface unit for *ISS* (Russian segment) |
| BVS-2 | On-board ventilation system used to dry the spacesuit |
| CCC | $CO_2$ control cartridge |
| CDR | Critical Design Review |
| CHX | Condensing heat exchanger |
| CNES | Centre National d'Etudes Spatiales |
| COF | Columbus Orbital Facility |

| | |
|---|---|
| COS | EVA SUIT 2000 Communication System |
| CPS | Chestpack subsystem for ESSS (earlier called EICM) |
| CPSU | Communist Party of the Soviet Union |
| CSC | Combined Service Connector |
| DC | Direct control mode, for SAFER |
| DCS | Decompression sickness |
| DM | Docking module |
| DMS | ESSS data management subsystem |
| DOS | Long-term orbiting station |
| DUK | Institute of Dirigible Technology, Moscow |
| DU-2 | Buran remote manual control unit of PLSS system |
| DV | Descent vehicle |
| EC | Emergency control mode, for SAFER |
| EICM | EVA Information Communication Module, for ESSS |
| EMU | Extravehicular Manoeuvring Unit, US EVA spacesuit for Shuttle |
| EOSC | Emergency oxygen supply and control (ESSS) |
| ESA | European Space Agency |
| ESEM | EVA Suit Enclosure Module, for ESSS |
| ESLM | EVA Life Support Module, for ESSS |
| ESO | *Buran* spaceplane experimental compartment |
| ESSS | European Space Suit System |
| ESTEC | European Space Research & Technology Centre, Noordwijk, The Netherlands |
| EVA | Extravehicular Activity |
| EWS | Emergency and Warning System |
| GCTC | Gagarin Cosmonaut Training Centre ("Star City"), Moscow region |
| GK NII | Air Force Scientific Research Institute |
| GKZh | Pressure cabin for animals |
| GN | Hydraulic weightlessness, water training suits (ORLAN-GN) |
| GNIIIA&KM | State Scientific Research Institute of Aviation and Space Medicine |
| GP | Pressure glove |
| GSh | Pressure helmet |
| GStM | Strela Cranes on *ISS* |
| GZU | Plume deflectors for thrusters on *ISS* |
| HCM | Hand controller module, for UPMK |
| HUT | Hard upper torso |
| IAM | Pavlov Red Army Institute of Aviation Medicine, Moscow |
| *ISS* | International Space Station |
| IVA | Intravehicular Activity |
| JSC | Joint stock company |
| KRT-10 | Antenna on *Salyut-6* |
| KVO | Russian water-cooled garment |
| LCD | Liquid crystal display |

| | |
|---|---|
| LCG | Liquid-cooled garment (ESSS) |
| LII | Gromov Flight Research Institute, later LII |
| LiOH | Lithium hydroxide |
| LP-2, LP-5, LP-7, LP-9 | Contaminant control cartridge for ORLAN-type spacesuit (with LiOH) |
| LPOV | Low-pressure oxygen ventilation (ESSS) |
| LSS | Life support system |
| LTA | Lower torso assembly |
| L3M | 1972 programme for a three-person lunar mission |
| L1 | Circum-lunar programme in the USSR |
| L3 | Lunar programme for landing of manned module in the USSR |
| MCS | Movement control system, for UPMK |
| MLI | Multilayer insulation |
| MOS | Moon orbiting spacecraft |
| MPS | Movable propulsion system |
| MS | Moon spacecraft for landing on the lunar surface |
| MSTP | ESA Manned Space Transportation Programme |
| MTFF | *Columbus* Man-Tended Free-Flyer |
| NASA | National Aeronautics and Space Administration |
| NASDA | National Space Development Agency (Japan) |
| NIIAM | Scientific and Research Institute of Aviation Medicine |
| NII | Science and Research Institute |
| NIKhI | See Tambov NIKhI |
| NOSC | Nominal oxygen supply and control (ESSS) |
| NPO | Scientific and production association |
| NPP | Scientific and production enterprise |
| N1-L3 | Lunar mission programme by OKB-1 using the N1 launcher |
| OKB-1 | Experimental Design Bureau, later Central Design of Experimental Engineering (TsKBEM), later RSC Energia |
| OKB-52 | Experimental Design Bureau, later Central Design Bureau of Machine Engineering (TsKBM), later NPO Mashinostroyenia |
| OKB-124 | Later Nauka Enterprise |
| OPM | US optic characteristic monitor and radiation sensor on *Mir* |
| ORK | Combined Service Connector, CSC |
| ORU | Orbital replaceable unit |
| OS | Orbiting station |
| OSCA | On-board suit control assembly |
| OTSST | Programme for orbiting heavy satellite/station |
| OV | Orbiting vehicle |
| PDR | Preliminary Design Review |
| PLSS | Personal life support system |
| PNO | Instrument and Science Compartment on *Mir* |
| PPK | Post-flight prophylactic suit |
| PPP | Portable power plant |
| PRSO-2 | *Buran* flight simulator (in NPO Molnia) |

| | |
|---|---|
| PVC | Polyvinyl chloride |
| PVF | Foamed polyvinylformal |
| PVU | Portable ventilation units |
| RD&PE Zvezda | Research Development and Production Enterprise "Zvezda" |
| RF | Russian Federation |
| RIR | Backpack life support system with injector ventilation (РИР in Russian) |
| RKA | Rosaviakosmos (Russian Aerospace Agency) |
| RPS-62 | $CO_2$ control cartridge for SKV and KRECHET |
| RSA | Rosaviakosmos (English abbreviation) |
| RSC | Rocket and Space Corporation |
| RSS | Reusable space system (e.g., *Buran*) |
| RV | Re-entry vehicle |
| RVR | Backpack life support system with fan ventilation (regeneration and ventilation backpack) |
| S | Full pressure suit |
| SAFER | Simplified Aid For EVA Rescue |
| SB | Bomber crew full pressure suit |
| SCLSS | Self-contained LSS |
| SES | Suit enclosure subsystem for ESSS (earlier called ESEM) |
| ShK | *Voskhod-2* airlock (Volga) |
| ShKK | *Buran* airlock |
| SI | Fighter pilot full pressure suit |
| SIS | Support and interface subsystem for ESSS |
| SK | Full pressure suit for *Vostok* system (SK-1 and SK-2) |
| SKB-AP | Special Design Bureau of Analytic Instrument Building, Leningrad |
| SKB-KDA | Special Design Bureau for Oxygen and Breathing Equipment, Orekhovo-Zuyevo |
| SKK | Russian material experiment (space environment exposure) on *ISS* |
| SKV | Prototype for semi-rigid EVA spacesuit |
| SM | Service module |
| SPSR | US *ISS* flight experiment |
| STS | Space Transportation System (Space Shuttle) |
| SZP-1 | On-board recharging/refilling and checkout system |
| T | Training (ORLAN-T) |
| Tambov NIKhI | Science and Research Chemistry Institute, Tambov |
| TBK | Thermal vacuum chamber |
| TDMA | Time division multiple access |
| TM | *Soyuz*-TM |
| TMG | Thermal micrometeorite (protection) garment |
| TMP | Thermal and micrometeoroid protection |
| TsAGI | Central Institute of Aerohydrodynamics |
| TsKBEM | Earlier OKB-1, today RSC Energia |

| | |
|---|---|
| TsKBM | Earlier OKB-52, headed by V.N. Chelomey, today it is "NPO Mashinostroyenia" |
| TzUP | Mission Control Centre, Korolev |
| UPMK | Cosmonaut Transference and Manoeuvring Unit |
| V | Ventilation (ORLAN-V) |
| VDC | Volt DC |
| VK | Vacuum chamber |
| VKS | High-altitude combined full pressure suit (full pressure suit plus anti-G suit) |
| VS | High-altitude full pressure suit |
| VSS | High-altitude rescue full pressure suit |
| WCG | Water-cooled garment |
| 3KV | *Voskhod* spacecraft programme |
| 3KD | *Vykhod*, later *Voskhod-2* spacecraft programme |
| 7K | Circum-lunar spacecraft |
| 7K-L1 | Manned spacecraft for the L1 programme |
| 7K-OK | Spacecraft for rendezvouz and docking in earth orbit |
| 7K-T | *Soyuz* spacecraft (transportation) for the *Salyut* programme |
| 21KS | Cosmonaut transfer and manoeuvring system (UPMK) for ORLAN-DMA (OS *Mir*) |

## RUSSIAN ACRONYMS

| Cyrillic | English | |
|---|---|---|
| АХСГ | AKhSG | *Buran* suit PLSS cooling/drying unit |
| БО | BO | *Buran* contamination control assembly |
| БСС | BSS | On-board spacesuit interface unit for space stations (on-board suit control assembly, OSCA) |
| В | V | Ventilation (ORLAN-V) |
| ВА | RV | Re-entry vehicle (NPO "Mashinostroyenia") |
| ВКС | VKS | High-altitude combined full pressure suit (full pressure suit plus anti-G suit) (vysotny kombinirovanny skafandre) |
| ВС | VS | High-altitude full pressure suit (vysotny skafandre) |
| ВСС | VSS | High-altitude rescue full pressure suit (vysotny spasatelny skafandre) |
| ГК НИИ | GK NII | Air Force Scientific Research Institute |
| ГН | GN | Hydraulic weightlessness, water training suits (ORLAN-GN) |
| ГП | GP | Pressure glove |
| ГШ | GSh | Pressure helmet |
| ДУ-2 | DU-2 | *Buran* remote manual control unit of PLSS |
| КО | KO | Design department |
| КБ | KB | Design Bureau |
| КРТ | KRT | Antenna on *Salyut-6* |

| | | |
|---|---|---|
| КВО | KVO | Russian water-cooled garment |
| Л3М | L3M | 1972 programme for a three-person lunar mission |
| ЛК | LK | Moon spacecraft (MS) for lunar landing |
| ЛОК | LOK | Moon orbiting spacecraft, MOS |
| НИИ | NII | Science and research institute |
| НПО | NPO | Scientific and production association |
| НПП | NPP | Scientific and production enterprise |
| ОКБ | OKB | Experimental Design Bureau |
| ОРК | ORK | Combined service connector, CSC |
| ОТССт | OTSST | Programme for orbiting heavy satellite/station |
| ПВУ | PVU | Portable ventilation units |
| РВР | RVR | Backpack life support system with fan ventilation (regeneration and ventilation backpack) |
| РИР | RIR | Backpack life support system with injector ventilation (regeneration and injector backpack) |
| С, СК | S, SK | Full pressure suit (skafandre) |
| СА | SA | Descent vehicle (DV) |
| СБ | SB | Bomber crew full pressure spacesuit (skafandre bombardirovschika) |
| СВМ | SVM | Super-module fibre |
| СЗП | SZP | On-board recharging/refilling and checkout system |
| СИ | SI | Fighter pilot full pressure suit (skafandre istrebitelia) |
| СКВ | SKV | Extravehicular Activity Space Suit (skafandre vykhoda) |
| СК-ЦАГИ | SK-TsAGI | TsAGI aviation full pressure suits |
| СМ | DM | *Buran* Docking Module |
| СМ | SM | Service module of the *Mir* and the *ISS* |
| СО | SO | Docking module of the *ISS* (DM) |
| СССР | USSR | Soviet Union |
| Т | T | Training (ORLAN-T) |
| ТБК | TBK | Thermal vacuum chamber |
| УПМК | UPMK | Cosmonaut Transference and Manoeuvring Unit |
| ЦКБМ | TsKBM | Central Design Bureau of Engineering industry, earlier OKB-52, headed by V.N. Chelomey, today "NPO Mashinostroyenia" |
| ЦКБЭМ | TsKBEM | Central Design Bureau of Experimental Engineering industry, earlier OKB-1, now RSC Energia |
| Ч-1 ... Ч-7 | Ch-1 ... Ch-7 | Chertovsky aviation full pressure suits |
| ШК | ShK | *Voskhod-2* airlock (Volga) |
| ШКК | ShKK | *Buran* airlock |
| ЭСО | ESO | *Buran* spaceplane experimental compartment |

# Figures

# Tables

# 1

# Introduction and background

In 1952, 51 years ago, the company Research Development and Production Enterprise (RD&PE) Zvezda was founded under the name "Plant No. 918" to develop, test and manufacture safety equipment for the crews of aircraft, including full pressure suits for Soviet military aircrews. With the start of Soviet manned space activities in the late 1950s Zvezda was directed to handle the crew's safety tasks, in particular rescue and spacesuit development.

The Russian authors of the book, all employees of Zvezda, directly participated in the development of spacesuits at Zvezda from the very beginning of the manned space programme. Over the years, many details of the programme have been lost to memory, and, thus, the authors have had to use archives as well as the memories of other participants in the programme.

Unfortunately, due to the climate of secrecy prevailing during the first decades after the Second World War, few open records and publications exist. Moreover, a large part of public archives has also been lost. Thus, certain dates and details are presented from the memory of the participants in those events. Therefore, they may differ from those that have already been indicated in publications of other authors writing about the history of Astronautics.

It is worth mentioning that this book is the first publication by any Russian authors on the history of the development of Soviet/Russian spacesuits, although certain facts may be indirectly covered in other publications on Astronautics.

The development of spacesuit technology in the Soviet Union did not start from scratch. In the initial phase, considerable experience was gained from the development of protective gear, pressure suits and life support systems (LSSs) for aircraft pilots and stratosphere aeronauts.

We should also mention some specific features of the development of Soviet spacesuit technology. To begin with, there was a strong dependence on approved or planned space programmes. Development of this or that hardware or LSSs in the Soviet Union occurred, as a rule, on the basis of resolutions of the Communist Party of the Soviet Union (CPSU) Central Committee or the Council of Ministers approval for appropriate space programmes. At the time of the "space race" the schedule from decision to planned launch was generally very short. Thus most development work was directly project-oriented.

Zvezda did, however, also run separate research and development work on its own initiative using budgetary funds released by the Aviation Industry Ministry (Zvezda at that time reported to this ministry).

The directions that space programmes took and dates for their implementation were often specified with the competition between the USSR and the USA taken into account. In his series of books titled *Rockets and People*, Boris Ye. Chertok (Chertok, 1994, 1996, 1997, 1999) describes this situation in detail. He also indicates the factors that influenced programme implementation as a consequence of transfer of activities between programmes and/or enterprises (e.g., delays in the N1-L3 programme and competition between the enterprises led by S.P. Korolev and V.N. Chelomey).

Many historical facts are given in the book *Rocket and Space Corporation Energia Named after S.P. Korolev*, 1946–1996 (Semeonov, 1996).

The experimental design bureau (OKB-1) (led by Chief Designer S.P. Korolev up to 1966), renamed over the years several times and now known as S.P. Korolev Rocket and Space Corporation (RSC) "Energia", was the prime contractor for manned space programmes in the Soviet Union and then in Russia. In this function Korolev and his company OKB-1 were the main customer for spacesuit development at Zvezda.

They were joined in this work on spacesuits and on-board LSSs by scientific and production association (NPO) "Mashinostroyenia" (then OKB-52) led by V.N. Chelomey. However, in retrospect, they did not have any practical results, as the programmes were terminated during preliminary design or early testing.[1]

Spacesuits and on-board LSSs for the *Buran* programme were ordered by NPO "Molnia" (Chief Designer G.E. Lozino-Lozinsky).

The suit concept to be used for a forthcoming space programme was often selected as a trade-off version, with the planned programme tasks and specified development time as principal requirements. Such an approach, as a rule, calls for the use of experience already gained in other programmes and the results of research work undertaken at that time coupled with less new basic research, due to the short development time. This cannot be considered a fully natural evolution of a suit family.

Such an approach was used for the development of suits and systems for the *Vostok* (SK-1 and SK-2 suits), *Voskhod-2* (BERKUT extravehicular activity—EVA—suit), *Soyuz-4/5* (YASTREB EVA suit equipped with the RVR-1P backpack LSS), *Soyuz* (the SOKOL-type suit) and *Buran* (the STRIZH suit) programmes (Figure 1.1). Probably the KRECHET and ORLAN suit family and

[1] Zvezda participated in two main programmes initiated at "Mashinostroyenia": a programme involving a flight around the moon, terminated during preliminary design, and the use of an ORLAN spacesuit at the *Almaz* Space Station, stopped during early testing of the suit in an airlock.

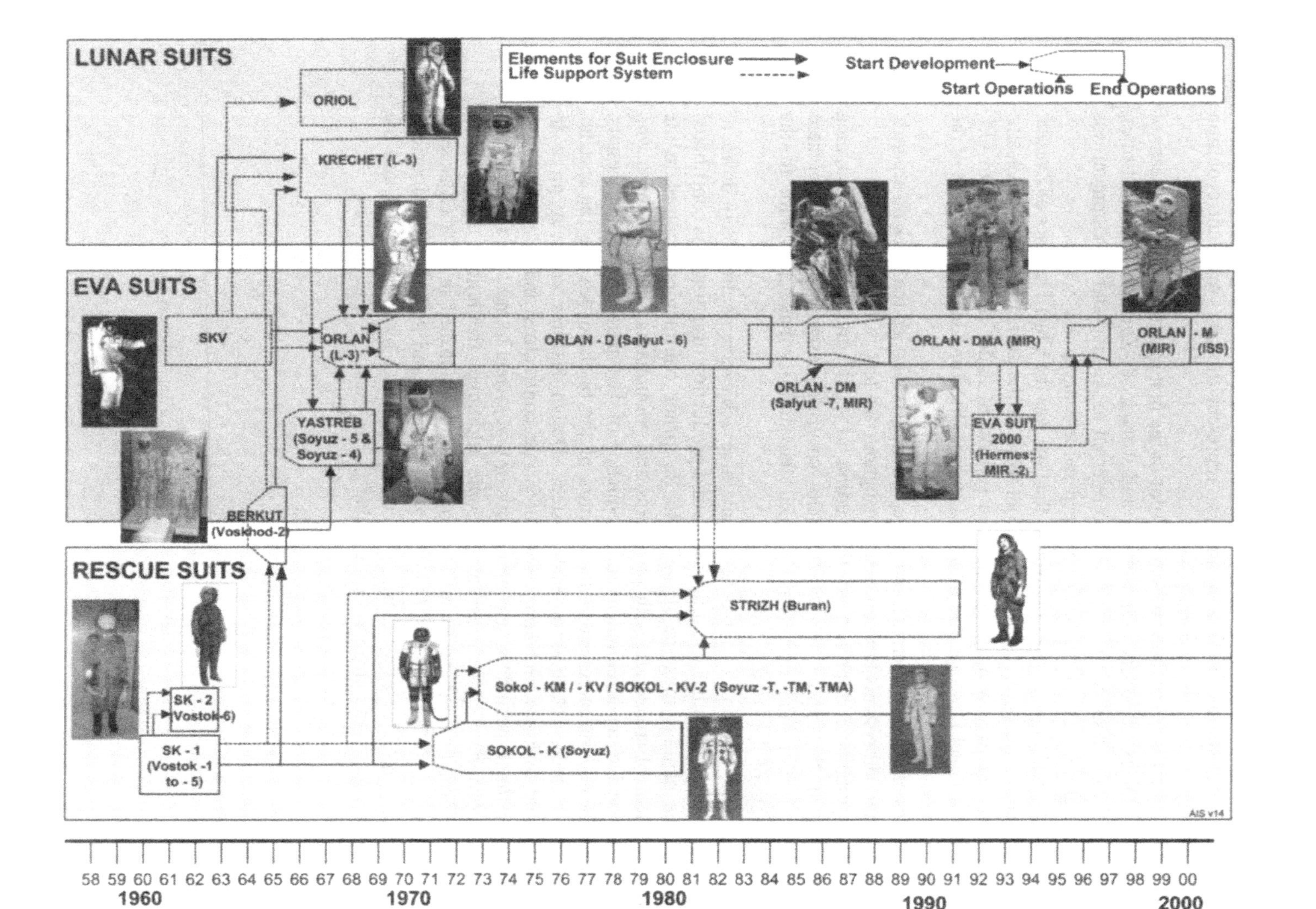


**Figure 1.1** The Soviet/Russian spacesuit family tree.

From Archive Skoog.

their systems—initially developed for the L-3 lunar programme and then continuously improved for further utilization in the *Salyut-6*, *Salyut-7*, *Mir* and international space station (*ISS*) programmes—represent the first, real, suit evolution (Figure 1.1).

The book describes the work carried out in the first instance by a large team of design engineers, scientists, test subjects and workers at Zvezda. Other institutes, organizations and subcontractors are also referred to, as spacesuit development was a well-integrated part of the Soviet manned space programme.

Before 1964, S.M. Alekseyev, who at the time was the Chief Designer of Zvezda, was in charge of general management of all work in the development of spacesuits. From 1964 up to now all work has been supervised under the leadership of G.I. Severin, the Chief Designer and now the General Director–General Designer of the enterprise.

Names of other key persons and project-managers who led the work on separate projects and subsystems are indicated throughout the text of the book.

The systematic approach to the setting of goals and the single ongoing management throughout the whole cycle—covering design work, development tests, production and operation—enabled the staff to make hardware to the highest state-of-the-art standards.

However, Zvezda do not believe that these achievements are solely down to Zvezda staff; many other scientific, design and production organizations contributed a lot. The plaudits must be shared with the prime contractors involved in the development of spacecraft, with physicians and cosmonauts, and with those who were involved in the development and production of the separate parts of spacesuits.

It was not the intention of the authors of the book to describe the specific, detailed, technical features of each system. Design features are only given to explain the positions and approaches of prime contractors and subcontractors, and the rationale behind the selection of this or that system to meet the requirements of the appropriate space programme. Some additional theoretical and practical design information can be found in two textbooks published in 1973 (Alekseyev and Umanski, 1973) and 1984 (Abramov et al., 1984) and in a number of articles and papers presented at symposia and conferences later by the spacesuit and life support designers and engineers of Zvezda.

# 2

# How full pressure spacesuits came about. Time period preceding the space mission era

## 2.1 FULL PRESSURE SUITS FOR FLIGHTS IN THE STRATOSPHERE

Full pressure suits for space use were preceded by full pressure suits for pilots of balloons and aircraft with open cockpits, flying at high altitudes where air is so rarefied that even breathing with pure oxygen under barometric pressure does not protect the pilot from oxygen starvation (hypoxia) and other sicknesses caused by high-altitude conditions.

The first known facts about development of high-altitude full pressure suits in the Soviet Union date from the early 1930s (Ivanov and Khromushkin, 1968).

The first full pressure suit, identified as Ch-1 [Ч-1] (Figure 2.1.1), was designed by the engineer E.E. Chertovsky of the Aviation Medicine Department of the Civil Air Institute of Scientific Research in Leningrad in 1931. It was a simple pressure-tight suit with a helmet fitted with a small visor. The suit did not have joints. Thus, when the suit was under pressure, substantial forces were required to flex arms and legs. With positive pressure under the enclosure, any work for the person wearing the suit was impossible.

On 30 January 1934 the Osoaviakhim-1 stratosphere balloon suffered a catastrophe.[1] The tragic event gave motivation to engineers and inventors of various backgrounds to look for means of personal protection for pilots to save them in the event of pressure-tight cabin decompression or to make use of unpressurized cabins.

The Ch-2 [Ч-2] full pressure suit by E.E. Chertovsky (developed in 1932–1934) used joints. They enabled the pilot to flex (to a certain degree) arms and legs. The

1 "At fifty-seven thousand feet the gondola broke loose from the balloon. The three aeronauts, Paul Fedoseyenko, Andrey Vasenko and Ilya Ousyskin, had parachutes but there was only a single hatch. Twenty-four bolts secured that hatch. In the wreckage it was found that they had only time to remove seven of the twenty-four hatch nuts. All three men were killed in the crash" (Payne, 1991).

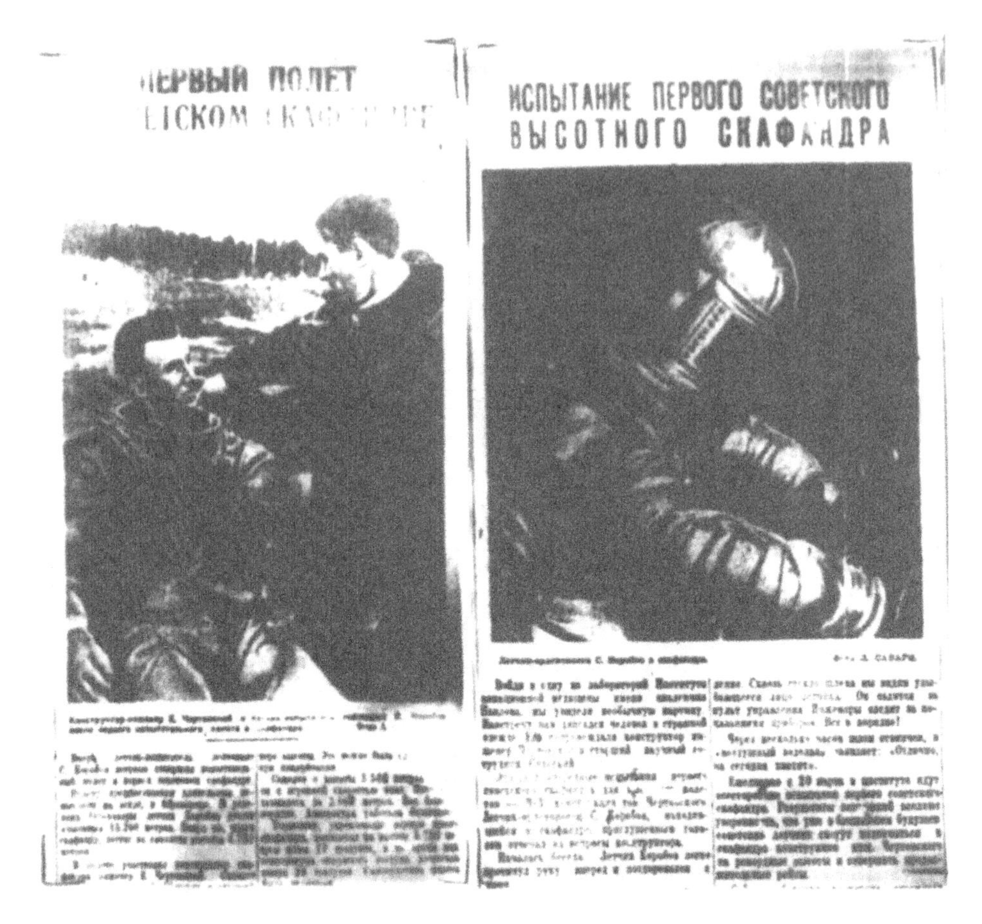
ИСПЫТАНИЕ ПЕРВОГО СОВЕТСКОГО ВЫСОТНОГО СКАФАНДРА

**Figure 2.1.1** Newspaper reports about the first Soviet high-altitude full pressure suit Ch-1 developed by design engineer E.E. Chertovsky (early 1930s).
From Archive Zvezda.

design of the next, Ch-3 [Ч-3], model developed in 1935–1937 already had all the basic elements of future full pressure suits and took into account physiological and hygienic requirements. Air Force physiologists V.A. Spassky and A. Apollonov of the Pavlov Red Army Institute of Aviation Medicine (IAM) in Moscow took part in the development of the Ch-3 suit as medical experts.

The Ch-3 suit (Figure 2.1.2) enclosure was made of rubberized fabric. With a positive pressure from 10 to 15 kPa, the pilot wearing the Ch-3 suit had sufficient mobility for piloting. After ground tests in a vacuum chamber, test pilot S. Korobkov performed flight tests of the Ch-3 suit at altitudes up to 12 km.

The years 1938 and 1939 saw the Ch-4 [Ч-4] and Ch-5 [Ч-5] full pressure suits followed by the improved Ch-6 [Ч-6] and Ch-7 [Ч-7] models in 1940.

In 1936, the TsAGI (Central Institute of Aerohydrodynamics) began developing high-altitude full pressure suits to support the increasing performance of aircraft (at

**Figure 2.1.2** Pilots wearing E.E. Chertovsky's Ch-3 suit.
From Archive Zvezda.

that time without a pressure cabin), as part of the race for new altitude records. A.I. Boiko and A.I. Khromushkin led the work. The latter led in particular the development of life support systems (LSSs) for full pressure suits. TsAGI put this work on a sound scientific basis, consistent with the high technological level already achieved by the world-known institute.

Development work started with the understanding and definition of the task and the concept of future design. The first and the most difficult part of the effort was to look for arm and leg joints of such a design that would enable the pilot to flex them without considerable effort. Many experiments with mock-up models were run to find the proper design.

Those soft joints that were developed were tested to ascertain not only their mobility characteristics but also the performance characteristics of load-bearing elements, fixtures and other components. For each joint, the capabilities of this or that material were investigated. Various commercially available rubberized fabrics were used as initial materials. However, the first phase of research work and the proper understanding of specific operating conditions revealed the necessity of having suit enclosure fabrics manufactured in accordance with special technical requirements.

Experiments with mock-up models generated a lot of valuable data for use in the development of the first pressure suit models. The first steps in the development of initial suit models revealed the problem of selecting a rational suit concept that would offer the highest operational characteristics. Manufacturing characteristics, comfort for the wearer (e.g., lined with cotton to absorb sweat, etc.), and performance characteristics were principal inputs for the development of full pressure suits.

The first SK-TsAGI-1 [СК-ЦАГИ-1] suit, a full pressure suit, was developed, manufactured and tested in 1937. The suit consisted of two parts, an upper torso or "shirt" and a lower part or "pants". The waist interface was made of metal. The lower removable part was integrated with a profiled elliptic ring made of aluminium, and the upper "shirt" incorporated a disconnect fixture with two locks located on both sides. The suit used joints of the accordion type. Initially, the impression was that such joints would offer the best mobility. However, thorough studies revealed that such joints had limited flexion angles and did not provide the suit with the required mobility. Moreover, such joints were very difficult to manufacture with strength properties sufficient to withstand high internal pressures. Pressure suits were then designed for the 30 kPa positive pressure.

The suit design included steel ropes working as longitudinal load-bearing elements. Many subsequent pressure suits used such load-bearing elements. Suit elongation under positive pressure was the problem that needed to be tackled from the very beginning. Therefore, design work on all full pressure suits started with identification of a force distribution pattern that would offer the best solution in terms of the reduced elongation problem.

In 1937, work began on the development of certain components of equipment for life support. Pressure suits were equipped with a self-contained oxygen supply system of the regeneration type (i.e., removal of $CO_2$ by means of a chemical absorbent) that made it possible to disconnect it from the on-board system when crew members moved from one place to another inside the aircraft or when it was used for parachute-jumping.

In 1938, two independent inventors Dr A.A. Pereskokov and the engineer Rappaport designed a full pressure suit with an LSS of the oxygen ventilation type.

In 1938, TsAGI developed the SK-TsAGI-2 [СК-ЦАГИ-2] full pressure suit. It was an experimental model designed for aircraft pilots. The suit incorporated a specially designed regeneration system sufficiently reliable for that time. First tests in a thermal vacuum chamber and flight tests aboard the TB-3 and modified R-5 aircraft were performed with the SK-TsAGI-2 suit. This model differed considerably from the SK-TsAGI-1 model. However, it imitated a diving suit in certain details. It is worth mentioning that the SK-TsAGI-2 suit was the best of those so far developed in the USSR. Pressure suits designed by Chertovsky, Pereskokov and Rappaport were already available at that time. The SK-TsAGI-2 suit was made as a one-piece suit (overall) with a removable helmet. Its enclosure was made out of a three-layer rubberized fabric. Therefore, the suit was unnecessarily heavy and bulky. The suit used soft, constantly creased joints. Such joints offered high mobility. However, they were made for this very suit in such a way that any change in joint length was impossible. The SK-TsAGI-2 suit also lacked height control capability. The shoulder joint was a type of Hook joint (two-degrees-of-freedom joint) with a metal ring glued to the enclosure. The suit was donned through the suit top opening, which was then covered with a soft top piece with a helmet opening. The top piece was sealed to the suit along its outline with two flat metal rings and wing nuts and accommodated the helmet fixed by means of a special, steel tape collar. The helmet, shaped like a truncated cone, was fitted with a single visor and provided rather limited visibility. Suit gloves were non-removable. A knitted overall worn under the suit provided internal thermal protection. The inner surface of the enclosure was covered with a single layer of flannelette for insulation and hands inside the suit were protected by thin woollen gloves. The person wearing the suit could stay for up to 4 hours at a temperature of −40°C. There was the option of active thermal protection, in which the pilot wore an electrically heated overall fitted with panels made of a special wire and cloth.

The SK-TsAGI-2 suit used regenerative life support equipment designed for 6 hours of operation. The equipment included many components of the already existing mine rescue oxygen equipment of the regenerative type. Hoses from regenerative equipment were attached to the suit through a disconnect lock. The lock could be disconnected with a single hand movement without any loss of pressure in the suit.

Test pilots P. Zozim, J. Stankevich and M. Gallay implemented the whole flight test programme for the suit. Laboratory and flight tests (Figure 2.1.3) of the SK-TsAGI-2 suit brought valuable experimental data and considerably affected the subsequent development of the SK-TsAGI-4 [СК-ЦАГИ-4] (Figure 2.1.4) suit, completed by the end of 1938.

On the basis of the earlier accumulated experience, the SK-TsAGI-5 [СК-ЦАГИ-5] and SK-TsAGI-8 [СК-ЦАГИ-8] full pressure suits were developed in 1940. The SK-TsAGI-8 was tested aboard the Soviet I-153 fighter, which did not have a pressure-tight cockpit or a protective cockpit canopy.

The Second World War saw the end of work on the development of "stratospheric" full pressure suits, and the post-war period was the time of rapid development of jet aircraft equipped with pressure-tight cockpits, ejection seats and other

**Figure 2.1.3** SK-TsAGI-2 during flight-testing in a TB-3 aircraft.
From Archive Zvezda.

technical innovations, which called for new approaches to the development of high-altitude protective gear for crew members.

## 2.2 AIR CREW FULL PRESSURE SUITS

As indicated above, TsAGI and LII (Gromov Flight Research Institute), were the first Soviet scientific aviation organizations involved in the development and testing of high-altitude full pressure suits.

TsAGI developed the full pressure suit technology for piloted flights in the period from 1936 to 1941, and LII was involved in similar work from 1946 to 1952.

**Figure 2.1.4** Full pressure suit Sk-TsAGI-4 developed by design engineer A.I. Boiko (1938).
From Archive Zvezda.

To develop and test full pressure suits, LII formed a department (High Altitude Laboratory), which included a design team, a production shop and a test laboratory equipped with a thermal vacuum chamber.

Four models of experimental, aircrew full pressure suits (VSS-01 [BCC-01], VSS-02 [BCC-02], VSS-03 [BCC-03] and VSS-04 [BCC-04]) were developed during 6 years of work (Figure 2.2.1 and Figure 2.2.2).

A.I. Boiko led the work researching the development of the suit, and A.I. Khromushkin and A.M. Gershkovich led the development of LSSs.

With the formation of Plant No. 918 (now Zvezda) in October 1952, certain people of the LII high-altitude department joined the new organization. S.M. Alekseyev, who also worked for LII as a manager of a design and manufacturing department, got the position of Chief Designer.

Plant No. 918 got down to the development of high-altitude, aircrew full pressure suits in line with existing rules for the development of aviation technology.

**Figure 2.2.1** VSS-04 full pressure suit.
From Archive Zvezda.

It formed several design teams, a suit assembly shop and a high-altitude test laboratory.

The most experienced design engineers with expertise in soft enclosure technology who graduated from the Institute of Dirigible Technology (DUK), and young specialists who graduated from the Moscow Aviation Institute were hired to work on full pressure suits.

Table 2.1 lists aircrew full pressure suits developed in the period from 1953 up to 1959.

An active part in the development of the pressure suits specified in Table 2.1 was

**Figure 2.2.2** VSS-04 full pressure suit with open helmet.
From Archive Zvezda.

also taken by the leading designers A.Yu. Stoklitsky and I.I. Derevianko (suit enclosures), A.M. Gershkovich and I.P. Abramov (LSS) and many others.

In 1959, work on full pressure suits for manned space flights started. However, work on full pressure suits for aircrews still continued. It included testing and final development of the S-9 [C-9] and Vorkuta full pressure suits.

In 1970, the Sokol aircrew full pressure suit was developed. It became a prototype of the cosmonaut SOKOL-K (cosmonaut cabin protection suit) suit. Ya. Rubashkin and A. Fadeyev were project-managers for the aviation suit.

In 1985, the Baklan full pressure suit for crew members of long-range aircraft

**Table 2.1.** Pre-space era development activities at Zvezda.

| Year of completion | Type (identification) of suit | Project-manager (leading designer) | Note |
|---|---|---|---|
| 1953 | VSS-04 | A.I. Boiko | Completion of the LII initiated project |
| | VSS-05 | S.P. Umansky | |
| | VS-06 | S.P. Umansky | |
| 1954 | VKS-1 (VSS-07) | A.L. Zelvinsky | |
| | VSS-04A | A.I. Boiko | |
| | VSS-04M | A.I. Boiko | |
| 1955 | SI-1 | S.P. Umansky | |
| | SB-2 | S.P. Umansky | |
| | SI-3 | S.P. Umansky | |
| 1956 | SI-3M | S.P. Umansky | |
| | SB-4 | S.P. Umansky | |
| 1957 | Vorkuta | A.I. Boiko | Start of Vorkuta activities |
| | SB-4B | S.P. Umansky | |
| | SI-5 | S.P. Umansky | |
| 1958 | SI-5 | S.P. Umansky | |
| | S-9 | S.P. Umansky | Start of S-9 activities |
| 1959 | Vorkuta | A.I. Boiko | Development and tests |
| | S-9 | S.P. Umansky | |

*Legend:*
VSS High-altitude, rescue full pressure suit (*Vysotny Spasatelny Skafandre*).
VS High-altitude full pressure suit (*Vysotny Skafandre*).
VKS High-altitude, combined full pressure suit (full pressure suit plus anti-G suit) (*Vysotny Kombinirovanny Skafandre*).
SI Fighter pilot full pressure suit (*Skafandre Istrebitelia*).
SB Bomber crew full pressure suit (*Skafandre Bombardirovschika*).
S Full pressure suit (*Skafandre*).

was developed with G. Paradizov as the project-manager. It goes without saying that the design of full pressure suits was improved from one model to another. Competition between specialists (project-managers, team leaders and design engineers) contributed a lot to the design improvement process.

The directions to be taken in the development of the full pressure suit design were specified by the requirements imposed (i.e., the envisaged use).

It is known that the full pressure suit completely isolates the wearer from the environment. The suit must provide the wearer with conditions adequate for life and activity in case the vehicle cabin is decompressed at altitudes higher than 7–8 km. Thus, the suit is the only personal protective gear that enables the pilot to continue the mission for a long time without the need to descend to a lower altitude.

All aircrew full pressure suits are similar in terms of their operational pattern.

The cockpit air-conditioning system supports pressure in the suit and provides ventilation for the pilot, and a special system provides the pilot with oxygen. During the flight with the cockpit pressurized, a positive pressure of 10–20 hPa (100–200 mm of water column) is supported in the suit to smooth out the suit enclosure and "remove" the suit weight from the pilot's body. If the cockpit is decompressed at altitudes higher than 10 km, the suit is fed with the absolute pressure corresponding to 10–11 km altitude. Depending on the maximum flight altitude (up to 20 km), the operating positive pressure in the suit could be from 150 to 220 hPa. By breathing pure oxygen that pressure level was sufficient and at the same time it provided the mobility needed for aircraft-piloting.

Below are the main requirements that aircrew full pressure suits had to meet:

1. The suit had to support the vital functions of the crew member and piloting capability in case of cockpit decompression.
2. The suit had to provide the crew member with cockpit escape capability and safe parachute descent.
3. The suit had to provide the crew member with a safe splashdown or landing in adverse climatic conditions.
4. With the advent of ejection seats, the suits had to protect the crew member from exposure to the incoming airflow in the ejection process and protect the head from the impact effect.

To meet these requirements, the suit had to have adequate mobility, both with positive pressure and without it, provide a good field of vision, protect the crew member's eyes and provide the crew member with good hygienic conditions through proper ventilation. Moreover, the suit had to be easy to don, have the height adjustment capability and preserve the height adjustment with the suit inflated. Furthermore, the suit had to provide the crew member with the proper thermal protection, have a buoyancy capability and have a minimized weight.

All aircrew full pressure suits were of the so-called ventilation type. That meant that, in case of emergency, the ventilation air, coming to the suit from the aircraft pressurization system, and spent oxygen were removed from the suit through a pressure control valve (with the helmet closed). Such a ventilation system is known as an open system, in contrast to a closed system (a regenerative suit).

There were two types of aircrew full pressure suits that were developed for many years under parallel programs and in competition with each other. They are the so-called mask and maskless (with separate ventilation) suits.

With a suit of the first type, the pilot used an oxygen mask in the helmet. The mask was fed with pure oxygen or an oxygen/air mixture for breathing. Air for ventilation of the suit came from the aircraft's pressurization system.

With a suit of the second type, the helmet volume (respiration zone) was separated from the remaining suit volume by a neck pressure-tight partition. The helmet functioned as an oxygen mask with a continuous oxygen (or air/oxygen mixture) supply.

In the long run, the maskless suit arrangement became the winner in the competition, since it offered the best operational and hygienic conditions for the crew member, although it required an increased amount of oxygen for breathing.

All full pressure suits of the VSS line (1953–1954) as well as the SI-1 [СИ-1], SB-2 [СБ-2], SI-3 [СИ-3], SI-3M [СИ-3] and SB-4 [СБ-4] (1955–1956) are of the mask type. The suits, beginning with the SI-5 [СИ-5] model (1957), are of the maskless type. The Vorkuta suit (1957) was developed in both configurations.

Plant No. 918 developed the first full pressure suits of the VSS [ВСС] line, namely VSS-04A [ВСС-04А], VSS-04M [ВСС-04М], VSS-05 [ВСС-05], VS-06 [ВС-06], VSS-07 [ВСС-07] (later VKS-1 [ВКС-01]) on the basis of the VSS suits developed by LII. Despite the differences in separate elements, the enclosure of these suits had the same basic design features (Figures 2.2.1–2.2.3):

1 A single enclosure made of two- or three-layers of rubberized cotton fabric performed the functions of the bladder and restraint enclosures.
2 A spatial (large size) helmet with a soft enclosure at the back of the head integrated with the suit body (except for the VSS-05 suit). A plastic visor was made either as a single front piece (Figure 2.2.3) or as a three-piece (front, side and upper) element (Figure 2.2.1).
3 A helmet disconnect was made in the form of two hinged semi-frames. The upper semi-frame was fixed at the lower edge of the visor and the lower semi-frame was attached to the suit body neckband. The disconnect made it possible to open the helmet like a purse and move the visor behind the head (Figure 2.2.2).
4 All operational disconnects were sealed with a rubber membrane stretched on the "knife" of a contacting part (disconnects of the helmet, cuffs and gloves).
5 The front opening used for donning/doffing was sealed with a wide front sleeve (so-called appendix) made of rubberized fabric. The appendix was glued to the enclosure along the opening and the helmet base. When the suit was donned, the appendix was twisted into a rope, fixed with a rubber or cotton lace and put under the enclosure. The latter was then laced up.
6 Soft joints (for bending inflated arm and leg enclosures) were made in the form of orange peels (specially tailored inserts in transverse cuts of the enclosure). The only exception was the VS-06 suit, which had crimped shoulder joints (Figure 2.2.3).
7 The restraint system was formed by cotton cords running along neutral axes of all enclosure parts (body, arms and legs).
8 The arms were fitted with removable airtight cuffs (Figure 2.2.2). In 1954, airtight gloves worn above the cuffs were developed (Figure 2.2.3).
9 The suits were fitted with a group flow line inlet in the form of short hoses fixed on the body enclosure and terminated by a single quick disconnect.
10 All these suits featured a low-operating positive pressure, from 150 to 220 hPa, which resulted in insufficient mobility of the enclosure, especially for the arms.

Starting with the SI-1 model (Figure 2.2.4), the suits featured a removable helmet

**Figure 2.2.3** VS-06 full pressure suit.
From Archive Zvezda.

and a modified restraint system for the body. But the design concept of the enclosure and its materials were as those used in the suits of the VSS line.

From 1953 through 1959, design engineers involved in the development of aircrew full pressure suits concentrated their efforts to:

- tackle the problem of suit elongation resulting from the effect of positive pressure;
- improve suit mobility to increase operating pressure;
- ensure matching of the suit with the seat and the helmet with the seat headrest;

**Figure 2.2.4** SI-1 full pressure suit.
From Archive Zvezda.

- develop a suit-sizing system and identify the size adjustment range;
- protect the head during ejection and select the helmet type;
- extend permissible time of wearing the suit under operating positive pressure;
- find the most convenient place for the location of the pressure control valve on the enclosure; and
- improve operational characteristics.

It is worth considering some of the listed aspects.

Experiments demonstrated that the elongation effect was caused by elongation of the longitudinal restraint cords of the body and, to an even higher degree, by the change in the shape of the buttock part of the enclosure. As one of the options of solving the elongation problem, a restraint system with a wide waistband cut in the back was developed. The band was connected to the longitudinal cords through a roller system. When the suit was pressurized, the band was designed to draw in the longitudinal cords. However, due to the high friction between the waistband and the enclosure, the system proved to be inefficient and did not solve the elongation problem. It was only used for suit height adjustment. Such a system was used in the SI-1, SI-3M, SB-4B and SI-5 full pressure suits (Figures 2.2.4–2.2.5).

Introduction into the design of a so-called front adjustment strap (i.e., a restraint strap adjusted with a special buckle that connected the neck ring with the groin zone of the enclosure) happened to be the most effective way of eliminating the enclosure elongation problem. For the first time, the front adjustment strap was used in the SI-3M suit (Figure 2.2.5). Later on, it was used for all soft full pressure suits.

As to the improvement of suit mobility, the most difficult element for specialists to develop was a shoulder joint. This joint with two degrees of freedom ensured arm abduction/adduction as well as arm bending/straightening in the longitudinal plane (forward/backward movement).

The VSS-04, SI-1 and SI-3M suits (Figures 2.2.1, 2.2.4 and 2.2.5) used a two-degrees-of-freedom soft joint (of the Hook type) with orange peels located in mutually perpendicular planes. The VS-06 suit (Figure 2.2.3) used a crimped joint with a sliding restraint cord inside the enclosure.

The best results were obtained with the SI-5 suit, for which shoulder airtight ball bearings designed by Zvezda specialists were used for the first time. However, effective as they were with the suit under positive pressure, the bearings caused discomfort to the wearer in the donning/doffing process, while wearing it without positive pressure in the seat equipped with the restrain system and during parachute descent.

Examples of various arrangements of the group flow line inlet unit are shown in Figures 2.2.4 and 2.2.5.

Of significant importance was to ensure the proper matching of the helmet and the headrest. The soft helmet of the VSS and VS suits matched well with the headrest when the helmet was closed. With the helmet in the open position, the visor was behind the head and caused significant discomfort to the pilot (Figure 2.2.2). That was one of the reasons for switching over, beginning with the SI-1 suit, to a removable spherical helmet with a soft enclosure at the back of the head (Figure 2.2.4). The visor was open in the forward–downward direction (Figure 2.2.6). A removable helmet improved the operational properties of the suit. It simplified the donning/doffing process and wearing the suit on the ground before the flight and after landing. Such a spatial helmet ensured a good field of vision and comfort conditions for the pilot. However, it did not protect the pilot's head from impact during the ejection and parachute descent.

The famous parachute-jumper (and parachute-tester) P.I. Dolgov from the Air Force GK NII wore a SI-3M [СИ-3М] suit (Figure 2.2.5) when he made an attempt at a record-breaking parachute jump from the balloon *Volga* (Figure 2.2.7) on

**Figure 2.2.5** SI-3M full pressure suit.
From Archive Zvezda.

1 November 1962. The suit was modified for this mission: the oxygen mask with its oxygen demand regulator was removed from it and continuous oxygen flow was supplied directly to the helmet. The pressure suit had a spherical helmet with a flat visor. The helmet visor was made of a single 1.5–3-mm-thick layer of acrylic plastic (Figure 2.2.8). When Dolgov was about to leave the gondola for the parachute jump, he straightened himself up and knocked his helmet against some object protruding from the hatch area. The blow was so strong that the glass broke and the suit depressurized, resulting in Dolgov's sudden death. In the next modification of the

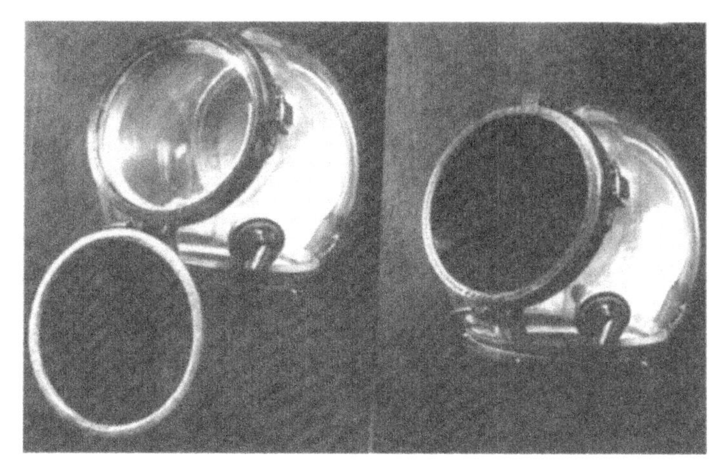

**Figure 2.2.6** SI-5 full pressure suit helmet.
From Archive Zvezda.

**Figure 2.2.7** The Volga gondola at the RF Air Force Museum in Monino on 11 May 1992.
From Archive Skoog.

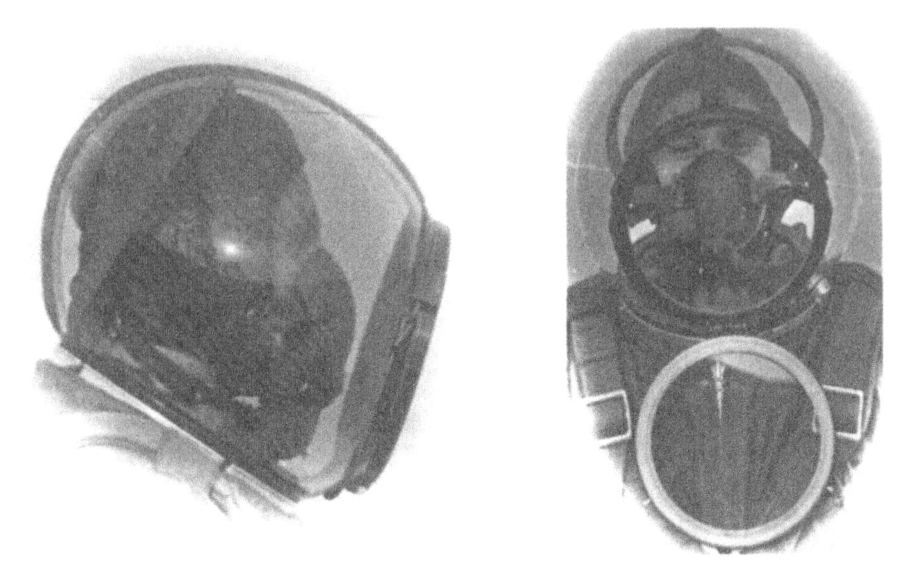

**Figure 2.2.8** Dolgov SI-3M helmet (this type was used without the mask).
From Archive Zvezda.

aviation maskless pressure suit—the SI-5—the helmet visor had two layers (see Figure 2.2.6, SI-5 helmet).

In this record-breaking attempt a second parachutist, E.N. Andreyev, performed an initial jump from a second compartment of the *Volga* gondola. Andreyev wore a mass-produced Air Force partial-pressure suit designed by Zvezda, in regular use at his Air Force unit. In this jump Andreyev left the gondola by means of an ejection seat. His jump was uneventful.

Beginning with the S-9 suit, the spherical removable helmet was replaced by a turning helmet with a metal casquet (Figure 2.2.9). The turning helmet was fixed on the head with a headset. An airtight bearing assisted side rotation of the helmet. Nod movement in the forward–backward direction was supported by a soft neck joint (Figure 2.2.9) fitted with two cords passed through the roller.

The first full pressure suits had arms sealed at the wrist with a pressure-tight cuff (Figures 2.2.1 and 2.2.2). When the suit was under positive pressure, the cuff additionally compressed the pilot's arm and sealed the suit. The pilot's wrists protruded from the suit and thus preserved his or her ability to work. However, the cuff did not protect the wrist from exposure to decreased pressure and therefore considerably limited the time of flight with the cockpit decompressed.

Therefore, beginning with the VS-06 and SI-3M models (Figures 2.2.3 and 2.2.5), suits were equipped with easily removed airtight gloves worn above the cuffs. The cuffs were modified and provided redundancy by means of cuff pressurization just in case the gloves were damaged. The limitation for time of wearing the suit at maximum altitudes was removed.

**Figure 2.2.9** S-9 full pressure suit.
From Archive Zvezda.

The SI-3M, SI-5 and then the S-9 suit models came with easily removed pressure-tight boots, available in many different foot sizes.

The pressure control valve located on the enclosure automatically maintained the required absolute pressure in the suit. The suit development process included improvement of pressure control valves and selection of the most convenient position for the valve on the enclosure (Figures 2.2.3 and 2.2.4).

The year 1959 is significant because it was this year that work on the SI-9 (S-9) and Vorkuta suits for Navy pilots began.

The S-9 suit was a follow-on model of the SI-5 prototype. The S-9 was the first suit to use a turning helmet with a metal casquet and a pressure-tight bearing. Therefore, the enclosure accommodated a soft joint located between the neck airtight bearing and the neck bent ring (Figure 2.2.9). The joint ensured nod movement in the forward–backward direction.

The suit enclosure was made of two-layer rubberized cotton fabric, as was done for the preceding suit models. To ease suit-wearing on the ground before and after the flight, a front transverse cut fitted with a zipper was made in the suit body. The zipper was in the open position to make walking easy. The pilot zipped the suit just before he occupied the seat in the cockpit. The closed zipper kept the suit in the "sitting position". Outerwear fitted with a soft aluminium thermal reflector (anti-fire coating) was fixed on the suit.

A pressure-tight inlet unit was located on the back to the left. A flotation rescue collar was fixed on the enclosure (see also Figure 3.2.3).

The Vorkuta suit model (Figures 2.2.10 and 2.2.11) was the first suit fitted with an enclosure, which consisted of two separate layers, instead of a single restraint enclosure and bladder. The outer layer worked as the load-bearing enclosure and the inner layer functioned as a bladder. Such an arrangement of the suit enclosure was a significant design achievement. It improved reliability, since a more durable technical fabric could be used for the restraint. It improved the elasticity of the enclosure particularly in joint areas. It simplified production techniques for the restraint (sewing instead of gluing) and shortened assembly time (parallel work on separate components).

The restraint of the Vorkuta suit was made out of the Russian polyester fabric "lavsan" and the bladder was made of foam rubber with a thickness of 3–5 mm. The inner enclosure also provided thermal insulation in case of an emergency splash-down. The bladders in the joints were made of thin sheet rubber. The suit's thermal insulation enabled the pilot to stay in cold water at a temperature from 0 to +10°C for up to 12 hours. The full pressure suits were manufactured in two versions, with and without a breathing mask (Figures 2.2.10 and 2.2.11).

In the first development phase, the restraint system was made up of a rope running along the body's neutral axes. The rope—through a system of rollers secured to the rigid liner on the body's sides—ended up at the waistband and was stopped at the front by a winch with a stopper. The winch was used to adjust the body length. And then, the now "classic" front adjustment strap system was introduced.

In various phases of the development, work on the Vorkuta suit, both spatial (fixed) and turning helmets were used. Subsequently, all Zvezda suits used separate enclosures (restraint plus bladder). The Sokol full pressure suit (1975) (Figure 2.2.12) was the last aircrew full pressure suit developed in line with such a design concept. The Sokol suit, other than its pure "high altitude" gear, could be fitted with a respirator that provided the pilot with protection before boarding and after aircraft landing in case the airfield was chemically contaminated.

The following sections will show that the experience gained from work on aircrew full pressure suits was widely used for the development of spacesuits, both

**Figure 2.2.10** Vorkuta full pressure suit, maskless type.
From Archive Zvezda.

for rescue (SK-1 and SOKOL KV-2) and EVA (extravehicular activities) suits of the KRECHET and ORLAN type for work on the Moon and in free space.

This was especially true for (1) the sealing systems of the operational disconnects of the gloves, helmet and backpack, which used a flat rubber profile sealing modified later into a hose sealing; (2) for the two-layer enclosure with its separate restraint enclosure and bladder; (3) for the design of load-bearing systems and soft enclosure joints; (4) for the design concept of gloves, pressure bearings and helmets; and (5) for the enclosure sizing system; etc.

All of this enabled specialists to develop the first space full pressure suits within the shortest possible time.

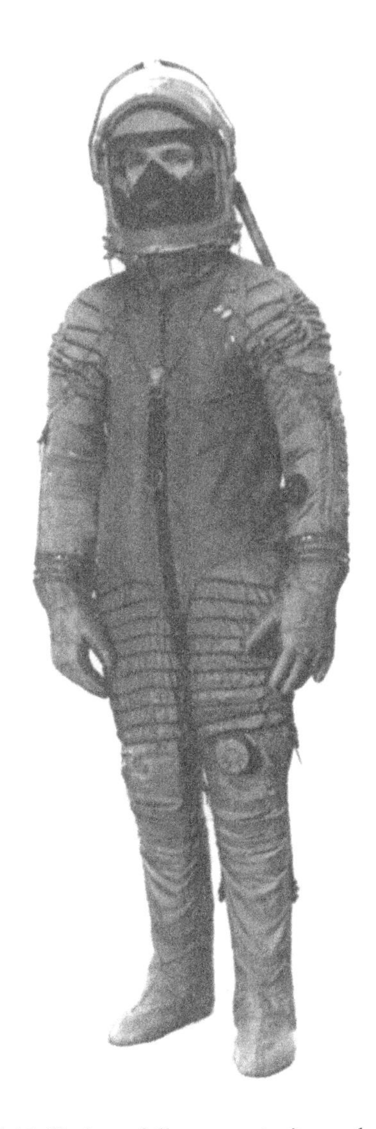

**Figure 2.2.11** Vorkuta full pressure suit, mask type.
From Archive Zvezda.

## 2.3 EQUIPMENT FOR THE INVESTIGATION OF THE VITAL FUNCTIONS OF ANIMALS DURING SPACE FLIGHTS

In the first years after its establishment, in parallel with aviation programmes, Plant No. 918 (Zvezda) also became involved in work on the safety of future space missions regarding life support systems for pilots and rescue means by high-altitude and high-velocity flights.

**Figure 2.2.12** Sokol aviation full pressure suit.
From Archive Zvezda.

Zvezda's design work in the emerging space programme resulted in all manner of equipment, including ejection sleds, oxygen equipment and full pressure suits for animals (dogs), developed and manufactured in 1953–1954 (Figure 2.3.1). The purpose of this work—carried out according to a resolution of the USSR Council of Ministers dated 6 February 1953 and an order of the USSR Academy of Sciences and the Scientific and Research Institute of Aviation Medicine (NIIAM)—was to study the possibility for living species to sustain altitudes up to 100–110 km after a rocket launch and then ejection from the rocket and the parachute descent.

**Figure 2.3.1** General view of the ejection sled mock-up with a pressure suit for a dog.
From Archive Zvezda.

The full pressure suit consisted of a pressure-tight enclosure made from a three-layer rubberized fabric and a removable helmet in the form of a plexiglas sphere. In the front portion of the suit there were two blind sleeves for the front paws of an animal. The rear top surface of the suit had a special opening with an appendix for putting the animal inside the suit. The suit had interior and exterior straps to restrain the animal. The suit was attached to a pull-out tray and inserted into a specially designed sled (Figure 2.3.2).

The suit's oxygen supply system, which was arranged on the rocket sled, incorporated three 2-litre cylinders with oxygen under 150 bar, a pressure reducer and a pressure gauge. The oxygen supply was designed for 2–2.5 hours of suit ventilation at a flow rate of $6\,\mathrm{l\,min^{-1}}$. At the launch of the rocket, oxygen supply to the suit was initiated and suit pressure rose to 400–440 hPa. This pressure was maintained automatically owing to an absolute pressure valve and a relief valve. Two sleds were installed in the payload compartment of the rocket. One of the sleds ejected at an altitude of 75–90 km and the other at 35 km. Then the parachute was forced to deploy at an altitude of 75–85 km in the first case and at 3–4 km in the second case. At the same altitude (3–4 km) a small port opened automatically to connect the inner suit cavity with the environment. A total number of 20 units of this equipment were manufactured. Six flights with the R-1D rocket were carried out between 1954 and 1956. Data on the following activities were gathered during those flights: animal behaviour (by film shooting), physiological functions under g loads, barometric pressure changes and total and partial weightlessness, which lasted about 220 s.

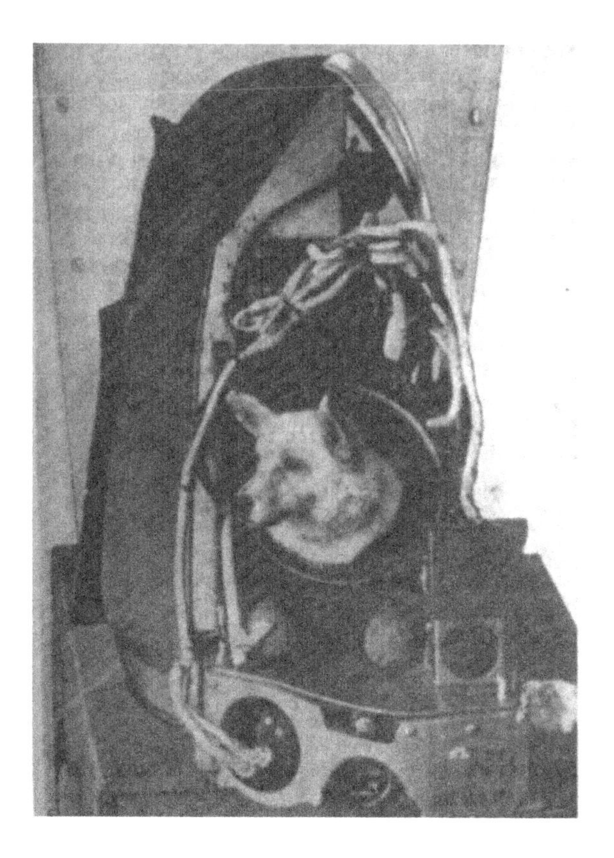

**Figure 2.3.2** Fully equipped ejection sled with the spacesuit attached to it (the helmet is removed).

In 1956–1957—as part of a continuation of work on the development of LSSs for pilots during high-altitude flights up to interplanetary space—a pressure cabin for animals (GKZh) (Figure 2.3.3) was developed by order of NIIAM. This work was crowned by the one-way flight of the dog Laika in the second satellite, *Sputnik 2*, on 3 November 1957. Animal behaviour under vibration, acceleration and longer duration of weightlessness was studied during this mission. The pressure cabin for animals was equipped with a system for gas atmosphere regeneration and the automated supply of food and water to the animals. This work was carried out under the management of the leading designer A.M. Bakhramov in cooperation with NIIAM.

The next stage of the work that Zvezda took part in—investigation of the vital functions of animals under space flight conditions—was the creation of the ejection cabin, which was used in the initial stage of development and evaluation of the ejection seat for the *Vostok* spacecraft (Figure 2.3.4). This cabin was designed for two animals and equipped with feeding and waste management systems for the dogs (Figure 2.3.5). The animals were ventilated with the spacecraft's own cabin air using

**Figure 2.3.3** Equipment of the animal pressure-tight cabin (the pressure-tight housing is not shown).
From Archive Zvezda.

**Figure 2.3.4** Accommodation of the pressure cabin in the ejection seat.
From Archive Zvezda.

**Figure 2.3.5** Pressure-tight cabin for animals (with the cover removed).
From Archive Zvezda.

**Figure 2.3.6** Pressure-tight cabin (*at rear*) and pressure cabin in ejection seat in Zvezda Museum.
From Archive Skoog and Zvezda.

special fans. The ejectable container with the dogs Belka and Strelka successfully returned to the Earth after the second *Vostok* spacecraft precursor flight on 19 August 1960 (Figure 2.3.6). The research work performed with the use of animals made it possible to obtain by 1960 a large amount of real data to judge the feasibility of a manned space mission.

The equipment that supported the vital functions of the animals operated normally. Between 1954 and 1960 there were several cases when dogs died, but they were all due to equipment failure rather than with LSS operations.

# 3

# The *Vostok* era

## 3.1 HOW IT BEGAN

Less than a year after the launch of the first satellite in 1957, the OKB-1 (Experimental Design Bureau 1) got down to initial studies aimed at development of a manned Earth satellite. With instructions from Sergey Korolev, the first concept for a manned space flight was prepared within OKB-1 under the leadership of Michail Tikhonoravov and Konstantin Feoktistov in August 1958 and documented in the OKB-1 report, *Materials on the Preliminary Work on the Problem of the Creation of an Earth Satellite with Humans on Board* (Semeonov, 1996).

A resolution of the Soviet government and a corresponding order from the Ministry of Aviation Industry with the instruction to start work on the preparation of a manned mission aboard an Earth satellite were issued in January 1959. On 17 April 1959, the OKB-1 issued performance specifications for the development and manufacture of a pressure suit with an emergency air conditioning system, and on 22 May 1959 the Soviet government issued a resolution, which specified the main contractors and subcontractors. Those documents initiated work on the design and development of spacesuits in the USSR.

During 1959, Zvezda came up with the conceptual design and shop drawings of the first spacesuit (identified as the S-10 [C-10]) and manufactured two operating models for laboratory tests (Figure 3.1.1).

It is worth mentioning that Zvezda did not have a medical department at that time (it came into being in March 1960). The main partner of Zvezda in physiological and hygienic tests was the State Scientific Research Institute of Aviation and Space Medicine (GNIIIA&KM). Zvezda provided the Institute with an S-10 model for joint tests. B.V. Mikhailov participated in the tests as Zvezda's representative and the Institute test team was led by A.M. Genin and L.G. Golovkin.

The suit design and the life support system (LSS) arrangement were developed to ensure safety of the wearer in new emergency situations such as cabin

**Figure 3.1.1** General view of S-10 spacesuit breadboard in 1959.
From Archive Zvezda.

depressurization in orbit, disturbances in the cabin gas content or rescue in the event of splashing down (including rescue of an unconscious person).

The S-10 suit enclosure was developed on the basis of the design of previous aviation pressure suits. The suit helmet had a new design and was equipped with a system for automatic closure of the visor. The system featured an integrated parachute suspension and seat-restraining system, used as the suit reinforcement element, and a special combined service connector. The suit was meant to operate autonomously with a regeneration system. Purchase orders for the development of part of the on-board equipment of the system were given to the OKB-124 (known currently as the Nauka enterprise) and to SKB-KDA (Special Design Bureau for Oxygen and Breathing Equipment) (for the oxygen loop) led by Chief Designer P.I. Zima.

The performance specifications required ventilation of the suit with cabin air for up to 10 days at a cabin pressure of 1 bar and at a flow rate of 50 to $150\,\mathrm{l\,min^{-1}}$ in case the open loop was used, and for up to 14 hours in case the closed loop emergency system was in operation. Special oxygen equipment supplied the pilot with oxygen in the de-orbiting phase both before seat ejection and afterwards. Since the cabin temperature could reach the +40°C level, the suit used a special ventilation system and a unique liquid cooling system that sprayed water on the cosmonaut's torso in an emergency. The KP-50 oxygen unit was developed to automatically purge the suit with oxygen when the cabin pressure dropped. Under development were various options destined to support the breathing process after landing or splashing down when the cosmonaut might be unconscious. The problem was considered to be very important since the effects of space missions on humans were unknown. Also under development were ways to use waste management systems in the suit under space flight conditions.

At the beginning of work on the S-10 project, within the framework of the design bureau led by A.M. Bakhramov, Zvezda formed a special design department for high-altitude equipment with A.L. Zelvinsky as the Department Chief and A.M. Gershkovich as the Deputy Chief. The department incorporated several design teams. They included a team for development of partial pressure suits and helmets led by M.N. Arkhangelsky, teams for development of suit enclosures led by A.Y. Stoklitsky and I.I. Derevianko,[1] a team for LSSs led by I.P. Abramov and a team for new materials led by Z.B. Tsentsiper.

The meeting of the special committee, chaired by Academician Keldysh, at the Presidium of the Academy of Sciences of the USSR held on 18 July 1959 had arrived at the decision to make Plant No. 918 the prime contractor for development of LSSs and rescue systems needed for manned space missions.

Early in 1960, Zvezda continued work on the development of the S-10 pressure suit. However, in February 1960, the OKB-1 issued new performance specifications to Zvezda for the development of a protective suit (instead of the pressure suit).

There were several reasons for such a decision. The main ones were the weight deficit of the spacecraft and the negative attitude of OKB-1 design engineers led by K.P. Feoktistov about the necessity of a full pressure suit. Their rationale proceeded from the idea that the probability of cabin depressurization was considerably lower than the occurrence of other emergency situations that could have catastrophic consequences.[2]

The protective suit (identified as the V-3 suit [B-3])[3] was under development up to the end of August 1960. The main purpose of the V-3 suit was to protect the cosmonaut after landing under cold conditions (especially splashing down in cold water) (Figure 3.1.2).

1 Two suit enclosure teams worked in parallel on space and aviation suits by different orders and concepts.

2 This controversy continued until the *Soyuz-11* accident in 1971.

3 Protective suits were at this time given the same designation as the spacecraft. The first manned version of *Vostok*-type spacecraft was the *Vostok-3* (or V-3), V-1 and V-2 were unmanned versions.

**Figure 3.1.2** Testing of the V-3 survival suit under winter conditions.
From Archive Zvezda.

A waterproof enclosure worn on top of a special thermal protection garment used elements found in an immersion pilot suit. The garment was fitted with a ventilation system fed by cabin air during the whole flight from a self-contained ventilation unit. The garment torso was made of quilted foam rubber and woollen jersey was used for the legs and sleeves.

Zvezda came up with a conceptual design for the protective suit, made shop drawings and then manufactured eight suits. Several suits were delivered to the GNIIIA&KM to support physiological tests, and several suits were delivered to the Gromov Flight Research Institute (LII) for parachute-jumping tests. Zvezda performed 12-h coldwater immersion tests in a pool and 2-day open-air exposure tests in winter conditions.

The S-10 pressure suit project and later the V-3 suit project were managed by A.M. Gershkovich. Enclosures were developed by the team led by A.Y. Stoklitsky and another led by I.P. Abramov developed LSSs (jointly with subcontractors—mainly OKB-124 and SKB-KDA).

A.L. Zelvinsky, the Design Department Chief, and V.V. Fomenko, the Deputy Chief Designer, participated actively in the development of the above suits.

Discussions on the full pressure suit did not stop when work on the protective suit was under way. The most persistent advocates of the full pressure suit were Air Force representatives (namely, V.A. Smirnov, the Chief of an Air Force Department Office, and S.G. Frolov, a Department Chief of the State Scientific Research Institute of the Air Force). Physicians and Zvezda specialists supported them.

Heated discussions had culminated by the summer of 1960. A full pressure suit with an LSS of the closed loop type was proposed again. Design engineers of the OKB-1 stated that there was no weight margin for that. Then S.P. Korolev personally got involved with the discussion. A meeting was held at Zvezda at the end of August 1960 with the participation of all parties involved: S.P. Korolev, K.P. Feoktistov, G.I. Voronin, V.A. Smirnov, A.M. Genin, L.G. Golovkin and Zvezda specialists led by S.M. Alekseyev.

Taking into account the time available up to the planned date of the first manned space flight, the meeting reviewed various full pressure suit versions, including the one that was connected to the cabin regeneration system. When G.I. Voronin stated that the on-board system for regenerative pressure suits that had a closed loop system for gas regeneration would be ready no earlier than the end of 1961, S.P. Korolev said that he would agree to make 500 kg available for the suit system people provided that the suit with the appropriate system was ready by the end of 1960.

Since the delivery time schedule was very tough, the meeting arrived at a compromise settlement to accept development of a simplified version of a self-contained pressure suit system by maximizing use of already developed elements of the S-10 full pressure suit and the V-3 protective suit, coupled with use of the experience gained during work on high-altitude full pressure suits.

In September 1960, performance specifications for a full pressure suit (identified as SK-1 [CK-1]) were finally signed. The SK-1 suit was only required to operate as an open loop system for 5 hours in a depressurized cabin using on-board compressed oxygen and air.

Of importance was the decision to put in place a self-contained system for the suit (as opposed to similar systems integrated with the on-board LSS, like those planned in the USA). Despite its greater weight, such a system offered certain advantages. It offered high reliability, since in the event of a failure of the on-board LSS, even in a pressure-tight cabin, there was the chance of switching over to the self-contained system. Moreover, the development of the suit and its system was independent of the development of the on-board LSS. This made it easy to distribute responsibilities between the contractors developing the suit and the on-board LSS (in this very case between Zvezda and Nauka, respectively). Whether a single system or a combined on-board and spacesuit system were selected, no single entity could be found responsible for any possible failure.

Components of the suit's self-contained ventilation and oxygen supply system were located partially on the Zvezda-developed ejection seat and partially in the descent vehicle and spacecraft instrument module (Figure 3.1.3). It is worth mentioning that the concept of a rescue suit self-contained system is currently used for *Soyuz* vehicles.

By December 1960, Zvezda had manufactured eight SK-1 suits for in-house tests, for tests at the GNIIIA&KM and for delivery to the OKB-1. In parallel with spacesuit manufacture at Zvezda, development of the LSS and its components was carried out.

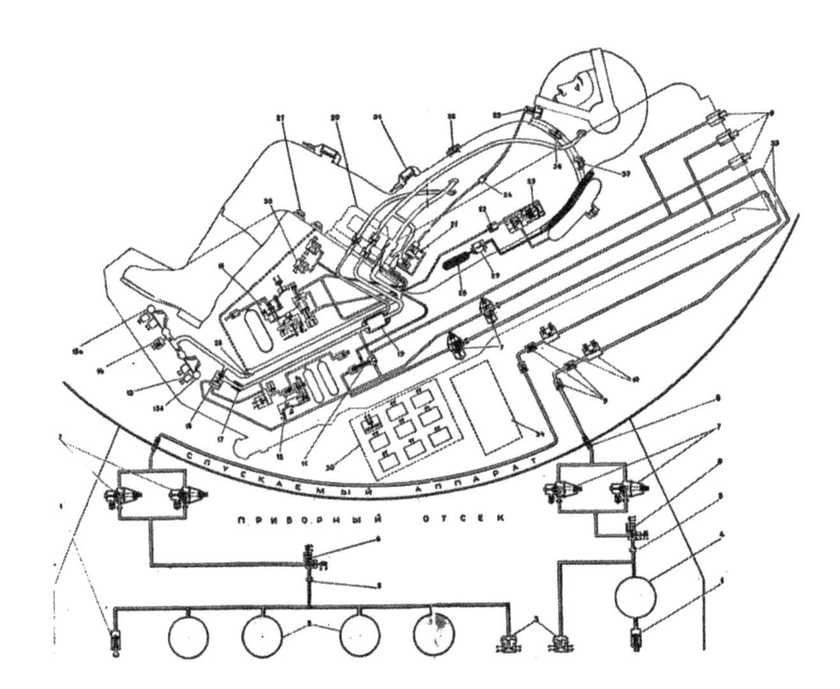


**Figure 3.1.3** Overall LSS for spacesuit, spacecraft and ejection seat (above: re-entry vehicle, below: instrument compartment): 1—charging connection; 2—air bottles; 3, 35—pressure sensors; 4—oxygen bottle; 5—filter; 6—electromagnetic valve; 7—reducer; 8—umbilical connector; 9—check valve; 10—pneumatic switch; 11—oxygen set; 12—oxygen and ventilation unit; 13—radiointerference filter; 14—switching device; 15a—primary fan; 15b—redundant valve; 16—shut-off valve; 17—noise suppressor; 18—parachute oxygen device; 19—additional leak-tight bag; 20—combined service connector; 21—oxygen injector; 22—relief valve; 23—collar inflation actuator; 24—connector; 25—breathing valve; 26—gasflow rate control valve; 27—switch; 28—$CO_2$ bottle; 29—starting head; 30—barometric relay; 31—absolute pressure regulator; 32—relief valve; 33—connecting hoses; 34—electric converter; 36—inhalation valve; 37—air inflow valve.

Original Zvezda.

## 3.2 FIRST SPACESUITS FOR THE *VOSTOK* SPACECRAFT

Yuri A. Gagarin undertook the world's first manned space flight[4] on 12 April 1961 aboard the *Vostok* spacecraft. Gagarin wore a Zvezda-developed full pressure suit identified as the SK-1 suit (Figure 3.2.1 and 3.2.2).

The SK-1 suit was also used by cosmonauts G.S. Titov, A.G. Nickolayev, P.R. Popovich, V.F. Bykovsky and V.V. Tereshkova (she wore the SK-2 [CK-2] suit) in their space missions.

[4] Contrary to reports in the West stating that other Russian cosmonauts preceded Gagarin in space, but perished in the attempt (maybe because of radio communication misinterpretations regarding the Ivan Ivanovich manikins), Zvezda can confirm that no one preceded Gagarin.

**Figure 3.2.1** SK-1 full pressure suit used by Yuri Gagarin (worn by Zvezda test engineer Yuri Orekhov).
From Archive Zvezda.

In combination with the LSS, the SK-1 suit ensured the following main requirements were met:

1 Normal hygiene conditions for the cosmonaut in the pressurized cabin for 12 days. A separate suit opening and the Vostok capsule waste management system were used (similar to the current system in *Soyuz*).
2 Safe occupation by the cosmonaut of a depressurized cabin for 5 hours in orbit and safe occupation of the descent module for 25 minutes.

**Figure 3.2.2** SK-1 spacesuit without the protective overall (worn by Zvezda test engineer Victor Yefimov).

From Archive Zvezda.

3 Protection of the cosmonaut in case of ejection at altitudes up to 8 km at a dynamic pressure up to 2,800 $kg\,cm^{-2}$.
4 Oxygen supply for breathing during parachute descent.
5 Survival of the cosmonaut in coldwater (after splashing down) for 12 hours (outside a rescue boat) and for 3 days under the temperature of −15°C after landing or in a rescue boat.

In case of cabin decompression, the suit operating pressure was maintained between 270 and 300 hPa, which corresponds to an altitude of 10 km.

The SK-1 suit LSS included a ventilation system, which used cabin air, as well as an emergency ventilation and oxygen supply system (see Section 3.3), developed in cooperation with the Nauka and SKB-KDA enterprises.

The suit system included the following components and units:

- enclosure (two layers with separate and pressure-tight enclosures);
- helmet with a dual visor and a device for automatic closure of the visor;
- removable gloves and cuffs;
- internal thermal protection suit with a ventilation system;
- protective coverall;
- boots adapted for parachute-landing;
- rescue float support collar with a $CO_2$ inflation system;
- group suit inlet interface unit;
- headset;
- emergency radio set;
- survival kit for survival in the wild (a pistol, a knife, a mirror and a shark repellant—fluorestine).

The suit mass amounted to about 23 kg. The SK-1 suit was designed, manufactured, tested and prepared for nominal operation within an extremely short time, almost 6 months. Involvement in the programme of all the leading specialists who were at that time working on high-altitude full pressure suits, utilization of experience gained in the previous 1.5 years in space apparel as well as the use of certain components from aircrew full pressure suits made such a hurried programme possible to implement.

The SK-1 spacesuit used the thermal protection suit and ventilation system developed for the V-3 full pressure suit. The SK-1 suit helmet was an improved version of the S-10 suit helmet fitted with a device for automatic closing.

The enclosure, a principal component of the suit, was similar to that of the Vorkuta aircrew full pressure suit (Figure 2.2.10). A.I. Boiko, the Vorkuta project-manager, was also involved in the development of the SK-1 suit enclosure. The enclosure consisted of two separate layers: an outer restraint layer was made of the strong polyester fabric "lavsan" and an inner pressure-tight layer was made of a natural rubber sheet 0.6 mm thick (in contrast to the Vorkuta suit that used a pressure-tight layer made of thick rubber with sealed pores).

Like the Vorkuta suit, the SK-1 suit used soft joints of orange peel type and the restraint system of sleeves and legs with length controllable cords. The body restraint system used a steel rope running from armpits to hips and then to a semi-rigid belt with grooves in the front and back. The rope terminated at a front-located drum with a ratchet-and-pawl gear. The drum was used to control the length of the cord during adjustment to fit the cosmonaut's body.

In the most critical phases of the ascent and descent of the spacecraft the suit had to be in sealed mode, the helmet closed and gloves worn. In the orbiting mode, with a normal cabin pressure, the cosmonaut could open the helmet and take off the gloves (the cuffs were not removed). In this mode, the cosmonaut could control all spacecraft systems, eat and answer calls of nature. To do the latter, the suit was fitted with a small appendix located in the lower part of the don/doff opening, through which a special receptacle of the waste management system was inserted in the suit. During the flight the appendix was tied and the opening laced up. In case of pressure drop the cosmonaut had to close the helmet and put the gloves on.

The cosmonaut would land after automatic ejection at an altitude of about 8 km, followed by parachute descent.

The SK-1 suit was of the maskless ventilation type. The helmet space was separated from the body space by a rubberneck partition fitted with an exhalation valve and an air inflow valve. When the visor was open the partition was automatically stretched out from the neck (a rope mechanically pulled the collar apart) to provide the cosmonaut with the necessary comfort.

Besides the "conventional" requirements imposed on a high-altitude suit helmet—such as good visibility, comfortable conditions for the head, protection of the head in the ejection process, ease of opening and closing the visor, no misting up of the visor—the SK-1 suit helmet had to meet additional requirements, such as automatic closing of the visor by an electric signal in case of cabin pressure drop and prior to ejection.

The spatial helmet designed for the VSS-type suits (see also Section 2.2) was found to be the nearest match to meet these requirements. This helmet was selected as a prototype for the S-10 and SK-1 space suits (Figure 3.2.1). The helmet was fitted with a large visor in the form of part of a sphere with its upper edge connected to a soft back-of-the-head element made of rubberized fabric. With the helmet open, the visor was positioned above the cosmonaut's head and the soft part was folded behind the head. The helmet interface was located along the lower edge of the visor. The interface incorporated two semi-frame joints coupled to the visor sides near the ears. The helmet was also fitted with two rigid casques, an inner one located between the head and the helmet's soft enclosure and an outer one located above the helmet enclosure. This design was selected to avoid the folded, soft part of the helmet being squeezed between the cosmonaut's head and the seat headrest. With the helmet open, the visor and helmet enclosure fitted between the casques. The casques also protected the cosmonaut's head during ejection and parachute descent. To improve reliability and prevent misting up, the visor was double-glazed with an 8-mm gap between the layers.

Since the suit was worn for the whole flight, the helmet's inner casque was glued to the top part of the bladder. The helmet interface was sealed with a sealing hose placed in the groove of the lower semi-frame. With the helmet closed, the hose was pulled taut on the mating "knife" of the upper semi-frame. The hose cavity was connected to the suit cavity and, thus, with the suit inflated, the hose was additionally pressed against the knife. Such a sealing method proved to be very reliable and viable. It was failure-free even when the mating parts considerably deformed. Subsequently it was used in the SOKOL-KV and ORLAN (backpack/body interface) spacesuits. In a closed position, the semi-frames were held by two quick disconnects fixed on the lower semi-frame.

The suit enclosure accommodated an absolute pressure regulator as well as a relief valve, which operated when positive pressure in the suit exceeded 285 hPa. An

**Figure 3.2.3** Testing of the SK-1 spacesuit with flotation collar inflated.
From Archive Zvezda.

electric inlet for communication and telemetry was positioned on the right side of the body next to the group inlet for pneumatic lines.

Suit gloves were easy to remove. They were attached to the rings on the arms with quick disconnects. The attachment ring was fitted with an airtight bearing that enabled the cosmonaut to rotate his or her wrists under positive pressure. Cuffs were also fixed on the rings to seal the suit when gloves were not worn. The glove consisted of four layers: the outer layer made of fabric was the restraint enclosure; the second layer was a bladder made of latex rubber; the third layer made of foam rubber 5–6 mm thick provided thermal insulation; and the fourth layer was a liner made of a slippery fabric. The boots were made of leather. They were worn above the restraint enclosure and were fitted with special buckles to avoid boots coming off during parachute deployment.

A decorative orange-coloured overall made of kapron (similar to nylon) was worn above the suit. The overall incorporated a flotation collar inflated from a $CO_2$ bottle. The collar could also be inflated by means of a special hose with a mouthpiece (Figure 3.2.3). The orange colour of the overalls was selected to facilitate the search for the cosmonaut in case of landing or splashing down in the wild. The overalls had pockets for a pistol, a knife, a radio set and a radiation dosimeter.

A thermal protection overall was worn above the underwear to protect the cosmonaut from excessive cold after landing or splashing down. The overall consisted of four layers including a silk outer cover, a layer of foam rubber, a layer of pure wool jersey and a silk lining. After landing, the suit could be taken off and the thermal protection coverall could be used by the cosmonaut as outer warm clothing, if that was necessary. The inner surface of the overall accommodated elastic, reinforced, perforated hoses as well as panels, which formed the suit ventilation system. Ventilation system air ducts ended in two (left and right) elastic manifolds made as flat reinforced hoses. While donning the suit, the manifolds were connected to the air loop inlet piece of the group connector through rubber couplings.

The headset, equipped with two microphones on adjustable holders and two earphones, provided communication support when the cosmonaut was suited up. The headset also incorporated two removable laryngophones, which were used at the time of entering orbit.

It is worth mentioning that implementation of the SK-1 suit programme within such a short time was facilitated by the fact that design, flight model manufacture and test activities ran practically in parallel. As a matter of fact, in November 1960, S.P. Korolev's OKB-1 got down to comprehensive tests of the suit's LSS jointly with *Vostok*'s spacecraft systems. However, only laboratory tests of the hardware had been carried out by that time and final tests for the suit were in the initial phase.

The test programme included strength tests (verification of the static strength margin), dynamic tests with simulation of the effects of an explosive decompression from the 1,013-hPa level down to the 41-hPa level, exposure to mechanical loads on a shaker and a centrifuge, tests in a vacuum chamber, flight jump tests with landing and splashing down, thermal tests including 11-day tests in the spacecraft cabin and exposure to cold water in a pool, tests on the operating life of the suit and full-scale sea tests under stormy conditions. Tests were carried out at Zvezda, Gromov LII, the GNIIIA&KM, Feodosia Air Force Base, the Central Institute of Aerohydrodynamics' (TsAGI) hydro lab and other organizations.

In the first quarter of 1961, before the first manned space flight, two flights with a dummy ("Ivan Ivanovich")[5] were carried out. For these flights, the suit and its on-board systems were readied to operate in the manned mode in case of cabin decompression.

The "Ivan Ivanovich" anthropometric manikin (Figure 3.2.4) with recording instruments was designed and manufactured in 1960. It was designed with the help of the Moscow Institute for Prosthetics. To facilitate spacesuit-donning and accommodation of the manikin in the *Vostok* seat, it was provided with flexible limbs (leg and arm prostheses with joints). The manikin body included a metal structure covered by leather and had a removable head made of metal with foam rubber glued on to it, on which was painted the "face". The head was connected to the manikin body through the open helmet after spacesuit-donning. The manikin was, as far as its mass and centre of gravity go, representative of separate parts of the human body. There were cavities inside the manikin containing instruments to record acceleration, angular rate changes, level of space radiation and radio communication checkout. Radio communication was checked by retransmission of radio station music that

[5] This was not the official name of the manikin and was invented by journalists describing the events.

**Figure 3.2.4** The Ivan Ivanovich manikin.
From Archive Zvezda.

was received in the capsule and picked up by a microphone. The retransmission of popular chorus folk songs was used for this purpose to avoid popular misinterpretation of "Ivan Ivanovich" being a real cosmonaut (Chertok, 1996).

The final phase of the flight, which included ejection and parachute descent, ran in the nominal mode to give a final demonstration of the proper operation of all systems (Figure 3.2.5). To avoid any misunderstandings the dummy (or manikin) was given a cloth carrying the word "mock-up" (*Maket*) over its face (Figure 3.2.6).

A team of Zvezda specialists led by F.A. Vostokov (Figure 3.2.7) was sent to the space launch centre to verify and prepare Zvezda-delivered hardware for the Gagarin

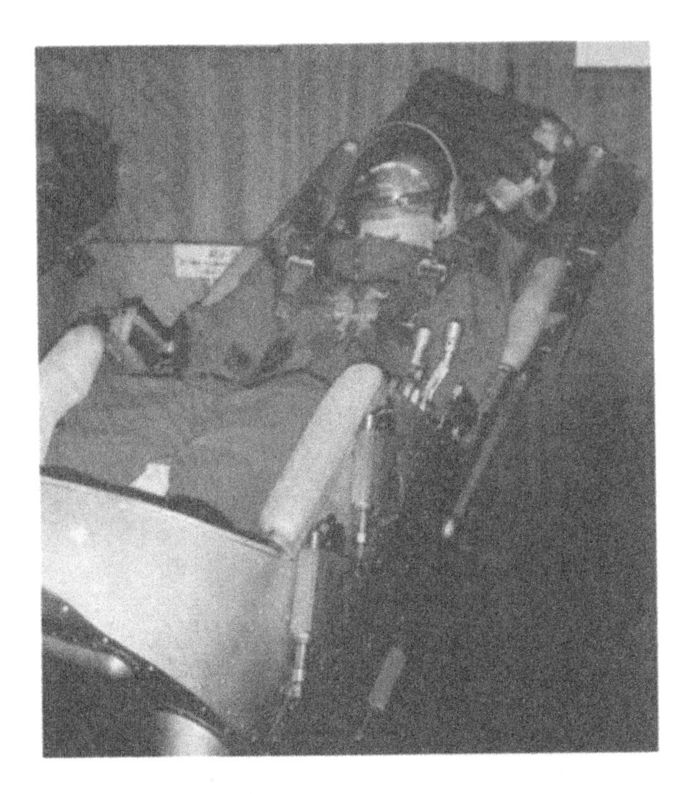

**Figure 3.2.5** Model of Ivan Ivanovich in the ejection seat in the Zvezda Museum. (One of the original Ivan Ivanovich manikins, which was flown in March 1961 to test the complete system of spacecraft, suit and ejection seat, is today in the RSC Energia Museum.)
From Archive Zvezda and Skoog.

**Figure 3.2.6** The sign *Maket* (puppet or mock-up) used by Ivan Ivanovich.

**Figure 3.2.7** The team that prepared the spacesuits and other Zvezda equipment for Gagarin's flight in Baikonur. *Front row (left to right)*: I. Skomorsky, I. Abramov, K. Yuryeva, F. Vostokov, N. Smirnov. *Back row*: N. Borisov, V. Svertshek, V. Davidyantz, M. Ikonnikov, G. Lebedev, A. Panov, V. Yelmanov, S. Zaytsev, G. Petrushin. Missing are N. Rogachiov, B. Firkin and G. Davydov.

From Archive Zvezda.

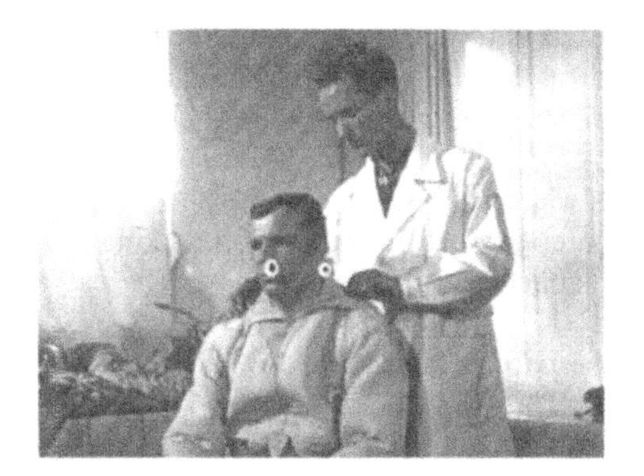

**Figure 3.2.8** Vitaly I. Svertshek assists Yuri A. Gagarin in spacesuit-donning, prior to the mission on 12 April 1961.

flight. The crew consisted of several support groups including a group for preparation of the suit system and LSS led by I.P. Abramov and a group for preparation of electric and radio equipment led by I.I. Skomorovsky. Vitaly Svertshek, a member of the first group, assisted and instructed Yuri A. Gagarin (Figure 3.2.8) and subsequently G.S. Titov in the donning process and performed suit leakage tests at the launch pad when the cosmonaut was aboard the spacecraft. Vitaly Svertshek is currently the First Deputy General Director and First Deputy General Designer of Zvezda. Georgy Petrushin was the main helpmate of V.I. Svertshek.

In the assembly and test building at the cosmodrome (at Site No. 2), special rooms were set apart for Zvezda-manufactured equipment. They were fitted with compressed air, oxygen (from stationary cylinders) and power supply lines. The rooms also accommodated Zvezda equipment for independent and preflight tests of the spacesuits and ejection seats. Next door to these rooms, there was a room specially set up for medical check-up of the cosmonaut and medical sensor installation.

In compliance with pre-developed plans and procedures the Zvezda team ran independent tests and pre-launch preparation of the flight equipment in these rooms. The Zvezda team also, jointly with specialists from OKB-1 and the cosmodrome, supported spacesuit/ejection seat testing together with the spacecraft's on-board electric and pneumatic systems.

It is important to remember that the performance capabilities of humans in a space mission had been inadequately studied in the period preceding the first manned space mission. Therefore the spacesuit was equipped with a number of automatic devices—used to close the helmet visor, open a breathing valve after landing, inflate flotation means and perform other functions, etc.

Nevertheless, the many specialists involved in the support of the mission were concerned about the landing phase. This subject was under discussion even right up to the last hours prior to the flight. The discussions resulted, for example, in a proposal to fit to the parachute suspension harness a plate bearing a sketch that explained how to open the suit helmet if a cosmonaut was not able to do so himself.

On the launch day, work in the Zvezda laboratory started 5 hours before the launch. Gagarin and G.S. Titov accompanied by their instructors and heads arrived about 4 hours before the launch. The cosmonauts underwent a medical check-up, and after that spacesuit-donning began.

Having put the suits on, the cosmonauts waited in test seats in the next room. This was when Yu.A. Gagarin gave his last interview before the mission, talking to S.P. Korolev, who specially came from the launch pad for that purpose. He also signed his first ever autographs.

Moreover, to identify the nation that the pilot belonged to, the decision was made to paint *CCCP* (USSR) on the helmet. This last-minute decision was made after Gagarin had already donned the spacesuit[6] (Figure 3.2.9). It was Zvezda test engineer Victor Davidyantz who painted *CCCP* on Gagarin's helmet at the cosmodrome. When Gagarin and the mission support team waited for a bus to drive them to the launch pad, S.P. Korolev came to the suit-donning room to excitedly give final instructions to Gagarin. Gagarin looked very confident and tried to calm Korolev down, assuring him that all instructions were firmly fixed in his memory.

As the hardware was being shipped to the space launch centre, the test

[6] The reason to paint *CCCP* on the helmet was that the very first manned space flight was not publicly known, and less than a year before Gary Powers crash-landed in the USSR after having been shot down in the famous U-2 incident. How would a farmer out in the steppe know that the pilot was a Soviet cosmonaut and not a US spy pilot in the event that the pilot was unconscious?

**Figure 3.2.9** Official photo of Yuri Gagarin, taken after the flight.
From Archive Zvezda and Skoog.

programme had still not been completed, and that necessitated making final corrections directly in the process of preparation of the hardware for the launch. This fact, in particular, explains why Gagarin's suit in photographs and movie shots taken several days before the launch day (e.g., Figure 3.2.8) at the launch centre differs outwardly from the suit in pictures shot on the launch day (e.g., Figure 3.2.10) and after the launch (Figure 3.2.11). Final adjustments to the suit on the results of sea tests in Feodosia were made a few days prior to the launch.[7]

A special bus to transport cosmonauts to the launch pad was prepared at Zvezda (Figure 3.2.12). It was a common passenger bus produced by the motor plant at Lvov. Its passenger compartment was modified to accommodate two workstations for suited cosmonauts. These workstations had air ventilation lines, with the air supplied from special bottles located at the rear of the bus.

In this bus Zvezda's support personnel, other cosmonauts and instructors usually accompanied a cosmonaut and his back-up. Spacesuit spare parts (gloves, headsets, etc.) were also kept in this bus.

[7] The ventilation valve below the helmet had to be modified. This was to prevent possible water penetration into the spacesuit. The original valve was white (Figure 3.2.8). The replacement had a white inner ring with an outer black cap (Figure 3.2.10).

**Figure 3.2.10** S.P. Korolev gives last instructions to Yuri Gagarin. In the middle is Marshal K.S. Moskalenko.

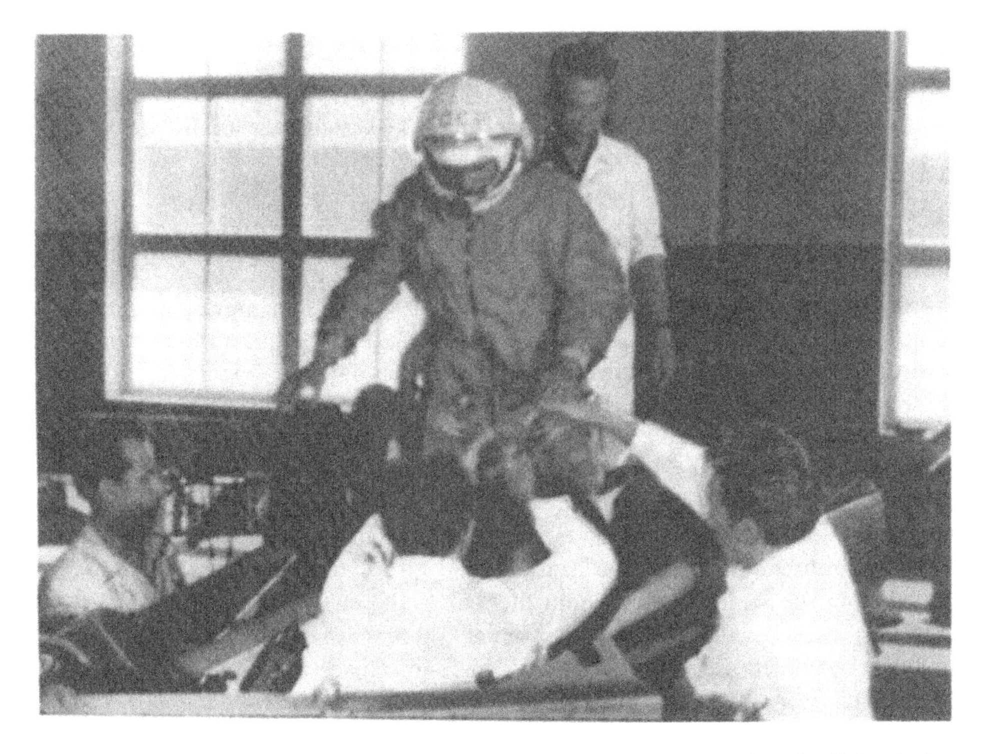

**Figure 3.2.11** This centrifuge test is a good example of a picture taken after the flight, during the shooting of a movie of the Gagarin flight. Although a similar suit is used, the scene depicts an event long before the launch, when the lettering *CCCP* would not have been on the helmet.

**Figure 3.2.12** Gagarin in the bus on the way to the launch pad. Behind Gagarin is Titov (in the spacesuit) and cosmonauts G.G. Nelyubov and A.G. Nikolayev.

Two hours before the launch the cosmonauts and their entourage made for the bus. The bus did not make any stop on the way to the launch pad (contrary to reports in the press that Gagarin got off the bus to answer a call of nature). On the launch pad the cosmonauts reported to the Chairman of the State Commission and made for the elevator. They were accompanied by F.A. Vostokov (Zvezda) and the Leading Designer O.G. Ivanovsky (OKB-1), who assisted Yuri Gagarin to get in his seat. After that V.I. Svertshek, N.S. Smirnov and B.S. Firkin, who were at the launch pad in front of the spacecraft hatches, ran a final check on the spacesuit and ejection seat systems.

Not long before the spacecraft launch, the bus with G.S. Titov and the specialists accompanying him left the launch pad, after which G.S. Titov was assisted in suit-doffing.

The *Vostok* flight only took 108 minutes. So, the Zvezda specialists awaited its outcome near the building where the mission control managers sat. Chief designer S.M. Alekseyev, who represented Zvezda in the State Commission, was also there. Prior to that event he participated in Commission meetings, making decisions on selection of the cosmonaut and flight certification for the spacecraft systems.

At the same time a team of specialists from the Ministry of Defence, OKB-1 and some other companies who were tasked with assuring Gagarin's safe descent and landing, were already in the area of the planned landing. Zvezda representatives A.M. Bakhramov and A.K. Malyshev were included in that group. As we all know, Gagarin's ejection and landing were successful.

**Figure 3.2.13** Yuri Gagarin and General Kamanin together with the Zvezda team responsible for all the suit equipment for Gagarin's flight, at the visit to Zvezda in Tomilino on 25 April 1961. *Front row (from left to right)*: F. Vostokov, F. Melikhov, Yu. Kilosanidze, Yu. Gagarin, S. Alekseyev, General N. Kamanin, V. Fomenko, General L. Goreglyad. *Back row*: A. Boiko, A. Miskaryan, B. Demyanovsky, V. Yelmanov, N. Borisov, I. Abramov, A. Malyshev, A. Istratov, I. Skomorovsky, A. Zelvinsky, G. Lebedev, A. Bakhramov, A. Panov, F. Sebrin, A. Grachiov, V. Svertshek, A. Stoklitsky.
From Archive Zvezda.

The cosmonauts did not give any post-flight critical comments on the space pressure suits. However, one should keep in mind that no emergency situation occurred during these missions, and thus the cosmonauts operated wearing unpressurized suits, with ventilation provided by cabin air. G.S. Titov was the first to use the waist management system. To do so, it was necessary to untie and afterwards to tie again the suit small appendices. During his flight, A.G. Nikolayev got free from his restraint system to hover above the seat. Critical comments were not given on the above operations.

The regular visits of cosmonauts to the main contractors after the manned flights became a normal procedure in the initial phase of the Soviet space programme. The visit to Zvezda, after Gagarin's flight, took place on 25 April 1961, at which time the usual group photo was taken (Figure 3.2.13).

Much of the test programme was carried out with simultaneously run test phases and called for many test models. Eight SK-1 suits were manufactured in 1960, twenty-three in 1961 and nine in 1962. This quantity included training suits and flight models worn by Yu.A. Gagarin and G.S. Titov in 1961, A.G. Nikolayev and P.R. Popovich in 1962 and V.I. Bykovsky in 1963.

After G.S. Titov's flight aboard *Vostok-2* (Figure 3.2.14), certain changes were introduced in SK-1 suit design to improve its reliability and operation. The changes primarily concerned the suit's LSS (see Section 3.3).

**Figure 3.2.14** Titov visits Zvezda in August 1961 after the flight. *From left to right*: E.F. Shvartsburg, V.V. Fomenko, V.T. Davidyants, B.V. Mikhailov, V.I. Svertshek, Yu.D. Kilosanidze, G.S. Titov, I.I. Skomorovsky, S.V. Zaytsev, G.S. Karavchuk, I.I. Chistyakov, N.S. Smirnov, S.M. Alekseyev, F.A. Melikov, A.K. Malyshev.
From Archive Zvezda.

## 3.3 LIFE SUPPORT SYSTEM FOR THE SK-1 SUIT

To support the vital functions of a cosmonaut wearing a full pressure suit, *Vostok* spacecraft were equipped with a system that assured the necessary physiological and hygienic conditions within the suit enclosure. The main components of the system were built in to the ejection seat and arranged on the suit. Compressed air and oxygen bottles to maintain the gas pressure in the suit and supply oxygen in case of emergency decompression of the cabin were installed outside the spacecraft in the instrument module (Figure 3.3.1). The nominal flight conditions required that only ventilation of the suit be provided to remove heat, sweat, $CO_2$ and other gas discharges of the human body. A fan supplied the suit with air through two separate lines, one feeding the helmet and the other bringing air from beneath the body enclosure to the body extremities. An elastic neck partition separated the helmet from the remaining lower suit cavity. A valve on the neck partition released air from the closed helmet, removing it to a cavity under the body enclosure. Air for ventilation came into the cabin through a pressure control valve installed on the enclosure.

Bottles located in the instrument compartment automatically supplied the suit (and, therefore, the cabin) with air in case of emergency depressurization of the cabin and decrease in ambient pressure down to 730 hPa. If the emergency air supply did not compensate for decrease in pressure due to air leakage from the cabin and the

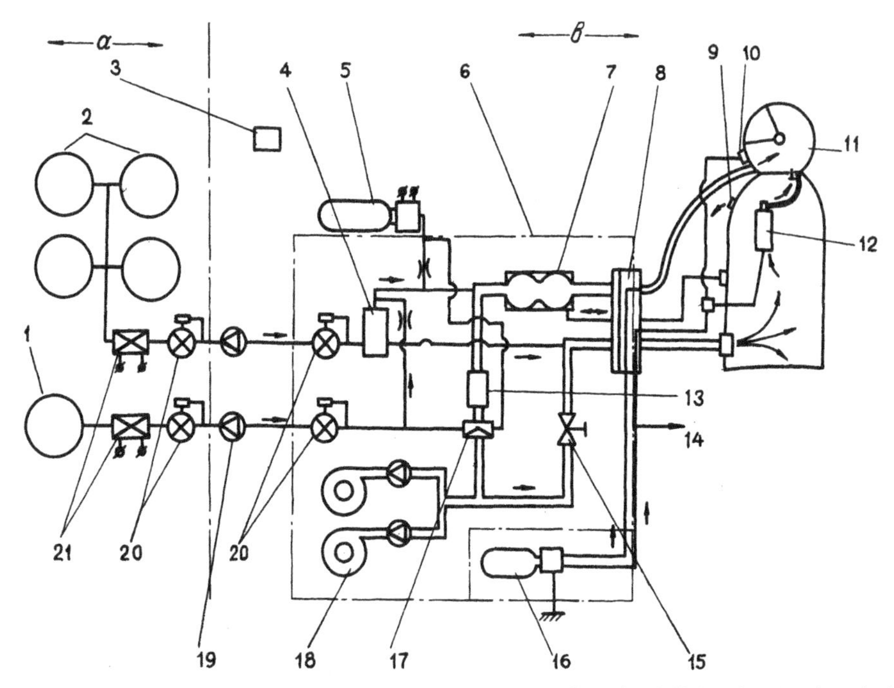


**Figure 3.3.1** Flow diagram of a *Vostok* cosmonaut wearing the SK-1 suit. *A*—detachable module; *B*—re-entry-capsule; 1—primary oxygen bottle; 2—air bottles; 3—barometric relay; 4—oxygen regulator; 5—oxygen bottle in re-entry capsule; 6—ejection seat; 7—reservoir; 8—interface; 9—SS pressure regulator; 10—breathing valve; 11—spacesuit; 12—life support unit (provides for breathing during and after splashdown); 13—silencer; 14—to floating means; 15—ventilation control valve; 16—oxygen parachute device; 17—pneumatic shut-off valve; 18—fan; 19—check valve; 20— reducer; 21—electropneumatic valve.

From Archive Zvezda.

cabin pressure fell to 600 hPa, the helmet visor would close at this pressure level and on-board oxygen bottles would start feeding the suit. In this case air ventilated the body enclosure, and an oxygen/air mixture, with the oxygen content increasing with increase of the barometric "altitude" in the cabin, ventilated the helmet. With barometric pressure less than 300 hPa (corresponds to the 9-km altitude), the helmet was fed with pure oxygen. With further decreases in cabin pressure, fans were switched off and positive pressure was established under the suit enclosure. For fire safety reasons, the total content of oxygen in the gas coming out of the suit through the pressure control valve did not exceed 40–45% in all operating modes of the LSS.

The oxygen supply line was equipped with an elastic reservoir (a rubber bag in a pressure-tight housing) connected to the suit. The purpose of the reservoir was to decrease respiratory resistance at deep inhalation of air.

The flow diagram (Figure 3.3.1) does not show the LSS components that operated after separation of the cosmonaut from the seat. These components will be described below. In the descent phase, separation of the instrument module (together with oxygen and air bottles) was followed by automatic actuation of the oxygen supply to the helmet from two bottles located in the seat. Each bottle contained 2 litres of oxygen under 20 MPa pressure. At an altitude of about 7 km, the cosmonaut ejected and parachuted. Oxygen entered the helmet from a 1-litre bottle (at 20 MPa pressure) installed in the profiled cover separated from the seat. When the oxygen bottle ran out, a helmet-mounted valve automatically opened to let ambient air into the helmet so that the cosmonaut could use it for breathing, before opening the helmet visor.

Special considerations were given to splashdown conditions involving strong wind and high waves. To counter such a situation, the suit was fitted with two different bladders. One was automatically inflated when the bladder's filler contacted water. The bladder was arranged on the rigid framework of the backpack with the parachute system. The other floating bladder was located on the suit and the cosmonaut manually actuated the filler after release from the parachute system. The shape, volume and location of both bladders ensured the cosmonaut would return from any initial position to lying on his back with his head above the water level. The test programme showed that when a test subject was exposed to strong wind and waves with steep crests (especially when the parachute turned into a sail and dragged him), splashes covered the helmet and water entered the suit through the opening of the respiratory valve. Therefore, the decision was made to extend (beginning with *Vostok-3*) the duration of the independent LSS of the cosmonaut in the sealed suit for as long as was nessary for the cosmonaut to release the parachute canopy, get rid of the suspension system and get in the emergency inflatable dinghy. Only when these operations were completed, was it time for the respiratory valve to be automatically open.

This resulted in installation of a small cartridge containing $CO_2$ absorber and an injector placed beneath the suit enclosure (Figure 3.3.2). At an altitude of 4 km, the oxygen flow directly entering the helmet was automatically switched over to the

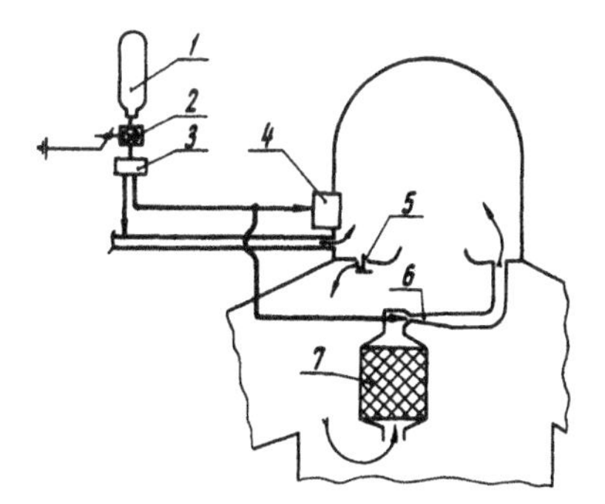


**Figure 3.3.2** Flow diagram of the *Vostok* cosmonaut LSS after separation from the ejection seat. 1—oxygen bottle; 2—starter; 3—mechanism for switching over oxygen flow from the helmet ventilation line to the injector at altitudes below 4 km; 4—valve automatically open after depletion of bottled oxygen to let ambient air in for breathing; 5—exhalation valve on the neck partition; 6—injector; 7—cartridge with $CO_2$ absorber.

From Archive Zvezda.

**Figure 3.4.1** Female version (SK-2) of the *Vostok* spacesuit for Valentina V. Tereshkova (here worn by Zvezda test engineer Svetlana Novak).
From Archive Zvezda.

injector. The latter provided gas circulation under the suit and the cartridge absorbed the $CO_2$ exhaled in the respiratory process. Such a flow arrangement made it possible for the cosmonaut to stay in a sealed suit for 30 minutes after ejection without any increase in the amount of oxygen stored in the bottle.

## 3.4 THE SK-2 SUIT

The SK-2 suit (Figures 3.4.1–3.4.3) was the model specially developed for the flight of Valentina V. Tereshkova, the first female cosmonaut.

**Figure 3.4.2** Valentina V. Tereshkova's official photo with a dedication to I.P. Abramov.
From Archive Zvezda.

**Figure 3.4.3** S.P. Korolev talking to Valentina V. Tereshkova before her flight. I.P. Abramov is in the background.

The SK-2 suit differed from the SK-1 suit mainly in the enclosure cut-out that took into account the specific features of a female body. The enclosure featured a decreased shoulder breadth, an increased hip girth and a decreased opening in the neck partition. In accordance with the decreased shoulder breadth, the restraint system of shoulder joints was modified to retain arm mobility.

Moreover, the position of the cord that kept the helmet at the front was changed. It was moved down from the breast. The gloves were also modified. The thickness of the thermal protection layer was decreased and mobility of the thumb improved. The lever for opening the respiratory valve and handles on the visor were modified to improve reach and make them handy. The required changes were also introduced in the design of the receptacle of the waste management system.

Final non-female-specific changes were also introduced in the last units of the SK-1 suit (used by Bykovsky). To support the test, training and flight programmes of V.V. Tereshkova, eight SK-2 suits were manufactured in 1962. G.I. Viskovskaya, a female Zvezda engineer, took an active part in the preparation of the suit for the mission and the donning of Tereshkova.

In their debriefs after their long-term flights, both V.F. Bykovsky and V.V. Tereshkova, who did not doff their SK-1 and SK-2 spacesuits during the flight, noted that sitting in the same posture for several days caused discomfort and pain due to suit element and medical sensor pressure on the body. Later rescue suits were only used in the most critical phases of the flight and caused no problems (see Chapter 7).

# 4

# Spacesuit and equipment for the world's first EVA

## 4.1 INTRODUCTION

The six manned missions carried out on board *Vostok* spacecraft between 1961 and 1963 proved the high reliability of the spacecraft and their equipment. This successfully implemented technology coupled with an actually carried out life science research programme clearly showed that the capabilities of *Vostok*-type spacecraft lent themselves to further use in manned space exploration.

Therefore, in 1963, the manufacture of four more *Vostok* spacecraft began, and an additional purchase order for further spacecraft of this type was under discussion. To expand the range of applications of the new lot of *Vostok* spacecraft, some new functions and technologies were introduced. Some of them concerned RD&PE Zvezda items. It was planned, in particular, to realize a "soft" landing of the descent vehicle and, thus, eliminate the need for the cosmonaut to eject, and a "space walk" from the spacecraft (Keldysh, 1980).

Rapid implementation of the plans was of significant importance for the forthcoming missions of the *Soyuz* vehicles, which were already in the production line.

The new spacecraft resulting from modifications to the *Vostok* spacecraft was named *Voskhod*. This was the space vehicle used in October 1964 for a space mission performed by a crew of three cosmonauts. A two-man crew used the *Voskhod-2* spacecraft in March 1965 when A.A. Leonov performed the first space walk (extravehicular activity—EVA) from it (initially this spacecraft had the trivial project name *Vykhod* (exit)).

*Voskhod* was the first spacecraft to incorporate shock-absorbing seats designed to reduce the possible g loads that crew members would be exposed to in case the soft landing system failed to operate.

It is worth mentioning that, due to the mass and space deficit that resulted from switching over to the three-seat arrangement for the *Voskhod* and later the *Soyuz* spacecraft, OKB-1 (Experimental Design Bureau 1) specialists made a decision not to use a full pressure suit as an emergency protective measure in case of cabin decompression. The arguments of the opponents were not considered, since OKB-1 design specialists pointed to the successful missions of the *Vostok* spacecraft (emphasizing the low probability of descent vehicle decompression).

As a result of the successful performance of the *Voskhod-2* mission it was planned to manufacture several additional *Voskhod* spacecraft including some for EVA. The newly developed backpack with a life support system (LSS) of the regenerative type and the Cosmonaut Transfer and Manoeuvring Unit (UPMK) were intended for use with these spacecraft (see also Chapter 9).

Later on, with the Moon programme gaining momentum and the *Soyuz* spacecraft under intensive development, further work with the *Voskhod* spacecraft was terminated.

## 4.2 EQUIPMENT FOR THE WORLD'S FIRST EVA

In early 1964 when the proposal to perform a manned exit into space (an EVA or space walk) from the *Voskhod-2* spacecraft was formulated, Zvezda specialists had already gained considerable experience in the development of aircrew and space full pressure suits. The SK-1/-2 full pressure suit with its LSS had been developed and successfully used by all *Vostok* crew members. Moreover, experimental work on advanced full pressure suits for lengthy space missions were on the way. Concepts and separate units for LSSs of the regenerative type were also under development.

However, a space walk from a spacecraft was quite a challenge for the specialists and called for development of new technologies to protect the cosmonaut from the hostile environment of free space. Furthermore, OKB-1 stated requirements such as provision for a space walk from the descent vehicle without its depressurization, use as far as possible of existing equipment with minimal of changes in the spacecraft design and use of the smallest equipment in terms of mass and dimension. These tasks, however, were made simpler by the fact that the first space walk was planned to be short.

No acceptable technical approach to undertaking a space walk from the *Voskhod* spacecraft was available at the beginning of the design work. The *Voskhod* cabin was not designed for lengthy operation in decompression mode, and there was no space available in the spacecraft to accommodate the special airlock—Volga. OKB-1 specialists studied various versions of a deployable airlock, while specialists at Zvezda, headed by Chief Designer G.I. Severin in January 1964, proposed an airlock with a soft inflatable shell. A soft inflatable airlock would make it possible to support a space walk with little change to the existing designs of the spacecraft and launcher fairing.

A governmental resolution giving the schedule for manufacture of *Voskhod* (3KV) and *Vykhod* (3KD) spacecraft was issued on 13 April 1964.

The Zvezda concept was finally accepted in April 1964 during a meeting between S.P. Korolev and his specialists at OKB-1. Participants in the meeting from Zvezda were G.I. Severin, N.L. Umansky and I.P. Abramov.

"УТВЕРЖДАЮ"
Главный конструктор ОКБ-I
ГКОТ
( КОРОЛЕВ )
" 9 " 6 1964г.

"СОГЛАСОВАНО"
Главный конструктор з-да 918
ГКАТ
( СЕВЕРИН )
" 9 " VI 1964г.

ТЕХНИЧЕСКОЕ ЗАДАНИЕ

на разработку и изготовление шлюзовой камеры
и системы шлюзования корабля "3КД"

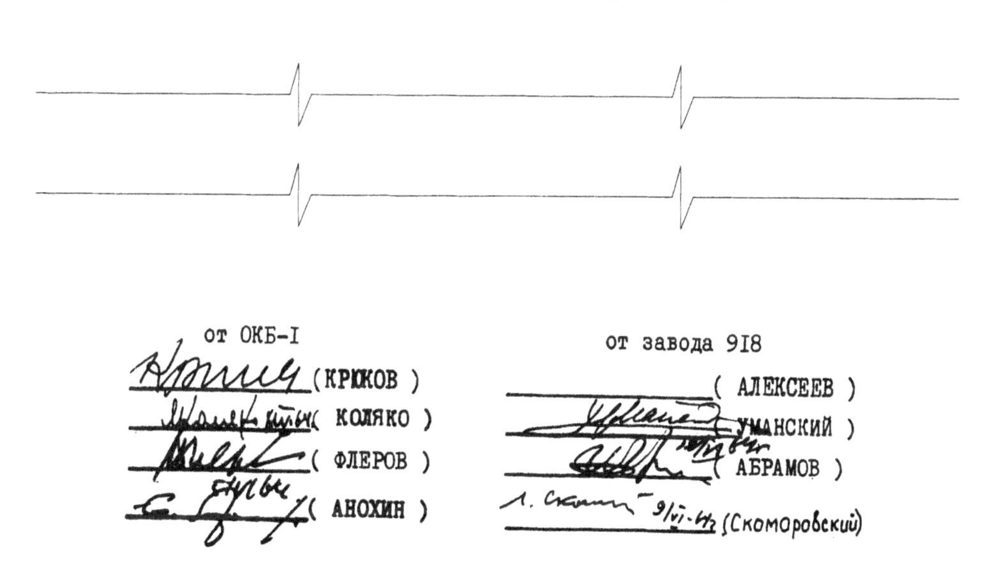

от ОКБ-I
( КРЮКОВ )
( КОЛЯКО )
( ФЛЕРОВ )
( АНОХИН )

от завода 918
( АЛЕКСЕЕВ )
( УМАНСКИЙ )
( АБРАМОВ )
9/VI-64 (Скоморовский)

**Figure 4.2.1** The title page of the performance specifications for airlock development approved by S.P. Korolev and S.M. Alekseyev, deputy of G.I. Severin. *Bottom of the page*: signatories to this document.

From Archive Zvezda.

Concept approval was followed by intensive studies of the space full pressure suit, the airlock and the airlock system. Results of the conceptual studies became the basis for the performance specification signed on 9 June 1964 (Figure 4.2.1) and for the corresponding government directive dated 8 July 1964 on the development of an airlock and a full pressure suit with a backpack LSS. The resolution specified Zvezda as the prime contractor of this equipment. Suppliers were instructed by the same directive to furnish Zvezda with the needed items within the shortest time.

The airlock Volga of the *Voskhod-2* spacecraft consisted of an upper rigid portion with an EVA hatch and a lower assembly ring fixed to the spacecraft flange. The upper and lower portions were connected to a pressure enclosure and a reinforcing framework, incorporating a system of longitudinal air beams in the form of inflatable rubber cylinders with a cover made of a high-strength fabric. The stowed airlock was secured on the outer surface of the descent module above the module exit hatch. The airlock housed systems that included deployment of the enclosure by means of inflating the air beams, an airlock pressure control system operating in the airlock process, a control panel, cosmonaut safety and fixation means for EVA, a system for separation of the airlock from the spacecraft on completion of the programme and some other components.

The arrangement of the airlock's main elements is shown in Figure 4.2.2 and a general view of the "soft" airlock is given in Figure 4.2.3. Between 1964 and 1965 seven sets of airlocks were manufactured. Two of them were used during the unmanned and manned missions of the *Voskhod-2* spacecraft. The remaining five sets were used in the airlock test programme at Zvezda and as back-up hardware. Currently, three of them are in the museums of Zvezda, RSC Energia and in the Memorial Museum of Cosmonautics in Moscow. Of the other two airlocks, one is in the private museum of the Tess Fund (Denver, USA) and the other is in another private collection outside the Russian Federation (the actual location is not known for sure).

With the aim of selecting the design and operating concepts of the full pressure suit and the LSS, Zvezda specialists analysed the experience already gained in development of the existing and the most promising concepts for full pressure suits and LSSs.

Because of the limited volume in the spacecraft, the decision was made to use the BERKUT spacesuit both in the rescue mode (in case of cabin decompression or LSS failure) and the EVA mode.[1] Assured reliability of hardware, cosmonaut safety and the creation of effective operating conditions for the suited cosmonauts were of primary importance.

In particular, to bring the selected concepts to fruition, the *Voskhod-2* full pressure suit used for the first time a back-up pressure enclosure (operating in case the main pressure enclosure was damaged) and a relatively high operating pressure (about 400 hPa) to prevent decompression sickness. It should be noted that at first the performance specification stipulated a spacesuit operating pressure of 270 hPa and a capability to switch over to 350–400 hPa (only in case of decompression sickness). But on the basis of man evaluated tests carried out at Zvezda, the decision was made to use 400 hPa as the main operating pressure mode and to have

1 In later projects for orbiting stations, two different full pressure suits were used for these purposes.

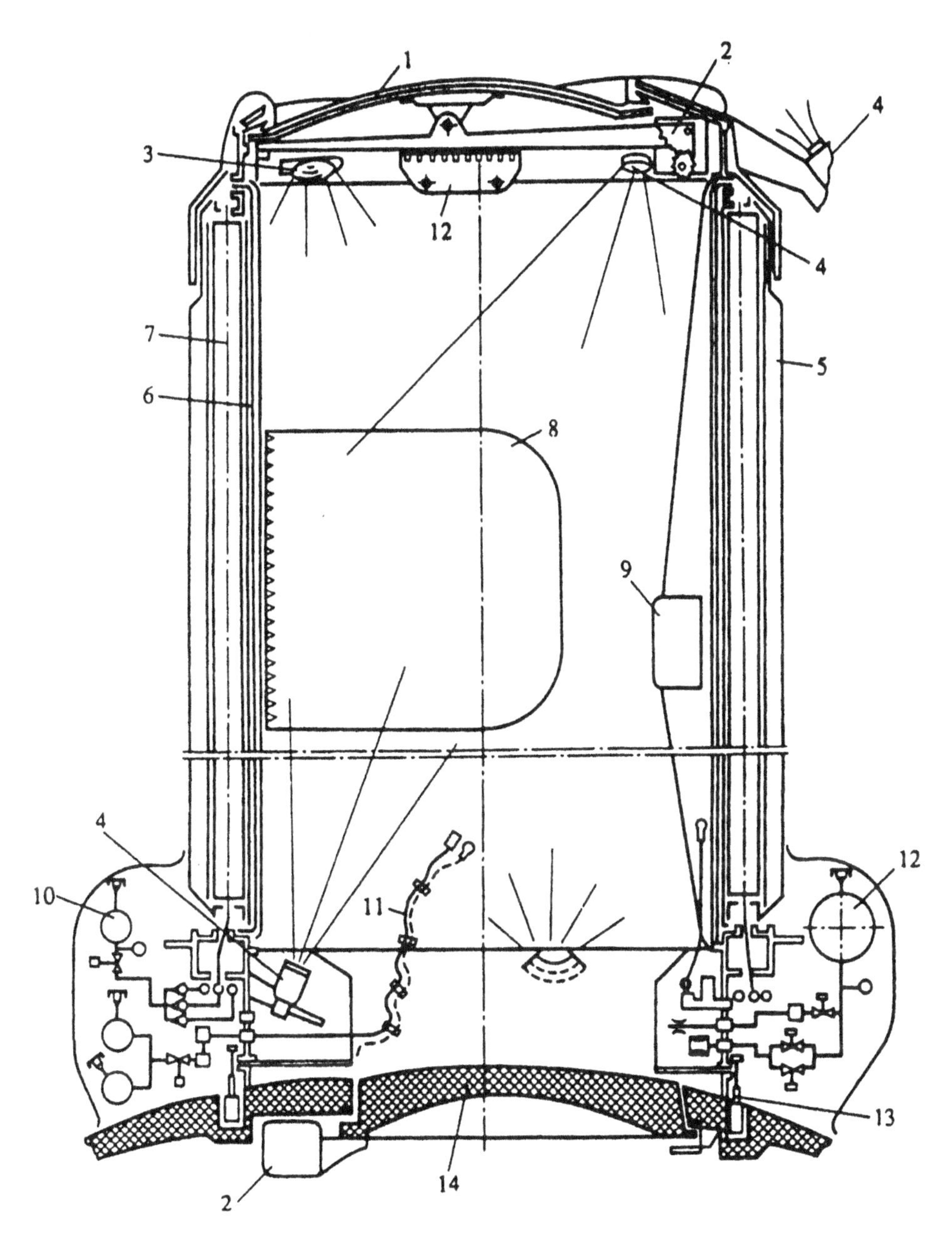


**Figure 4.2.2** Design solution for the *Voskhod-2* spacecraft airlock. 1—EVA hatch cover; 2—drives for hatch opening; 3—light; 4—cameras; 5—soft enclosure; 6—pressure bladder; 7—air beams; 8—elements to attach the equipment inside the airlock; 9—control panel; 10—air beam gas inflation system; 11—safety tether with oxygen supply hose; 12—airlock air inflation system; 13—mechanism to blast off the airlock after EVA; 14—hatch of the descent vehicle for *Voskhod-2*.

From Archive Zvezda.

**Figure 4.2.3** Airlock (without a thermal protection enclosure).
From Archive Zvezda.

the capability to switch over to the decreased pressure mode of 270 hPa for a short period of time in order to increase suit flexibility.

Recalling that the task was to develop EVA support means for the *Voskhod-2* mission within the shortest possible time and proceeding from the rather short operation of the suit in a self-contained mode, the decision was made to select a full pressure suit of the ventilation type with separate ventilation of the helmet and the enclosure.

Such an approach made it possible to arrange the on-board LSS in a similar way to that of the *Vostok* spacecraft (the only changes involved making the system applicable to a two-seater spacecraft), to utilize a number of the earlier developed units and to use the most simple and reliable self-contained backpack system operating in the open loop. The main portion of the on-board LSS components was located in the descent vehicle and arranged in two modules, one each to the left and the right of the seats (Figure 4.2.4). In contrast to the on-board system of the *Vostok* programme suit, the *Voskhod-2* on-board system contained gas for 3-hours operation in an emergency situation.

The basic part of the backpack system (the code name was KP-55) was developed and manufactured by SKB-KDA (special design bureau in the city of Orekhovo-Zuyevo). The EVA crew member donned the backpack (Figures 4.2.5 and 4.2.6) in the descent module prior to the EVA. The backpack was secured to

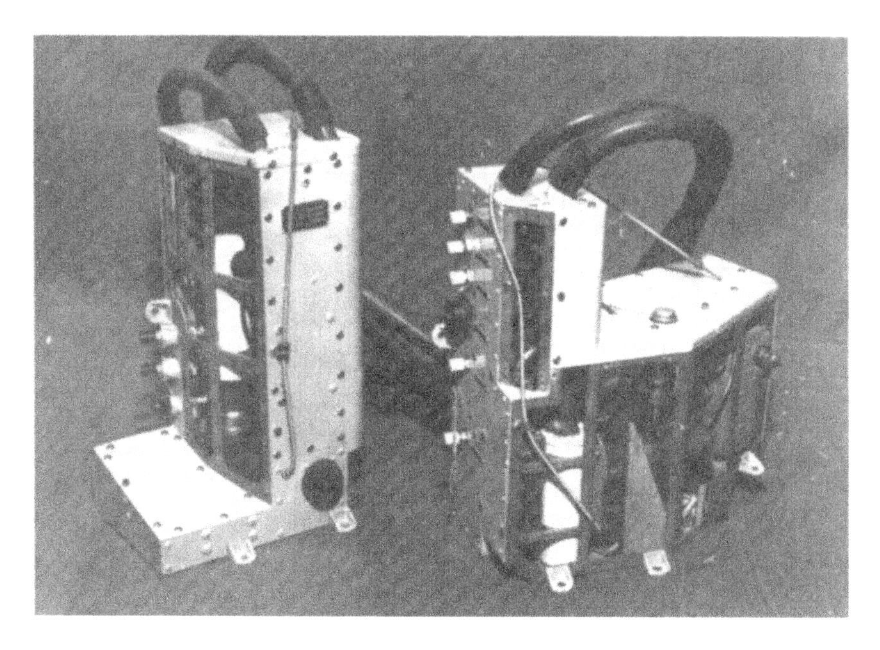

**Figure 4.2.4** Units with the on-board LSS assemblies for the BERKUT spacesuit.
From Archive Zvezda.

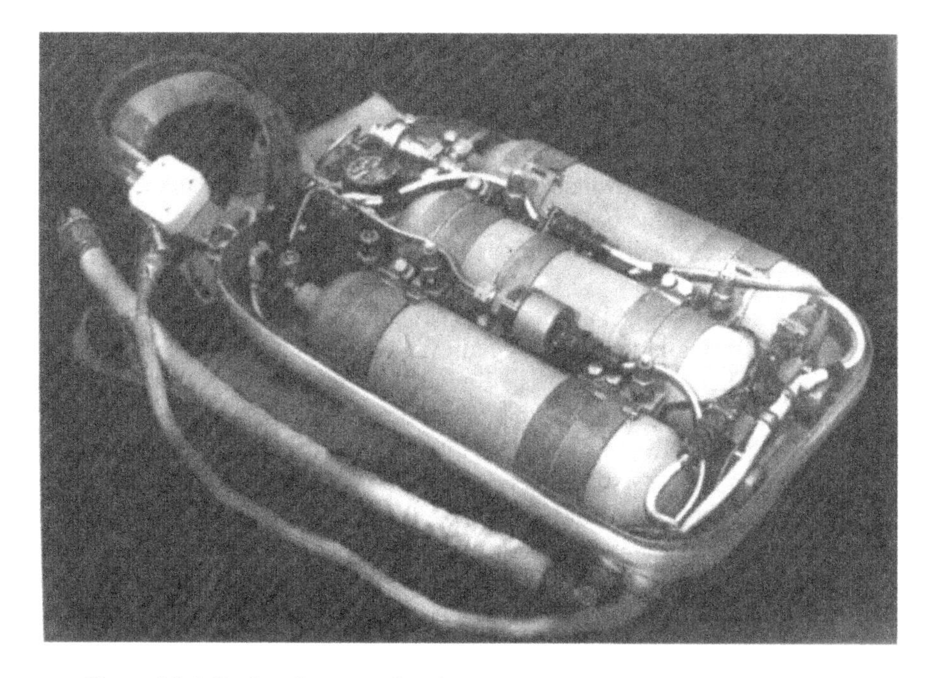

**Figure 4.2.5** Backpack system for the BERKUT spacesuit (cover removed).
From Archive Zvezda.

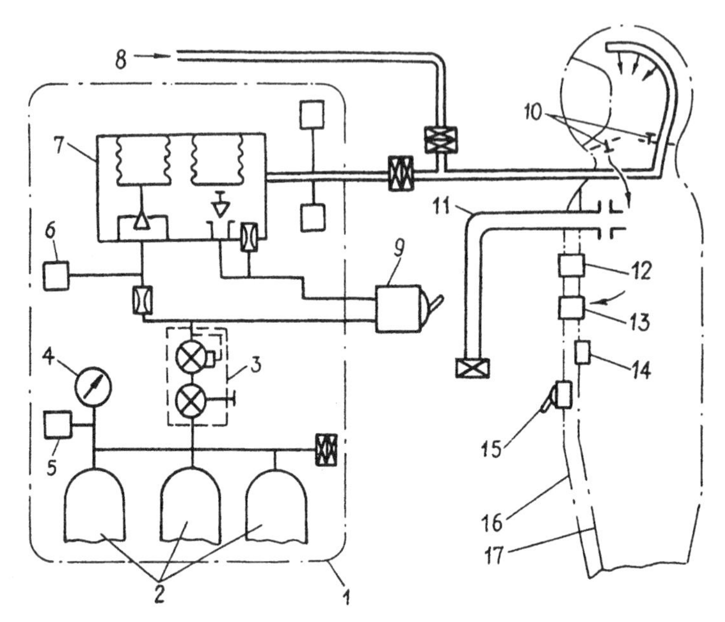


**Figure 4.2.6** Diagram of the BERKUT spacesuit's LSS. 1—Backpack with LSS components; 2—oxygen bottles; 3—reducer with shut-off valve; 4—pressure gauge; 5—pressure transducers; 6—emergency oxygen supply annunciator; 7—oxygen supply regulator; 8—emergency oxygen supply hose (from airlock system); 9—oxygen supply handle; 10—exhale/intake valve on the SS neck seal; 11—on-board oxygen supply hose; 12—relief valve; 13—suit primary pressure regulator; 14—suit low-pressure regulator; 15—suit low-pressure mode shut-off valve; 16—suit primary bladder; 17—suit redundant bladder.

From Archive Zvezda.

the suit by a suspension system. Oxygen was stored in three 2-litre bottles at 22 MPa pressure. The crew member switched on the oxygen supply by means of a remote control. All oxygen from the backpack was supplied to the helmet, then flowed into the suit enclosure and finally was dumped through a pressure control valve into space.

The oxygen flow from the backpack was designed to provide suit pressurization, oxygen for normal breathing and $CO_2$ removal for 45 minutes. As is known, the actual duration of Leonov's EVA was 12 minutes. The cosmonaut remained in the suit under vacuum conditions for about 23 minutes.

There were three modes of oxygen supply from the backpack: a nominal mode with an oxygen flow rate of 16–20 litres per minute (normalized), a mode during the airlock period with an oxygen flow rate of 25–30 litres per minute (with an ambient pressure of about 550 hPa) and an emergency supply mode with an oxygen flow rate up to 30 litres per minute. The emergency oxygen supply was switched on automatically when absolute pressure in the suit dropped below 240 hPa. The operation scenario also provided for a back-up oxygen supply to the suit via a hose from oxygen bottles located in the airlock.

The concept of the selected system and oxygen flow rates were verified in the test

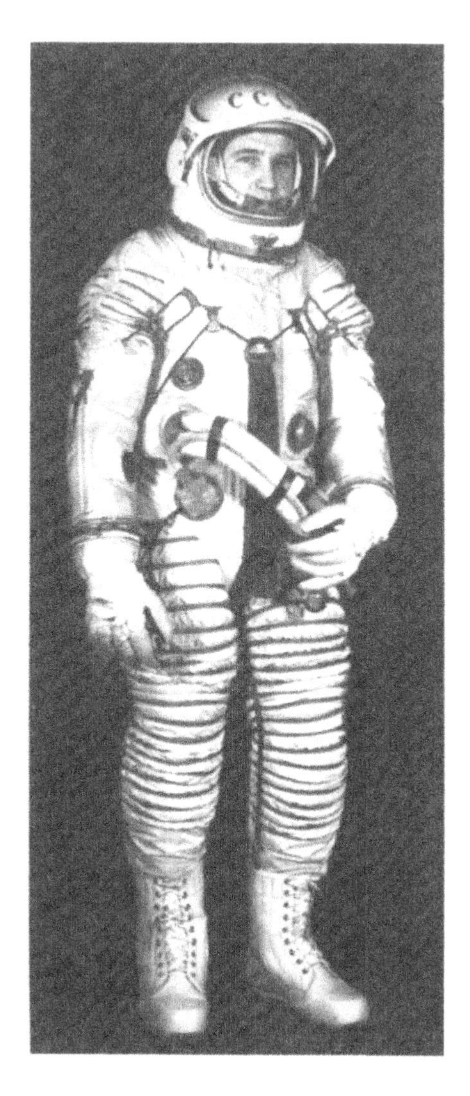

**Figure 4.2.7** BERKUT spacesuit (without outer thermal protection), worn by Zvezda test engineer Victor Yefimov.
From Archive Zvezda.

programme, which demonstrated that even with zero heat removal through the suit enclosure the cosmonaut would not lose the capability to work (i.e., body accumulated heat was acceptable for the given EVA duration).

The BERKUT spacesuit enclosure worn by A.A. Leonov used the design of earlier developed air and spacecrew protective outfits and, in particular, the SK-1 full pressure suit (the body enclosure with the restraint system, arms, gloves) and the S-10 experimental full pressure suit (the leg enclosure, internal ventilation system) (Figures 4.2.7 and 4.2.8).

**Figure 4.2.8** General view of the BERKUT spacesuit with the backpack in place (worn by Zvezda test engineer Victor Yefimov).
From Archive Zvezda.

The BERKUT suit enclosure consisted of four layers: a restraint layer made of high-strength nylon-type fabric, two pressure bladders (primary and back-up) made of sheet rubber and an inner nylon liner with an internal ventilation system. In contrast with the earlier used full pressure suits, the BERKUT suit was fitted with special outer clothing made up of multilayer screen vacuum thermal insulation.

The suit helmet was developed on the basis of the GSh-8 aircrew pressure helmet. It was an easy-disconnect non-turning helmet, which incorporated a metal casque and a movable visor. There was a manually controlled (by a special handle) light filter inside the helmet.

The crew member involved in the EVA was secured by a special 7-m umbilical that incorporated a shock absorbing device, a steel rope, a hose for emergency oxygen supply and electric cables to transmit medical and technical data to the spacecraft and support voice communication with the spacecraft commander.

The spacesuit of P.I. Belyayev, the *Voskhod-2* commander, was of the same design as that of A.A. Leonov. If the necessity arose, the commander could depressurize the spacecraft cabin, open the hatch and enter the airlock to help A.A. Leonov. In this case the vital functions of the commander would be supported by oxygen supply from the on-board LSS via an emergency hose.

The first space walk was preceded by a research-and-test programme, which included a number of novel computations, research work and tests, as well as a training programme for the crew. In particular, a procedure for thermal computational studies of the "man–suit" system was developed, the effects of high vacuum and other space environment factors on spacesuit materials were studied, procedures for the simulation of space conditions in ground facilities were developed and a significant number of tests in vacuum chambers were carried out.

As already indicated, the reliability of the equipment and the safety of the crew members were the important considerations that dictated design and development work. For instance, a great deal of work was done to ensure the proper deployment of the airlock, which was a difficult thing to assess under sea level conditions. To provide a uniform deployment of the airlock, the airlock enclosure structure was fitted with special strands that ruptured in the air beam inflation process. The air beams were subdivided into three sections. Their arrangement and operating pressure were selected in such a way as to ensure normal deployment and stability of the airlock with one of the sections failed. With changes in ambient temperature from −30°C to +40°C taken into account, the air beam operating pressure was specified to be at least 400 hPa (condition for airlock stability), but not higher than 800 hPa (dictated by strength).

The airlock system also provided a means of monitoring many parameters and manual back-up of the most critical operations, such as opening/closure of the airlock exit hatch, operation with pressurization valves and pressure dump valves. Under consideration were certain emergency situations: the descent vehicle becoming depressurized in any mission phase, the crew member having to enter the descent vehicle with the spacesuit under positive pressure (e.g., when the airlock exit hatch fails to close) or the crew member undertaking the space walk needing assistance from the crew member in the descent vehicle. Probable enrichment of the airlock atmosphere with oxygen exiting from the spacesuit was also assessed. These situations were important considerations in the selection of concepts, schematics and design solutions for the airlock, spacesuit and their systems. They were included in the equipment test programme and in the cosmonaut training programme.

A wooden, airlock breadboard model was built in July–August 1964 in order to make preliminary verifications of airlock dimensions, arrangement of the equipment in the descent vehicle and crew member passage through the hatches. These verifications were carried out as ground tests together with a *Voskhod* capsule. Initial tests were carried out in early August by Test Pilot S.N. Anokhin of OKB-1, wearing a BERKUT suit mock-up.

**Figure 4.2.9** Airlock in a stowed position during centrifuge-testing at Zvezda.
From Archive Zvezda.

The official Breadboard Review was held at OKB-1 as early as August 1964. It included fit check work in the descent vehicle and airlock mock-up of the following crew candidates suited in the BERKUT: A.A. Leonov, P.I. Belyayev, Ye.V. Khrunov and V.V. Gorbatko.

On 24–25 September 1964 a functional airlock mock-up and a *Voskhod-2* capsule were used in Baikonur for a demonstration of the system to Nikita Khrushchev during his visit to the cosmodrome. Guy Severin demonstrated the airlock and the BERKUT spacesuit and S.P. Korolev the *Voskhod-2* in the presence of the cosmonauts.

When the whole development and test programme for all systems had been completed, Zvezda issued a corresponding "statement" (flight certification) at the end of 1964. The programme included both technical (Figure 4.2.9) and man-evaluated tests and covered sea level tests, tests in thermal vacuum chambers, water flotation tests and strength and life tests. Special attention in the test programme was given to the operation of equipment under conditions simulating those in space, taken to the maximum degree possible. Thus, besides conventional tests, the programme provided opportunities to verify:

- tests of separate assemblies and units as well as non-metallic materials used in the pressure suit and airlock after exposure to high vacuum (down to $1 \times 10^{-8}$ torr) to verify material compatibility with a vacuum and friction levels in moving parts exposed to a vacuum;
- special cold tests of the pressure suit and those components that have multilayer screen vacuum thermal insulation;
- tests of the suit and systems in a new vacuum chamber (TBK-30) at Zvezda that

provided pressure conditions down to $2 \times 10^{-2}$ torr; they included the first man-evaluated vacuum chamber tests.

The man-evaluated tests with the test subjects wearing BERKUT spacesuits had been run in the descent vehicle thermal mock-up at OKB-124, and the comprehensive inter-agency tests of all the systems had been run in the TBK-60 thermal vacuum chamber at the Air Force Scientific Research Institute (GK NII) by February 1965.

A large group of Zvezda specialists led by G.I. Severin participated in the development of the suit and airlock: S.M. Alekseyev, V.G. Galperin, I.P. Abramov, O.I. Smotrikov, M.N. Doodnik, A.Yu. Stoklitsky, V.V. Ushinin, I.I. Derevyanko, D.V. Kuchevitsky, G.S. Paradizov, I.I. Skomorovsky, I.I. Chistyakov, as well as V.Ya. Tereshchenko (from the Design Bureau in Orekhovo-Zuyevo).

A significant contribution to the implementation of the airlock and suit test programme was made by B.V. Mikhailov, V.I. Svertshek, D.I. Sendick and others. Thermal analysis of the airlock and suit was made by A.N. Livshits, B.M. Bliyev and G.T. Sharapov. N.P. Strekozov and A.A. Klintsov carried out the strength analysis.

Responsible for the selection and tests of non-metallic materials were Z.B. Tsentsiper, D.S. Stoklitskaya, D.S. Abramova, V.I. Streltsova and others. It is also important to mention the Department of Aerospace Medicine headed by A.S. Barer. Of special importance was the work of department specialists on selection of the suit operating pressure with the aim of preventing cosmonaut decompression sickness.

Figure 4.2.10 shows a simplified organizational diagram of the suit and airlock development work. It is worth mentioning that, due to the lack of time for the development of hardware, a special structure for operating the management of work was set up that differed from the standard development management structure (Figure 14.2.1).

OKB-1 specialists carried out the modifications to the *Voskhod-2* spacecraft that were dictated by new tasks and developed the upper and lower assembly rings and the sortie hatch. Deputy Chief Designer P.V. Tsybin headed the OKB-1 activities on the *Voskhod-2* programme. S.P. Korolev continuously monitored the whole process of work on the programme. Execution of the scheduled work was regularly checked by the Ministry of the Aviation Industry.

Flight tests, which involved a short-term simulation of weightlessness conditions aboard the Tu-104 flying laboratory (Figure 4.2.11), and cosmonaut training were performed at the Gromov Flight Research Institute (LII) under the leadership of Leading Engineer E.T. Berezkin.

The whole mission scenario and step-by-step procedures for the airlock and EVA period were developed (Figure 4.2.12). Given below are the main phases of the mission and the operational procedures.

**1** The initial position (launch and orbit with the descent vehicle pressurized):

- two suited crew members are in the flight seats, suit helmets are closed or open, nominal ventilation of the pressure suit from the on-board system is on (similar to the operating mode used aboard the *Vostok* spacecraft);

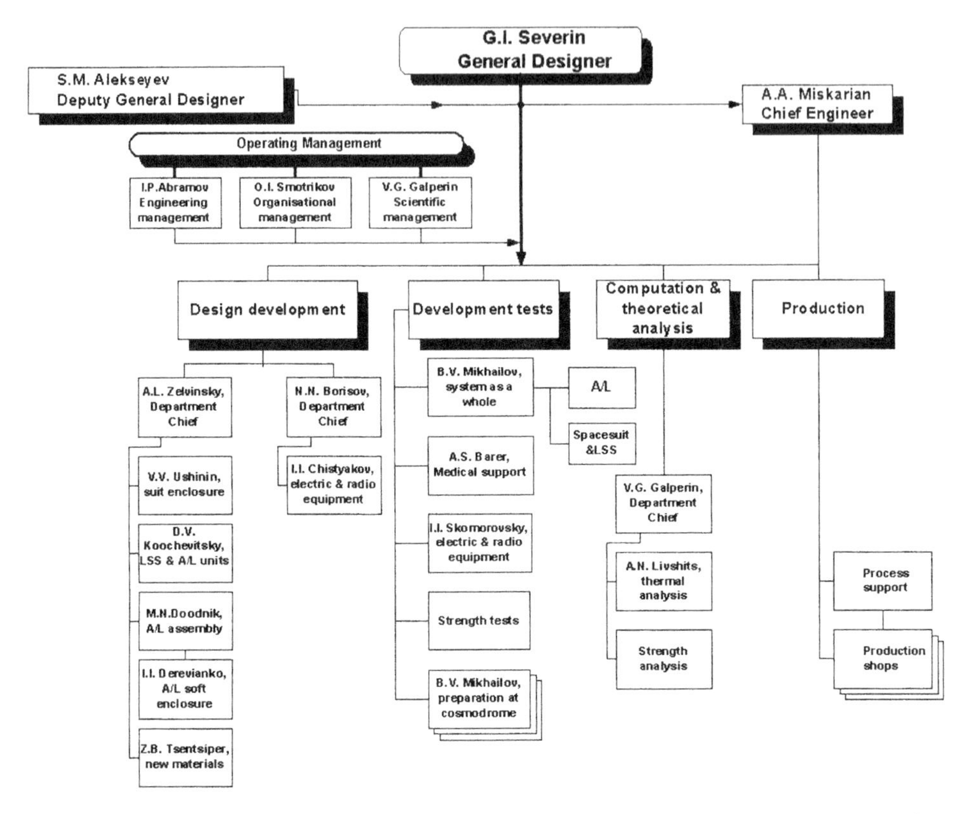


**Figure 4.2.10** Simplified organizational diagram of the *Voskhod-2* suit and airlock design, development and production at Zvezda.
From Archive Zvezda.

**Figure 4.2.11** Training in movement with the BERKUT spacesuit donned, under the weightlessness conditions in the Tu-104 flying laboratory.
From Archive Zvezda.

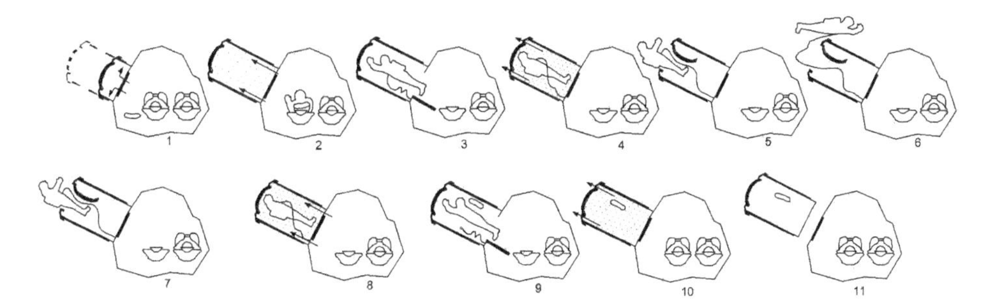


**Figure 4.2.12** Operation sequence for the *Voskhod-2* spacecraft airlock.
From Archive Zvezda.

- the airlock is in the stowed position (the airlock's internal cavity is connected via a special valve to open space to exclude positive pressure inside the stowed airlock).

Prepare for EVA:

- verify the initial state and switch on the power supply to the airlock systems;
- deploy the airlock and inflate the air beams (by sending command signals to pyrolocks, which keep the airlock in the stowed position, and activation of the air supply to the air beams).

**2** Pressurize the airlock and don the backpack:

- pressurize the airlock by opening the valve between the internal cavities of the descent vehicle and the airlock (to equalize pressure in the cavities at the level of 700 hPa), and, if the amount of air is insufficient, activate the airlock pressurization system and increase pressure to the 700 hPa level. Such a pressure level offers an adequate airlock enclosure load mode (strengthwise) and does not require oxygen supply for crew members;
- don the backpack.

**3** Open the descent vehicle/airlock hatch, connect the airlock oxygen system hose to the pressure suit of the EVA crew member and actuate the airlock oxygen system:

- disconnect the suit from the descent vehicle's on-board system:
- check the suit for leakage with the suit fed from the AL oxygen hose.

**4** Transfer to the airlock and close the descent vehicle/airlock hatch:

- depressurize the airlock;
- open the dump valve;
- switch over to the oxygen supply from the backpack.

With the airlock depressurized, the necessary positive pressure is automatically maintained in the suit.

**5, 6** Open the airlock sortie hatch and carry out the space walk:

- supply the suit with oxygen from the backpack;
- back the oxygen supply up via a hose from the airlock oxygen bottles (approximately designed for 80 minutes of operation).

**7, 8** Enter the airlock, close the sortie hatch and pressurize the airlock:

- the time required to pressurize the airlock, by air from the airlock bottles, is between 7 and 9 minutes (a safe pressure level that is equal to the suit pressure of 400 hPa is reached after 1.5–2 minutes);
- the airlock is equipped with a relief valve that blows off (under pressure) between 800 and 850 hPa and, thus, counteracts the situation when the pressurization valve fails to close;
- equalize the pressure in the descent vehicle and the airlock.

**9** Open the descent vehicle/airlock hatch and enter the descent vehicle:

- take the backpack off (it is left in the airlock);
- open the descent vehicle/airlock hatch;
- connect the suit to the on-board LSS.

**10** Close the descent vehicle/airlock hatch (after the airlock hose has been disconnected), and depressurize the airlock.

**11** Separate the airlock from the descent module, and jettison the airlock.

**12** Increase the pressure in the descent module, and activate the air supply from air bottles of the suit's self-contained on-board LSS, located on the instrument module. Air passes through the suit and enters the cabin, increasing the pressure to the level equal to that on Earth.

After that the air supply is switched off, and the suits start to operate in normal mode in the pressurized cabin.

The total duration of the above operations, from the moment gas leaves the descent module for the airlock up to the moment the descent vehicle's hatch is closed, equals approximately 1 hour and 40 minutes.

In the descent phase, after separation of the instrument module, the suits are fed with oxygen (in case of descent vehicle decompression from a special bottle located in the descent module). Oxygen supply from the bottle is switched on automatically. To make the landing safe, oxygen stored in the descent vehicle's bottle is released into the ambient atmosphere at an altitude of about 1.5 km.

Preflight preparation of Zvezda-developed systems at the Baikonur cosmodrome (Figure 4.2.13) was carried out by a large team of specialists led by B.V. Mikhailov.

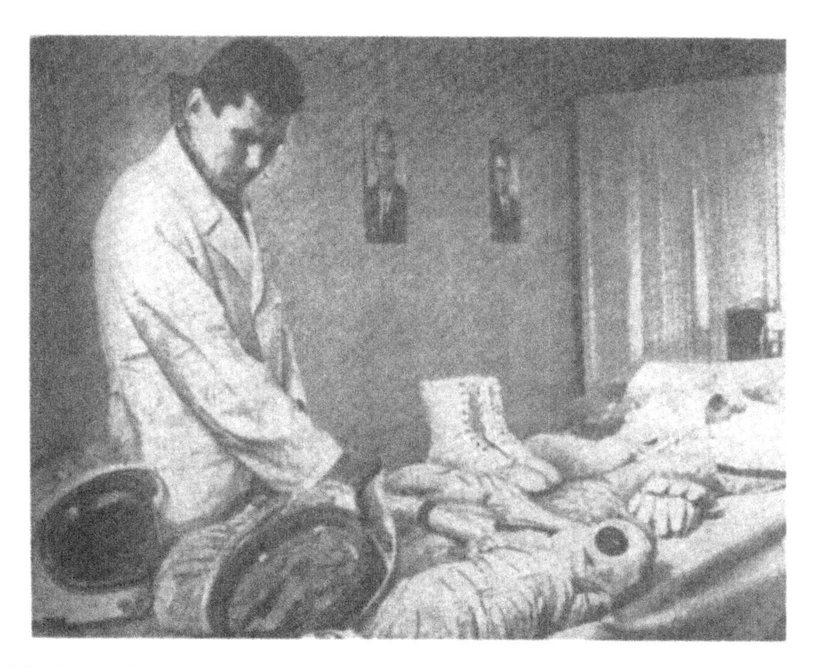

**Figure 4.2.13** G.S. Petrushin, one of the team members, prepares the BERKUT spacesuit for the *Voskhod-2* mission in Baikonur.

## 4.3 THE *VOSKHOD-2* FLIGHT

Some significant contingency situations, directly related to the Zvezda-furnished equipment, occurred in the preparatory period of and during the *Voskhod-2* mission. During ground-testing at Energia in January 1965 an airlock was accidentally dropped. Checks showed that the airlock was not damaged and fully functional. This airlock was then used in the test flight on board *Cosmos 57*. However, this incident had no influence on the future schedule of airlock-testing.

The *Voskhod-2* mission was preceded by the flight of an unmanned spacecraft (*Cosmos 57* on 22 February 1965), which carried a soft airlock and a BERKUT suit simulator. One of the tasks of the unmanned flight was to completely simulate the operation of the airlock and the airlock system and pressurization of the suit by command signals from the Earth. Several days before the launch, while verifying the airlock flight model at the launch centre, test specialists revealed that the airlock's sortie hatch, under conditions of no pressure differential, might not be closed tightly, indicated by the open contacts of the device used to monitor airlock closure. As a result, an error could occur in the airlock operation control programme (indicating the hatch failed to open). Zvezda specialists reported the event to S.P. Korolev, who straight away held a meeting of the involved specialists. To be on the safe side, the proposal was to give an additional close-the-hatch command from one of the command and measurement stations (far east of Russia) and duplicate it from the next neighbouring station. The decision was made on 19 February 1965

**Figure 4.3.1** Yu.A. Gagarin sees P.I. Belyayev and A.A. Leonov off on the bus to the launch pad.

despite objections of some representatives of mission control, who were wary of introducing any changes in the programme just a few days before the launch.

During the test flight of *Cosmos 57*, midway through the airlock operation programme, communications with the spacecraft were lost. Subsequent analysis of the telemetry data revealed that, accompanying the same simultaneous close-the-airlock-hatch command signals given to the spacecraft from the two ground control stations, another command signal—emergency destruction—had been initiated. Despite the fact that the flight programme was not complete, the bulk of operations with the airlock and the suit were found to perform normally, confirming the proper operation of the equipment. A decision was also made to check the airlock jettison procedure, since this operational procedure was not checked during the *Cosmos 57* test flight. This was done on an airlock mock-up installed on another *Cosmos* vehicle already prepared for flight (*Cosmos 59* on 7 March 1965). On completion of the above verifications, the *Voskhod-2* mission was authorized.

On the morning of 18 March 1965, P.I. Belyayev and A.A. Leonov were assisted in suit-donning in the Zvezda laboratory at Baikonur (Figure 4.3.1) and driven to the launch pad for boarding the spacecraft.

The first ever space walk was planned to take place on the very first pass over Soviet territory after entering orbit on 18 March 1965.

For S.P. Korolev this event was of the utmost importance, and, in order to promptly solve problems (if any), he requested that G.I. Severin together with I.P. Abramov were in the launch pad bunker, in the room next to the spacecraft launch

**Figure 4.3.2** A.A. Leonov during the first space walk ever, carried out on 18 March 1965.
Painting from Archive Zvezda.

control station. G.A. Tyulin, the Chairman of the State Commission, was there at the same time.

It should be noted that the Mission Control Centre (as we understand it now) did not exist at that time, and all important information coming from satellite tracking posts was immediately transferred to the mission managers in the form of reports.

The space walk was carried out in full compliance with the mission programme (Figure 4.3.2). It was noted from later telemetry analysis and the spacecrew report that there was a considerable increase in Leonov's heart rate in connection with the difficulties he had on his return to the airlock. Some journalists describing this situation mistakenly wrote about the suit ballooning. A spacesuit pressurized at 400 hPa has dimensions that remain the same both in vacuum and under normal Earth conditions. However, greater stiffness in the suit in combination with zero gravity caused difficulties in bending during the space walk and airlock re-entry. In order to facilitate his entry into the airlock, A.A. Leonov made the right decision in this situation: he decreased the pressure in the suit down to 270 hPa, which somewhat

**Figure 4.3.3** Meeting with A.A. Leonov and P.I. Belyayev at Zvezda. The first row of the platform (*from the left*): A.A. Miskarian, G.I. Severin, S.M. Alekseyev, A.A. Leonov, V.I. Kharchenko, P.I. Belyaev, N.P. Kamanin, M.L. Galay.

From Archive Zvezda.

reduced the force required for suit enclosure bending. As a whole, the difficulties encountered by A.A. Leonov may be explained by the fact that the procedures for returning into the airlock were not sufficiently developed under ground conditions, because the period of weightlessness lasted only a few tens of seconds during training in the aircraft and hydro lab training using the suit had not been introduced at that time.

Furthermore, as A.A. Leonov said many times after the mission, he tried to enter the airlock head first rather than legs first (contrary to his training at the ground facilities). As a result, he had to perform the difficult task of turning over inside the airlock to enter the descent vehicle.

During the space walk A.A. Leonov found himself unable to reach the starting cord of the photo camera arranged on the spacesuit. It is clearly seen on the shots made by the movie camera installed on the airlock (Figure 4.2.2). A.A. Leonov dismantled the movie camera and returned it to Earth. This inability may be explained by the combination of the previously unknown conditions of free space and weightlessness, which could not be tested on the ground. Leonov also reported that it was difficult for him to enter the hatch holding the camera with one hand.

Another contingency situation occurred on completion of the space walk. One of the crew members, while moving inside the descent module, accidentally switched on the air supply to the pressure suit from a self-contained gas supply source, which resulted in a considerable increase in the cabin pressure. People on the ground were in a state of panic until they could determine the cause.

And a further contingency situation: due to a malfunction in the orientation system, the crew members had to land the descent module manually and, as is well known, got stranded in the snow-covered taiga. The crew actually used the suits and the survival kit to keep themselves alive for 2 days, after landing in the wild.

The world's first space walk mission was successfully carried out from the *Voskhod-2* spacecraft. It was an outstanding achievement, which gave a new impetus to further studies aimed at the development of space walk technology. In particular, precious data on the locomotor activity of humans in zero gravity were obtained. These data were taken into account during crew-training for the next missions.

It is appropriate to mention the energy and enthusiasm with which the whole Zvezda staff worked on the design and experimental development of the systems designed to support the first ever space walk. Only 9 months passed between the day the performance specification for the airlock and the suit was signed on 9 June 1964 and the day A.A. Leonov space-walked on 18 March 1965. This fact was duly recognized at the ceremonial welcome to cosmonauts at Zvezda soon after the mission (Figure 4.3.3).

# 5

# EVA suit for the *Soyuz-4* and *Soyuz-5* missions

## 5.1 INTRODUCTION

In line with the government resolution of 16 April 1962 and OKB-1 (Experimental Design Bureau 1) performance specifications issued as early as November 1961, Zvezda had already in 1962 got down to research work on a rescue suit, a waste removal system and a lightweight ejection system for the 7K vehicle designed for a circumlunar flight (the *Soyuz* project). Zvezda ran mock-up studies and developed the conceptual design.

In September 1962 the space vehicle's tasks were redefined, and, instead of the circumlunar mission, the 7K-OK vehicle was given the task of developing technologies related to rendezvous and docking in near-Earth orbit. OKB-1 then issued new performance specifications, which fundamentally changed the direction of the work by requiring the development of a special flight garment rather than a rescue suit. The 7K-OK vehicle inherited the *Soyuz* name.

As a result of combined efforts between OKB-1 and the Chelomey Company on the development of the circumlunar flight system (the L1 programme), the 7K-L1 manned vehicle model was outlined in 1965. This vehicle was similar to the 7K-OK vehicle as far as Zvezda was concerned. A corresponding government directive was issued on 27 April 1966.

## 5.2 THE YASTREB SUIT FOR THE *SOYUZ* PROGRAMME

Late in 1964, S.P. Korolev made the decision to carry out in-orbit docking of two *Soyuz* vehicles, and early in 1965 the Scientific and Technical Advisory Council at the Ministry of General Engineering Industry accepted the Korolev proposal and made a decision to use the 7K vehicles for basic orbiting missions. However, according to the OKB-1 design documentation, no rescue suits to protect crew

(*a*) (*b*)

**Figure 5.2.1** General view of the YASTREB spacesuit with: (*a*) the RVR-1 and (*b*) RVR-1P backpacks (worn by Zvezda test engineers V. Byshkov (*a*) and V. Logvinov (*b*)).
From Archive Zvezda.

members in case of cabin decompression were planned in these missions (despite concerns expressed by Zvezda and Air Force experts).

The successful space walk by A.A. Leonov on 18 March 1965 was followed by a new mission objective for the *Soyuz* vehicles: facilitating the transfer of crew members from one vehicle to another through free space. Such an operation was required for one of the circumlunar flight plans (L1) when crew members were to move from the *Soyuz*-type transportation system in near-Earth orbit to the lunar vehicle. To do this, performance specifications were issued by OKB-1 to Zvezda in June 1965, and a formal governmental directive was issued on 18 August 1965.

The pressure suit planned for the *Soyuz* docking project was the YASTREB (Figure 5.2.1a, b). New requirements for the YASTREB pressure suit (in comparison with those for the BERKUT pressure suit used on the *Voskhod-2* spacecraft) revolved around use of the suit solely for space walks with suit-donning aboard the orbiting vehicle and increase in the duration of the suited mode by up to 2 hours.

As was usual then, the suit had to be developed within a short time (it was originally planned to use the suit in 1966). Therefore, development work started with evaluation of a life support system (LSS) of an open loop type, similar to that used for the *Voskhod-2* spacecraft.

It is worth indicating that, simultaneously with the development of the pressure suit for the *Soyuz* project, Zvezda ran experimental work on other pressure suits and systems for projects related to high-altitude flights. They included, in particular, a rescue pressure suit with a regeneration system for a rocket plane (1962) that was under development at OKB-52 (Chief Designer V.N. Chelomey), the SK-III full pressure rescue suit (Figure 5.2.2) for the Yastreb reconnaissance aircraft (1962–1963), and the SKV full pressure suit with a regenerative system and extra-vehicular activity (EVA) capability for a heavy satellite (1961 through 1965) (see Section 6.1).

In 1965 work also started on the development of hardware for the new lunar programme (L-3) with the proposed landing of a manned module on the Moon surface (described in Section 6.2).

By making use of the results of already completed development work on the layout and components for advanced systems, in addition to the open-type flow diagram, three other versions were considered for the *Soyuz* space walk suit: a closed-loop system with a liquid cooled garment and two closed-loop systems cooled by ventilation gas with gas circulation provided by a fan or an injector.

The open-loop system, which had an operating time of about 2 hours, proved to be unacceptable due to its very high mass, since it required that oxygen flow should be considerably increased (in comparison with the LSS of the *Voskhod-2* spacecraft) to provide the necessary thermal conditions for the crew member. From this point of view, the most promising version was the liquid-cooled garment. However, it required a considerable development effort over a longer period of time. Thus, because of the relatively short duration of the suited mode, the decision was made to postpone the application of such a version until the L-3 programme. Selected for further development were the remaining two versions (backpack systems) with heat removed from the human body by closed-loop ventilating gas systems.

The backpack LSS called RVR-1 [РВР-1] that used a fan as the source of gas circulation (Figure 5.2.3a, b) was developed by Zvezda, while the injector version called RIR [РИР] (Figure 5.2.4) was mainly developed by SKB-KDA in Orekhovo-Zuyevo.

Mock-up models of both backpack systems were manufactured and tested. On the basis of the analysis that was carried out, the RVR-1 version was accepted as the base line system since it was more promising in terms of possible further extension of operating time. The RIR version offered a simpler schematic, but required a large amount of oxygen, which could only be supplied to the suit from on-board bottles through a hose. However, oxygen bottles were not provided for on board the *Soyuz* vehicle, and therefore usage of such a schematic could not be implemented.

**Figure 5.2.2** SK-III full pressure rescue suit for the Yastreb reconnaissance aircraft (1962–1963) (worn by Zvezda test engineer V. Starikov).
From Archive Zvezda.

To develop the pressure suit and the backpack with the LSS, a number of new problems had to be solved. Primary among which was reduction of suit and backpack gas loop drag, because, by using the closed-loop LSS, the gas loop almost doubled in length. It was necessary to carry out a trade-off study of the ventilation gas flow direction: to the helmet or to the suit enclosure. Both options had their pros and cons associated with, for example, prevention of the pressure

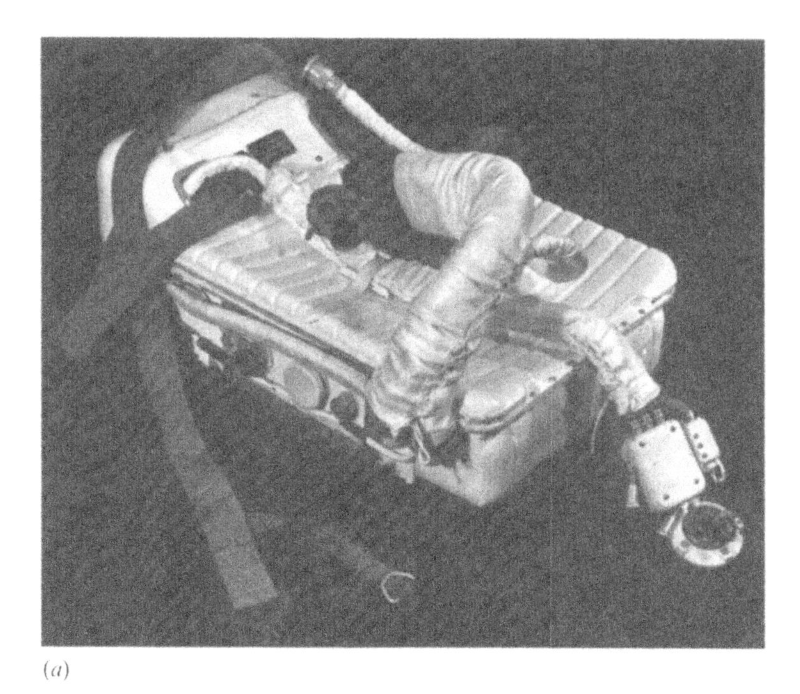

(*a*)

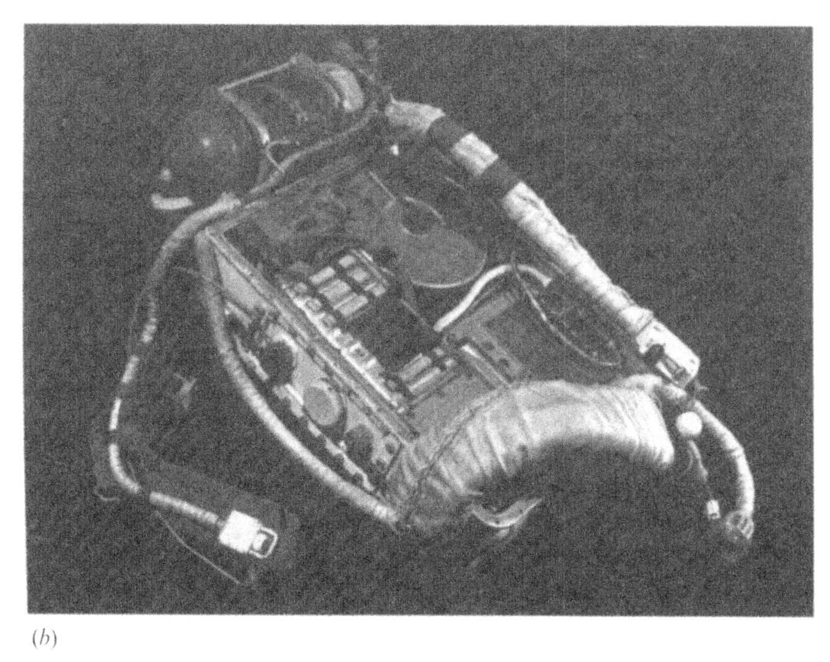

(*b*)

**Figure 5.2.3** RVR-1P LSS of the backpack type: (*a*) general view; (*b*) view of the backpack with the cover open.

From Archive Zvezda.

**Figure 5.2.4** RIR LSS of the backpack type.
From Archive Zvezda.

helmet visor misting up and the removal of human gaseous waste products from the gas ($CO_2$, etc). The change to the closed-loop LSS also called for improvements in suit gas-tightness, provision for suit-purging in order to remove nitrogen and create an oxygen atmosphere in the suit. Certain studies and tests, including man evaluated ones, were required to select the appropriate procedure for changing the gas composition and the adequate oxygen flow rate. Another problem was the first utilization (by the USSR) of bottled oxygen at 420 MPa pressure and corresponding hardware to handle it (bottles, valves, reducers and charging equipment). A new design, manufacturing processes and finally certification had to be implemented.

Problems related to the gas ventilation and regeneration system concerned a newly developed cartridge for removal of $CO_2$ and harmful impurities and the development of a centrifugal fan operating in the suit's oxygen atmosphere and under conditions of high humidity and significant changes in ambient pressure. To make the fan fireproof, Zvezda used a commutatorless motor developed by the All-Union Scientific Research Institute of Electromechanics in accordance with Zvezda-issued performance specifications. Also under study was a special version of a motor with brushes.

Hydrated lithium hydroxide (LiOH) compressed in the form of blocks was selected as the $CO_2$-absorbing substance. In comparison with other known potential candidates, LiOH offered the best performance-to-mass ratio and the

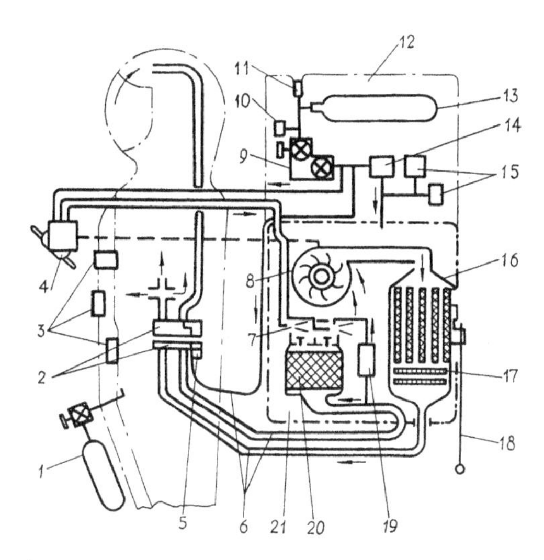


**Figure 5.2.5** LSS flow diagram for the YASTREB spacesuit: 1—emergency oxygen bottle; 2—combined connector; 3—suit pressure regulators; 4—remote control panel; 5—purge device; 6—spacesuit/pack interface tubings; 7—nominal injector and emergency injector; 8—centrifugal fan; 9—reducer with shut-off device; 10, 15—pressure transducers; 11—charging connection; 12—unpressurized compartment; 13—primary oxygen bottle; 14—emergency leakage compensator; 16—flash heat exchanger; 17—moisture collector; 18—heat exchanger valve switch; 19—monitoring unit; 20—$CO_2$ and contamination control cartridge; 21—pressurized compartment.

From Archive Zvezda.

lowest heat generation. There was no point in using oxygen-generating substances since, in any case, it was necessary to continuously supply the suit with oxygen to maintain a positive suit pressure. The $CO_2$-absorbing cartridge was developed in accordance with Zvezda-prepared performance specifications first by NII-404 (Scientific and Research Institute 404) in the city of Electrostal, and later by NIKhI in Tambov.

The selection of a heat control concept for the suited crew member was also a problem of major importance. Potential candidates under investigation within the specified period were all in-house-developed ways and means of heat removal from the human body. In particular, they included heat removal by a suit-mounted heat sink, cool storage, various cold panels that caused the evaporation of moisture in vacuum, an evaporating heat exchanger or sublimator and even application of a semiconductor refrigerating unit. Taking into account that external thermal conditions change depending on whether the crew member is in the shade or on the sunlit side, a decision was made to use vacuum shield insulation (like that used in the BERKUT suit) that minimizes heat flow into and out of the pressure suit under any conditions. In the long run, because of the relatively short EVA period planned and the expected low physical loads on the crew members, the final decision was to use an evaporating heat exchanger located in the backpack for heat removal (Figure 5.2.5).

**Figure 5.2.6** Evaporating heat exchanger during testing.
From Archive Zvezda.

The development of an evaporating heat exchanger capable of operating in zero gravity conditions and having the smallest possible dimensions called for extensive experimental work and thorough analysis. This was especially true for the separation of liquid and vapour in weightlessness conditions and the elimination of conditions that would result in liquid freezing at the evaporator outlet on the vacuum side (Figure 5.2.6). The evaporator concept was proposed and developed by R.Kh. Sharipov and a team of engineers.

To simplify the LSS units of the RVR-1 backpack and their interfaces and to improve their thermal conditions, the proposal was made to arrange the majority of units and components inside the pressurized backpack, whose internal cavity was connected to the suit atmosphere. This design concept is similar to that used for the SKV suit backpack (see Section 6.1). However, such a design concept of the backpack required certain measures to be taken to make electric equipment inside the oxygen-pressurised backpack fireproof. The pressurized section of the backpack accommodated a special measurement package designed to monitor the temperature at several points in the suit and the $CO_2$ content in the circulating gas. The Leningrad SKB-AP (Special Design Bureau for Analytic Instrument Building) in line with Zvezda-prepared performance specifications developed the measurement package. It is important to indicate that the arrangement of LSS components in the backpack's pressurized cavity turned out to be such an excellent design solution that Zvezda used it in subsequent projects.

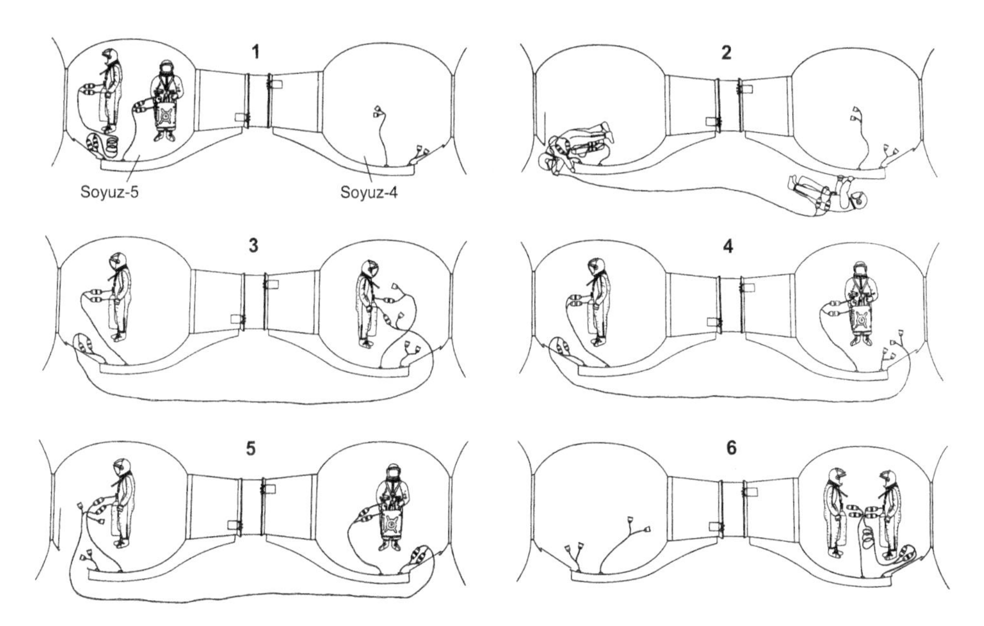


**Figure 5.2.7** Scheme for reconfiguration of the electric umbilicals and crew member transfer from one *Soyuz* spacecraft to another through free space (the sequence of crew actions are shown in 1–6).

From Archive Zvezda.

Electric power supply, radio communication and the monitoring of telemetry parameters were supported by the space vehicle's on-board systems via an electric umbilical. The electric circuit of the suit, backpack and umbilical made it possible to maintain continuous radio communication even with the electrical connector disconnected from one umbilical and connected to the other one instead. Moreover, a special electric connector was developed that enabled the suited crew member to handle it with the suit under positive pressure. As a matter of fact, a most original scheme for reconfiguration of the electrical connectors was developed, which made it possible to have just a single long umbilical for the transfer of two crew members (Figure 5.2.7).

Fire safety and reliability of the system were the main considerations in the development process. Therefore, all vitally important elements of the suit and LSS subsystems were given redundancy (i.e., became back-up systems in case of failures). In particular, the suit used a dual bladder and a dual helmet visor. Ventilation redundancy was provided by the injector. Redundancy was also introduced in the electric circuits. An emergency oxygen system was provided for.

Design work was supported by means of a comprehensive test programme, which involved more than 10 test models. Due to the novelty of the base line system, a back-up open-loop version of the LSS that could only be used with on-board oxygen supplied to the suit via a hose was also under development. The reconfiguration scheme, when the same hose was disconnected from one crew

**Figure 5.2.8** YASTREB spacesuit-donning (the test engineer is A.Ts. Elbakyan of Zvezda). From Archive Zvezda.

member and connected to the other, was similar to that used for connection of the electric umbilical.

The YASTREB suit enclosure (Figures 5.2.8 and 5.2.9) was designed along the lines of the BERKUT suit, but it featured specific differences resulting from the experience gained during Leonov's space walk, the application of a closed-loop LSS and changes in the way in which the suit was used. The YASTREB suit was designed as a pure EVA suit to be donned by crew members under zero gravity conditions in the spacecraft's living compartment, which was very small. Therefore, the suit enclosure was tailored for the "standing" position, the ventilation garment (Figure 5.2.10) was fixed to the suit enclosure, arm pressure cuffs as well as gloves were made removable and soft shoes were used instead of leather boots. Utilization

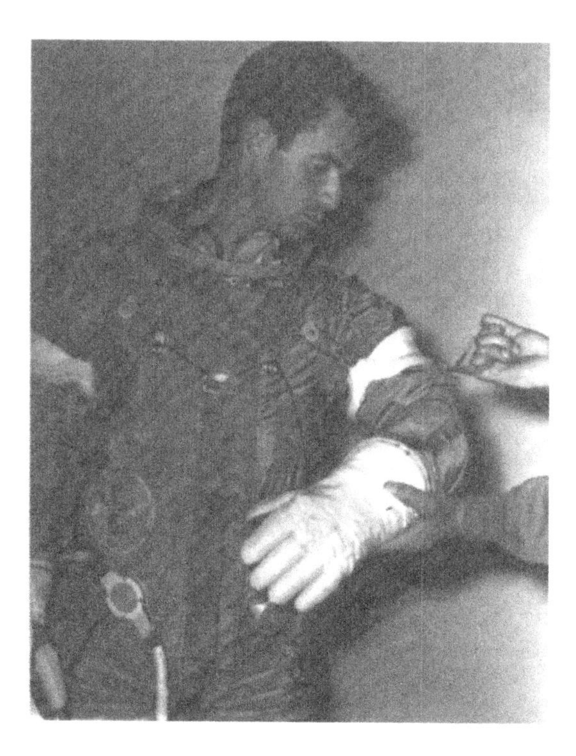

**Figure 5.2.9** YASTREB spacesuit arm length adjustment (the test engineer is A.Ts. Elbakyan of Zvezda).
From Archive Zvezda.

of the closed-loop LSS with gas circulation provided by a fan called for removal of the neck partition (i.e., the helmet and the suit enclosure formed a single volume).

There were also some additional changes. They included installation of a newly developed light filter outside the helmet (in contrast with the BERKUT's internal light filter), improvement of arm mobility, utilization of a newly developed, small-size service line connector (installed on the suit enclosure), installation of an emergency oxygen bottle on the suit body, etc. The main operating pressure in the suit was 400 hPa. As with the BERKUT suit, there was a possibility of switching over to a decreased pressure mode (270 hPa). However, pressure regulators were arranged on the suit enclosure in a different way.

Suit development was supported by a comprehensive test programme, with a large number of the tests and cosmonaut training (Figure 5.2.11) carried out in the Zvezda TBK-30 vacuum chamber (Figure 5.2.12, *top*), in the Air Force Scientific and Research Institute's (GK NII) thermal vacuum chamber (Figure 5.2.12, *bottom*) in conjunction with a mock-up of a spacecraft's living compartment (it was also used as the airlock) and aboard a Tu-104 flying laboratory (Figure 5.2.13). To facilitate the passage of suited crew members through the exit hatch (it was about 600 mm in diameter), it was expedient to fix the backpack with the LSS on the crew member's

**Figure 5.2.10** Ventilation version of the YASTREB spacesuit (unlaced from the spacesuit enclosure; worn by test engineer V. Bychkov of Zvezda).
From Archive Zvezda.

legs, at the front, rather than on the crew member's back. Such an arrangement required changes in the backpack suspension system. The modified backpack (Figure 5.2.1) was given the RVR-1P [PBP-1П] identification name (P[П] stands for the Russian word for "waist"). A total of 69 high-altitude manned experiments were carried out using the YASTREB spacesuit.

In accordance with the governmental directive dated 27 April 1966 mentioned in Section 5.1, YASTREB pressure suits fitted with RVR-1P backpacks were ordered for the L-1 programme (in-orbit transfer of crew members from the 7K-L1 trans-

**Figure 5.2.11** Ye.V. Khrunov and A.S. Yeliseyev (suited up) with Yu.A. Gagarin, V.M. Komarov and V.F. Bykovsky during a training session.
From Archive Zvezda.

portation system to the lunar vehicle). A new directive issued later in the same year specified the manufacture of YASTREB pressure suits with the RVR-1P backpack for the manned space vehicle (developed by the Chelomey Company) for the circumlunar mission and then, later on, for the *Almaz* orbiting station. Certain experimental activities and YASTREB suit fit checks in those spacecraft were carried out. However, with the advent of the L-3 programme, interest in the utilization of pressure suits for the L-1 programme disappeared. Use of the YASTREB suit in the *Almaz* programme was never realized, since it was replaced by the ORLAN suit in 1969. Between 1965 and 1969 a total of 24 YASTREB spacesuits and 18 backpacks for experimental work, tests, training and flights were manufactured.

## 5.3 THE FLIGHT OF *SOYUZ-4* AND *SOYUZ-5*

Final preparation of the YASTREB suits with the RVR-1P backpack for the *Soyuz* mission was carried out at the Baikonur space launch centre by a Zvezda team. Everything had been prepared for a mission by April 1967. Two spacecraft were under preparation for launch. The first spacecraft was to be piloted by one cosmonaut. The crew of the second spacecraft consisted of three members. Two of

**Figure 5.2.12** Spacesuit development testing in the vacuum chamber: (*above*) in the TBK-30 chamber at Zvezda; (*below*) in the mock-up of the *Soyuz* spacecraft's living compartment in the GK NII thermal vacuum chamber.

From Archive Zvezda.

these crew members, wearing YASTREB suits, were to transfer to the first spacecraft through free space.

Due to serious problems aboard *Soyuz-1*, revealed during the flight on 23 April 1967, the decision was made to descend the spacecraft ahead of schedule and cancel the launch of the second spacecraft. As is known, cosmonaut V.M. Komarov tragically perished aboard *Soyuz-1* on 24 April 1967, in the descent phase of the

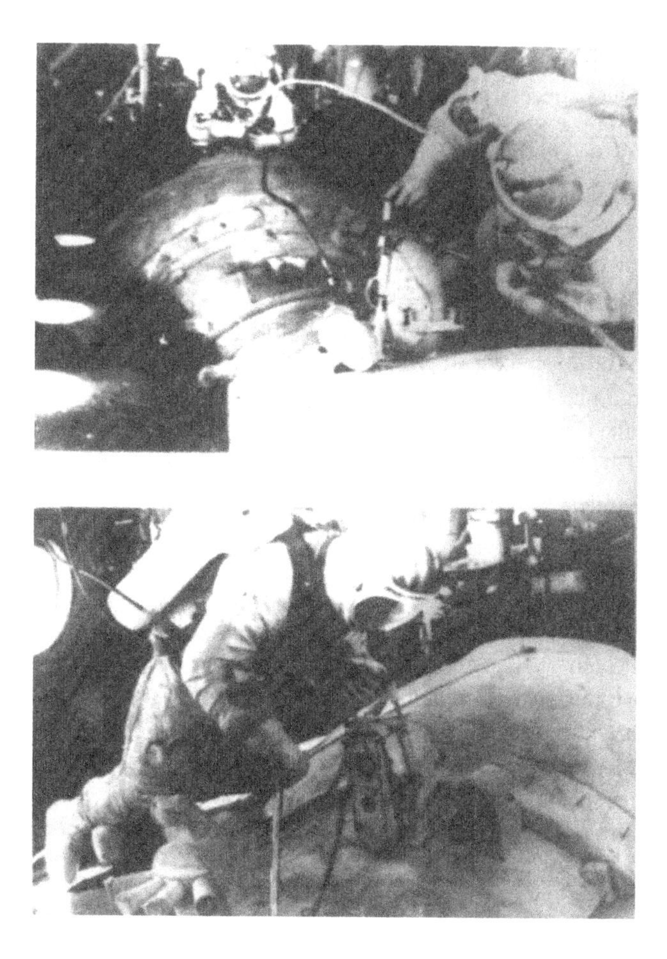

**Figure 5.2.13** Development of suited (spacesuit version with a backpack) crew member transfer from one *Soyuz* spacecraft to the other in the Tu-104 flying laboratory at the LII (Flight Research Institute).

From Archive Zvezda.

mission due to the failure of the parachute to open completely. After removal of the causes of the catastrophe, several flights of unmanned *Soyuz* spacecraft were carried out (six in the *Kosmos* series). Only after G.T. Beregovoy's mission aboard *Soyuz-3*, which docked with the unmanned *Soyuz-2* between 25 and 30 October 1968, were the *Soyuz-4* and *Soyuz-5* spacecraft prepared for a group mission. The group mission programme included the earlier planned, but not fulfilled experiment for the transfer of crew members from one spacecraft to the other through free space.

The preparation of suits and backpacks (as well as spacecraft) for the mission was carried out at a spare site at the Baikonur space launch centre under intensely cold weather conditions. The temperature in the room for the preparation of the hardware was close to 0°C. Personnel had to use a powerful airdrome blower with a heater to warm up the room and prevent freezing of the water used in the backpack heat exchanger.

The mission was successfully completed between 14 and 18 January 1969. On 17 January cosmonauts A.S. Yeliseyev and Ye.V. Khrunov transferred from *Soyuz-5* (commanded by B.V. Volynov) to *Soyuz-4* (commanded by V.A. Shatalov) and then returned to Earth aboard their new spacecraft. The space walk lasted 37 minutes. This time the management of the operations on spacecraft docking and the space walk was carried out from the Mission Control Centre, located near the city of Yevpatoria (Crimea). The mission managers and a group of specialists, including Zvezda's representatives, had transferred there just after the *Soyuz-5* launch. Telemetry data about spacecraft systems and spacesuit functioning were concentrated at the same place.

Transfer of the crew from *Soyuz-5* to *Soyuz-4* took place in accordance with the programme, but was carried out a little later than expected. As explained by the cosmonauts after the mission, the time deficit had arisen due to the fact that it was more difficult to carry out some operations during the mission under zero gravity and intensification of emotional tension during the space walk.

Moreover, the space walk was 11 minutes behind schedule, because, at the time of connecting the long umbilical, Ye.V. Khrunov, who was the first to go outside, mistakenly connected A.S. Yeliseyev's suit to the on-board system instead of his own. Ye.V. Khrunov had to return to the living compartment/airlock to remate the connectors after finding that ventilation became low.

Cosmonauts, moving along the spacecraft's external surface, could restrain themselves using special rigid handrails. Later, this method was also used for EVAs from orbiting stations, though in some cases the cosmonauts reported hand fatigue and difficulty in making delicate coordinated movements. Some difficulties with the *Soyuz-4* hatch closure were also encountered at the end of the EVA, due to intrusion of badly secured floating elements (straps, tethers, connectors, etc.). During the EVA from *Soyuz-5* to *Soyuz-4* the portable camera was not properly secured and drifted away from the cosmonauts. This is the reason that no photos from this EVA are available.

The mission provided valuable, additional experience in the way in which cosmonauts work in free space and evaluated the performance characteristics of a suit with a regenerative-type LSS under real conditions. Moreover, the transfer of two crew members through free space had demonstrated both the reality of such a method of transfer from one spacecraft to the other for programmes like the N1-L3 programme and the possibility of carrying out rescue operations in free space.

The YASTREB suit and the RVR-1P backpack were developed under the leadership of the project-managers I.P. Abramov (LSS), A.Yu. Stocklitsky (suit) and I.I. Chistyakov (electrical) (specialists of the design departments), with active participation of specialists of the test department led by B.V. Mikhailov, V.I. Svertshek and I.I. Skomorovsky.

After the successful *Soyuz-4/5* mission, the cosmonauts paid the usual visit to the Zvezda plant in Tomilino (Moscow region), meeting all the employees on 29 January 1969 (Figure 5.3.1).

**Figure 5.3.1** Meeting with cosmonauts Ye.V. Khrunov and A.S. Yeliseyev at Zvezda. First line (*from left to right*): N.P. Kamanin, Ye.V. Khrunov, E.A. Ivanov (speaker), G.I. Severin, B.V. Volynov, V.A. Shatalov, A.S. Yeliseyev.

From Archive Zvezda.

For the later flights of *Soyuz-7* and *Soyuz-8* (together with *Soyuz-6* in October 1969), no space walks were planned and hence no suits were manufactured by Zvezda.

# 6

# Spacesuits for the Soviet Moon programme

## 6.1 THE SKV EXPERIMENTAL SPACESUIT

The acronym "SKV" [CKB] stands for the "extravehicular activity (EVA) spacesuit" in Russian. SKV was a prototype for semi-rigid spacesuits of the KRECHET and ORLAN types, developed later by Zvezda for the L-3 Moon programme and the long-term orbiting stations *Salyut*, *Mir* and *ISS* (the International Space Station).

Preliminary design and development work for realization of the experimental EVA spacesuit were initiated at Plant No. 918 under the government resolution "On creation of powerful launchers, satellites, spacecraft and space pioneering in 1960–67", dated 23 June 1960.

Plant No. 918 (RD&PE Zvezda) initiated a special programme to carry out this work. The above-mentioned governmental resolution and further orders by the Ministry of Aviation Industry determined 1965 as the year for completion of the spacesuit's preliminary design. In April 1961 A.Yu. Stoklitsky, the chief of a design team, was assigned as acting lead designer for this programme. Research work into the life support system (LSS) for this suit were managed by I.P. Abramov.

During 1961 the work concentrated on the determination of reference design data for the spacesuit and its systems and making these data more precise. Draft performance specifications were received in December 1961 from Experimental Design Bureau 1 (OKB-1), the prime contractor for the OTSST programme (OTCCт = orbiting heavy satellite/station).

These performance specifications determined the major spacesuit characteristics: spacesuit operation time of 4 hours in self-contained mode and 8 hours of life support from the mothercraft, maximum spacesuit mass of 85 kg and maximum spacesuit-donning time of 5–10 minutes. The spacesuit had to be equipped with the cosmonaut transfer and manoeuvring unit (UPMK), enabling, of course, cosmonauts to transfer and manoeuvre in free space (see Chapter 9).

LSS development experience gained at Zvezda at that time showed that a closed-loop regenerative-type system would have the best compliance with the requirements set forth. Preliminary thermal analysis of the spacesuit and the LSS components was carried out, and the creation of test facilities for the spacesuit and its subsystems was initiated. The following facilities were developed: a bench for testing the stress level of soft enclosure materials, a facility for determination of the thermal resistance of soft materials and a facility for evaluation of spacesuit soft joints.

Since a number of totally new issues, of which Zvezda had no experience, were to be solved, a few subcontractors were involved in these activities. Zvezda started to prepare performance specifications for these subcontractors, in particular for:

- evaluation of radiation and meteorite protection;
- research into and selection of spacesuit materials;
- electric and radio equipment;
- development of $CO_2$ and toxic impurity absorbers; and
- development of the elements for the propulsion unit.

The company SKB-KDA (in the city of Orekhovo-Zuyevo) was traditionally responsible for development of oxygen equipment.

In 1962, selection of the optimum design solution for the SKV spacesuit and spacesuit mobility were the focus of the activities.

As far back as 1920 K.E. Tsiolkovsky, the founder of Russian scientific Cosmonautics, in his fantastic novel *Beyond the Planet Earth* (Tsiolkovsky, 1958) put forward the idea of using a spacesuit for undertaking a space walk from a space station. He describes his ideas by means of a character in his novel, who explains the spacesuit design:

> The time will come when we shall have to land on a planet ... in an atmosphere unsuitable for us to breathe: unsuitable either because of its special composition or because it is extremely rarefied. In order to keep alive in vacuum, or in a rarefied or unsuitable gas, we need special suits—the same kind will do for either contingency. Here they are. The suit covers the whole body from head to foot; it is impervious to gases and vapours; it is pliant, not too bulky, and does not impede the movements of the body. It is strong enough to withstand the internal pressure of the gases surrounding the body. The headpiece is fitted with special flat, partially transparent plates, to enable the wearer to see. It has a thick, heated lining, which is pervious to gases and vapours. There are fittings for containing urine and so forth. The suit is joined to a special box from which a sufficient quantity of oxygen is continuously fed to the wearer. The carbon dioxide, water vapours and other waste products of the body are absorbed in other boxes. Gases and vapours are kept in constant circulation under the suit in the permeable lining by means of special automatic pumps. A man needs one kilogram of oxygen. There are sufficient supplies for eight hours. Suit and oxygen supplies together have a mass not greater than ten kilograms; but of course nothing here has any weight at all.

Nine years later, in 1929 another famous scientist of Cosmonautics, the German Hermann Oberth, presents his version of a rigid spacesuit in his book *Wege zur Raumschiffahrt*.

Study of the specific features and operational conditions for the EVA spacesuit, analysis of the available literature (translations of US publications on spacesuit research and design, and the works of Herman Oberth) as well as almost 10 years of Zvezda experience (from 1952) in design, manufacture, testing and operations of aviation suits suggested that adoption of a spacesuit of the current classic design (soft full pressure suit plus a removable backpack with the LSS) outside the spacecraft would not be the optimum solution.

Among the drawbacks to the classic pressure suit design, the following were the major ones:

- considerable difference in dimensions and torso shape of the pressurized and unpressurized suit;
- difficult donning/doffing and sealing of the suit enclosure;
- suit elongation (so-called "ballooning") under positive pressure due to fabric deformation and distortion of the suit enclosure shape;
- limited range of suit enclosure adjustment to fit different human shapes;
- necessity of the backpack restraint system and difficulty of its operation;
- external communication lines between the suit and the backpack; and
- difficulty of attaching suit controls to the soft torso enclosure.

The preferred design concept was a suit with a hard upper torso (HUT), having a built-in LSS with controls attached to the torso, and arms with gloves and leg enclosures made of soft materials. Since Zvezda already had a lot of experience in "soft"-type suit development, it was decided to focus work in the initial stage on elaboration of an optimum design concept for a suit with a HUT.

As is known from the literature, the rigid suit projects of the Litton Company (USA) and the above-mentioned Professor Oberth were of no use for this task.

During 1962 the first breadboard model was manufactured to be used to define the shape and overall dimensions of the HUT (including entry hatch location and dimensions). Several versions of joints, both rigid and soft, and major pieces of soft enclosure parts were manufactured, and their testing initiated (Figures 6.1.1 and 6.1.2).

Along with the work on the suit enclosure, the suit's LSS and some of its components were developed. Possible designs of the self-contained LSS were analysed. They differed in their methods of heat removal from the human body (by ventilating gas or water-cooled garment), in their methods of gas circulation inside the suit (by a fan or an injector), and in their type of $CO_2$ absorber (regenerative or non-regenerative materials). As a result of these analyses, an LSS having the following three standard loops (Figure 6.1.3) was selected for further development:

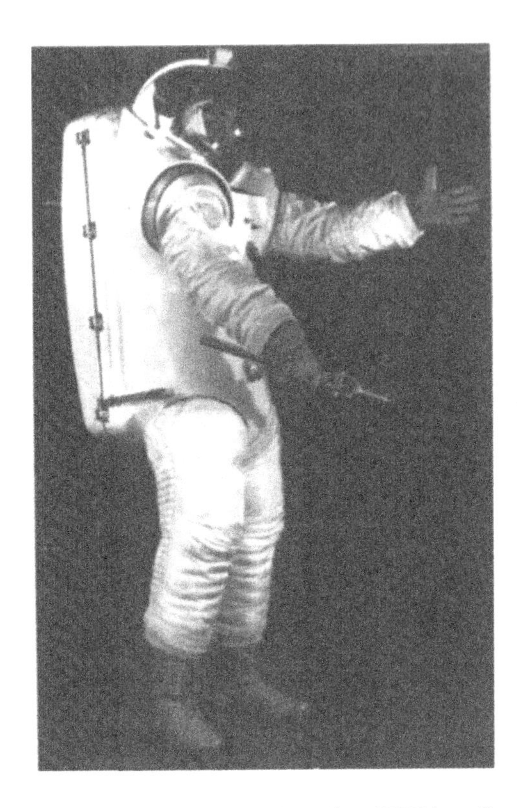

**Figure 6.1.1** General view of one of the first SKV breadboards.
From Archive Zvezda.

- one loop to supply oxygen for breathing and leakage compensation;
- a ventilation loop to remove moisture and $CO_2$ from the in-suit cavity, as well as some metabolic heat; and
- a cooling loop including an evaporating heat exchanger, water pump, water receiver and water-cooled garment.

A development test programme was run in conjunction with industrial partners for some self-contained LSS components like the fan, the $CO_2$ and contamination control cartridge and the oxygen unit.

The oxygen unit provided continuous oxygen supply to the suit with a normal and an emergency flow rate of 1.5 and 24.5 standard litres per minute, respectively.

An evaporating heat exchanger, the inter-tube space of which was filled with blocks filled with a porous material with high water-absorbing capability (trade name of polyvinyl formal—PVF), was developed. The heat exchanger featured a vapour vent valve that maintained the absolute pressure of about 6 mm Hg in the PVF zone. In case of fan failure, the injector was automatically initiated due to inlet pressure decrease. The injector provided helmet ventilation of $40\,l\,min^{-1}$ with

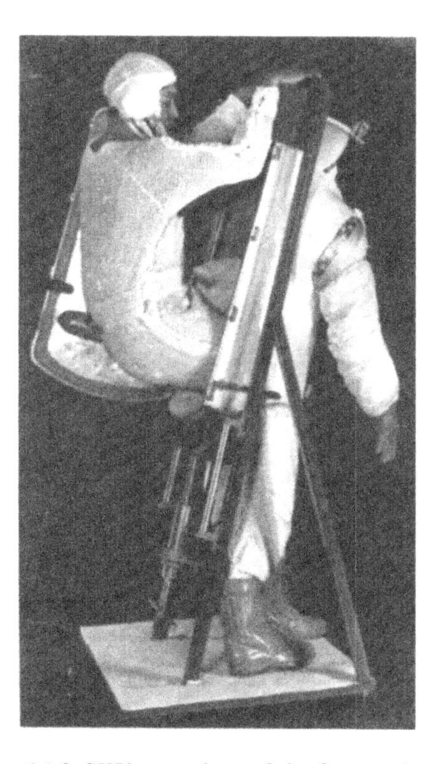

**Figure 6.1.2** SKV entry (one of the first versions).
From Archive Zvezda.

an oxygen flow rate of three standard litres per minute. This was sufficient for adequate $CO_2$ removal. The pressure-tightness check and suit-purging were foreseen.

Using the performance specifications developed by Zvezda, the Special Design Bureau of Analytic Instrument Building (SKB-AP) developed a measuring system that provided telemetry monitoring of the following data: oxygen and $CO_2$ content in the suit atmosphere, relative humidity, ventilating gas flow rate and temperature, and the temperature at two points on the suit surface.

It was planned to include electrical equipment in the suit assembly that would take bearings of the EVA cosmonaut in relation to the mothercraft (together with the on-board equipment), on top of its ordinary functions.

In 1963 final approval for the design concept of the SKV suit enclosure was given: a rigid torso made as a single component with the helmet (HUT) and having a rear entry hatch; a backpack with the LSS components located inside the suit served as the hatch cover; and soft enclosures for the limbs.

Note that the proposed unique design of the suit entrance turned out to be highly successful. It enabled suit operators to considerably simplify the process of suit-donning. This design has been used in all Soviet/Russian EVA suits right up to the present day.

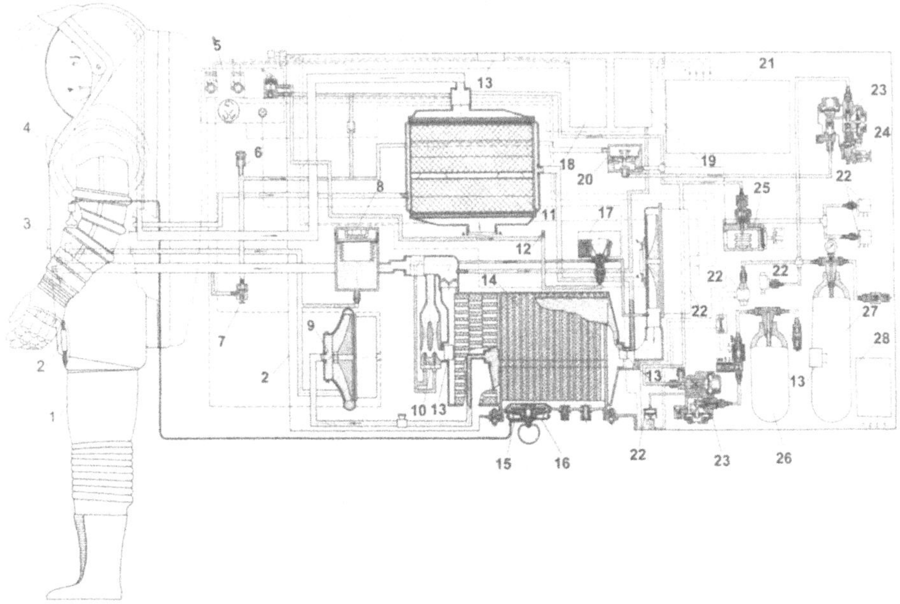

**Figure 6.1.3** Principal flow diagram of the SKV SCLSS (self-contained LSS) (original drawing): 1—spacesuit; 2—relief valve; 3—absolute pressure regulator; 4—pressure gauge; 5—control panel; 6—on-board system connection union; 7—purge valve; 8—hydraulic accumulator; 9—water pump; 10—injector assembly; 11—contamination control cartridge; 12—calibrated orifice for $1.5 l min^{-1}$; 13—temperature sensors; 14–16—evaporation heat exchanger, actuation pin and valve; 17—oxygen switch; 18—group of amplifiers; 19—$CO_2$ sensor; 20—locking device; 21—storage battery; 22—pressure annunciators and sensors; 23—reducer; 24—manual oxygen actuation handle; 25—leak compensator; 26, 27—redundant and primary oxygen bottles; 28—telemetry system module.
From Archive Zvezda.

In the initial version of the suit design, the whole suit body was rigid (from the groin to the helmet), and the shoulder and hip joints were of the so-called rigid spherical type (Figure 6.1.1).

The suit/backpack interface was located in the vertical plane. It included two milled frames made of aluminium alloy (AMG-6): the HUT and the backpack frames that were built into the HUT were made out of a composite material based on glass fibre. The HUT frame had a U-shaped cross section profile, where a sealing rubber hose of rectangular cross section was glued in. The backpack frame had an edge (knife) that pressed against the surface of the hose on the HUT frame interface and sealed the interface. The hose had holes connecting its inner cavity with the suit cavity, so that when the suit was pressurized the hose was further pressed to the knife, improving pressure-tightness.

The closure frames were connected on the left side by four hinges that enabled the backpack to rotate, like a door, opening the entrance hatch (Figure 6.1.2). There were six rod-type latches on the frames on the upper, bottom and right sides (four) of the closure. The latches were locked/unlocked by the cosmonaut in the suit, using a lever-type handle located on the right side of the HUT below the waist. To lock the latches the cosmonaut rotated the handle anti-clockwise and secured it in the upper position.

By recommendations of radiation protection experts, the HUT was made of a material with low atomic weight—glass-reinforced resin—in order to reduce secondary radiation resulting from electron impact on the suit during space walks in the Earth's radiation belts (according to performance specifications the orbit altitude was 450 and 36,000 km).

Development of production procedures for the glass-reinforced resin HUT and its manufacture were carried out at the Science and Research Institute of Plastics using Zvezda's drawings, tools and fixtures.

Technical testing was carried out of the oxygen equipment, the $CO_2$ control cartridge (RPS-62), the electric motor and the fan (in collaboration with the cartridge designer, a branch of NII-104 [Science and Research Institute 104]).

In the same year Zvezda presented a paper "Spacesuit for EVA sortie from a spacecraft into space" at a meeting of the Inter-Agency Scientific and Technical Council on Space Investigations of the USSR Academy of Sciences, chaired by K.D. Boushouyev, Deputy General Designer of OKB-1. The paper stated major technical problems and the status of spacesuit development. The Council approved the proposed suit design featuring the HUT. Later, such a spacesuit (HUT and soft enclosures for the limbs) was called "semi-rigid".

In 1964, taking the test results of the first HUT breadboard model into account, three samples of the second version of the HUT model (see Figure 6.2.2, *right*) were designed and manufactured. Two samples were tested for strength and pressure-tightness, and the third one was used for the manufacture of a fully functional mock-up of the suit. The work on the joints and other suit enclosure subassemblies, the LSS components, propulsion unit, materials testing and creation of the necessary test benches were continued. The second version of the evaporating heat exchanger with a moisture-absorbing agent was developed and tested. The regeneration-type LSS system underwent laboratory tests and fireproofing tests in an oxygen atmosphere. The latter were given a great deal of attention, because the adopted design—the semi-rigid suit—has its major LSS components, including electric equipment, located in the backpack pressurized cavity, which is connected to the oxygen-filled inner cavity of the suit enclosure.

In 1965 the second fully functional SKV mock-up for ground tests was designed and manufactured. The first ever water-cooled garment, some functional mock-ups of the LSS components and a breadboard of the propulsion unit were also designed and manufactured. A great amount of the work on calculations and studies into some subassemblies and subsystems of the suit (joints, inner ventilation system) and LSS (a version of the sublimation-type heat exchanger, the fan, the measuring system, etc.) were also carried out.

The results of the work carried out on the SKV suit and its systems were presented in the preliminary design report "Spacesuit for a sortie from a spacecraft into space", which was sent to the customer (OKB-1) in December 1965.

Creation of an alternative soft version of the EVA spacesuit named ORIOL under the leadership of S.P. Umansky, a lead designer, was continued at Zvezda in parallel with the work on the SKV suit (see also Section 6.3).

## 6.2 KRECHET AND ORLAN SPACESUITS UNDER THE L-3 PROGRAMME

In the middle of 1964, along with the circumlunar mission programme OKB-1 started to rescope the lunar mission programme in light of the availability of the N-1 launcher (N-1–L-3 programme). On 3 August 1964 the corresponding government resolution was issued. Later the government directive dated 10 February 1965 and orders by the Ministry of Aviation Industry confirmed this document.

On the basis of these documents Zvezda was commissioned to develop and supply the customer (OKB-1) with spacesuits that had a backpack LSS both for the cosmonaut undertaking the lunar landing and for the spacecraft commander. These activities also included a waste management system and on-board equipment for spacesuit operation support.

The L-3 complex was designed for a two-man mission to the Moon. It included two habitable vehicles: a lunar lander, landing on and taking off from the Moon with one cosmonaut, and a Moon-orbiting spacecraft, remaining in near-lunar orbit after separation of the lunar lander until its docking on completion of the work on the Moon. The spacecraft commander had to stay in the Moon-orbiting spacecraft while the other vehicle was on the surface of the Moon (Figure 6.2.1). This called for two types of spacesuits with different objectives and performance requirements:

- one for lunar surface activities at reduced gravity and long autonomous operation time; and
- one for in-orbit operation at zero gravity with an umbilical for short operation time.

In the course of preliminary mock-up activities, two versions of the lunar surface spacesuit were considered: a soft suit with a removable backpack having the in-house code ORIOL, and a semi-rigid suit with a built-in self-contained LSS (SCLSS). Further tests were needed to show which version was the best. The lunar semi-rigid spacesuit and its backpack with SCLSS were respectively named KRECHET and KASPIY [КАСПИЙ]. The orbital semi-rigid spacesuit and its backpack with SCLSS designed for the spacecraft commander were respectively named ORLAN and SELIGER [СЕЛИГЕР].

Realization of the lunar spacesuit called for the solution of some new problems. In particular, this suit should enable an unassisted man to perform the following tasks:

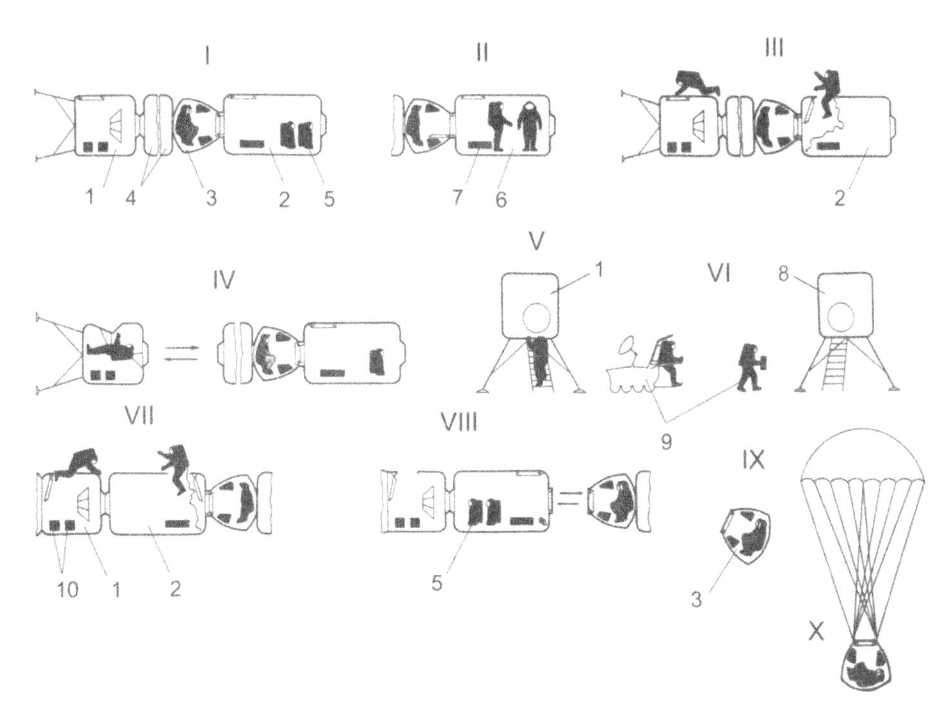


**Figure 6.2.1** Sequence of crew actions under the L-3 programme (drawn from the original sketch): 1—Moon lander; 2—Moon-orbiting spacecraft; 3—descent vehicle; 4—instrument and apparatus modules; 5—spacesuit arrangement in stowage; 6—spacesuit-donning and checkout; 7—system for spacesuit gas atmosphere change; 8—redundant Moon lander; 9—rover drive or walk; 10—spacesuit on-board system and SCLSS replaceable elements.
From Archive Zvezda.

- to fly the lunar lander and land it on the Moon;
- to walk on the lunar surface; and
- to protect a man against hostile environmental conditions, more adverse than those in an Earth orbit.

Therefore the Moon suit had to have the required leg mobility, stronger thermal protection and a second light filter (for shady conditions). High requirements were put on suit reliability, especially as only one cosmonaut was to land on the lunar surface.

Self-contained suit operation had to be increased to 10 hours and a continuous 52-hour suited stay in the lunar lander was considered. This requirement called for provision of drinking water, urine collection and removal and caused further difficulties at suit development and testing. Because of the high metabolic rates of a cosmonaut, it was necessary to adopt a new method of conductive heat removal from a cosmonaut's body using a water-cooled garment.

Development of the KRECHET suit barely commenced in 1966 due to a delay in transfer of performance specifications from OKB-1 (Experimental Design

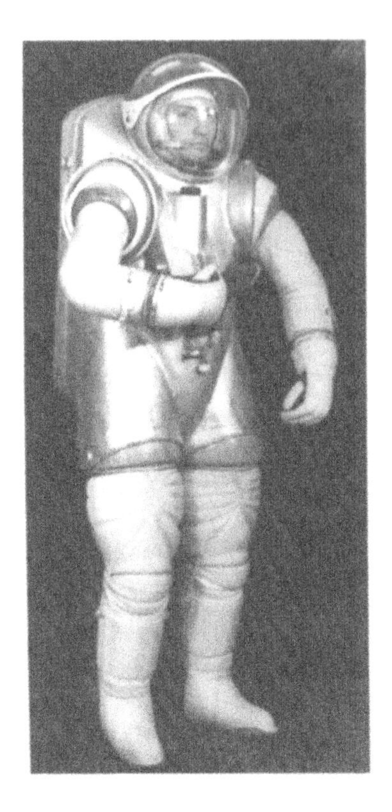

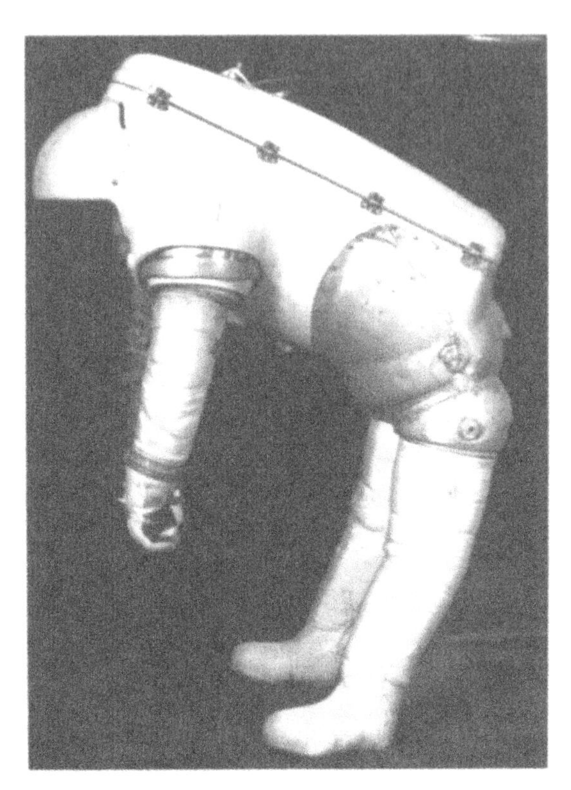

**Figure 6.2.2** Fully functional mock-ups of the KRECHET and SKV spacesuit (on the right is the SKV mock-up with the waist joint).
From Archive Zvezda.

Bureau 1). This suit was the modified SKV suit, the fully functional mock-ups of which had been manufactured and tested earlier. The necessity to modify the SKV suit was dictated both by experience obtained during suit manufacture and testing and by new requirements of the new performance specifications.

The HUT design had undergone a significant change. Instead of a composite material used for SKV, the KRECHET's HUT shell was made out of 1.2-mm-thick sheets of the aluminium alloy (AMG-3), which made it possible to reduce suit mass by 7 kg. This change became possible because radiation conditions on the Moon did not require a material of low atomic weight for cosmonaut protection.

Moreover, by substituting AMG-3 alloy for the composite material, HUT structure reliability and manufacturability were sharply improved owing to methods of shaping aluminium alloy parts and their joining by welding (developed in the aviation industry and widely used at Zvezda), instead of gluing the HUT shell as happened when using a glass-reinforced resin.

In 1966 three fully functional mock-ups of the KRECHET suit were developed and manufactured. They had a completely hard body and hard spherical shoulder

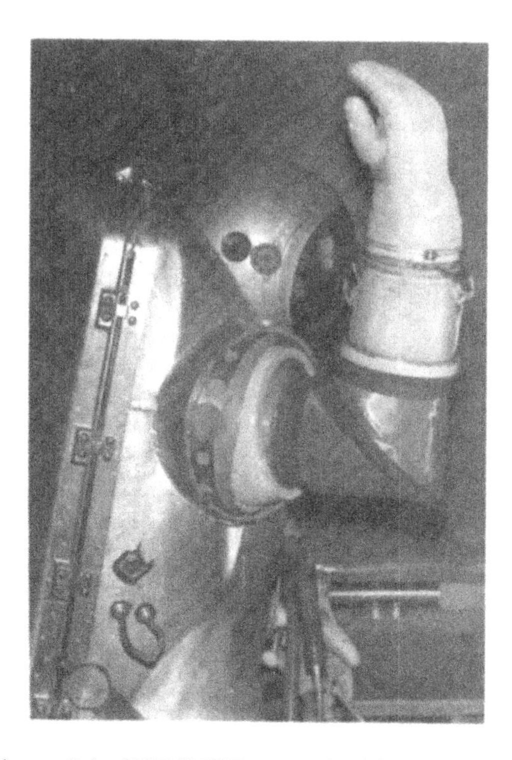

**Figure 6.2.3** Mock-up of the KRECHET spacesuit with the elbow pressure bearing.
From Archive Zvezda.

and hip joints (Figure 6.2.2, *left*). A fully functional mock-up of the KASPIY backpack was designed and manufactured for laboratory physiological tests.

Initial tests were conducted into walking while wearing the suit, multiple fit checks with the participation of test subjects of various heights and sizes and fit checks in the lunar lander mock-ups, when performance and the ergonomic characteristics of the suit were evaluated. Under the SKV and KRECHET programmes, a lot of experimental joints, both hard and soft, were designed and tested; in particular, the above-mentioned hard spherical joints with one (hip, waist) and two (shoulder and wrist) degrees of freedom, a hard elbow joint with an inclined "slanting" pressure bearing (Figure 6.2.3), soft joints "with orange peels" (shoulder and elbow), soft convolute joints (arm and leg enclosures) and soft joints for gloves (wrist and fingers). A special instrument for soft joint investigations—a kinemometer—and another for studying muscle forces in human joints were designed and manufactured as early as 1964. Extensive research work into soft joint mobility were carried out using these instruments facilities (Figure 6.2.4).

The tests and fit checks using the first KRECHET suit mock-ups had shown that its design needed further modification. Therefore, in 1967, new documentation was worked out and production of the KRECHET-94 modified lunar suit began (index "94" meant that this suit was designed for the lunar lander, which had a project code number "11F94").

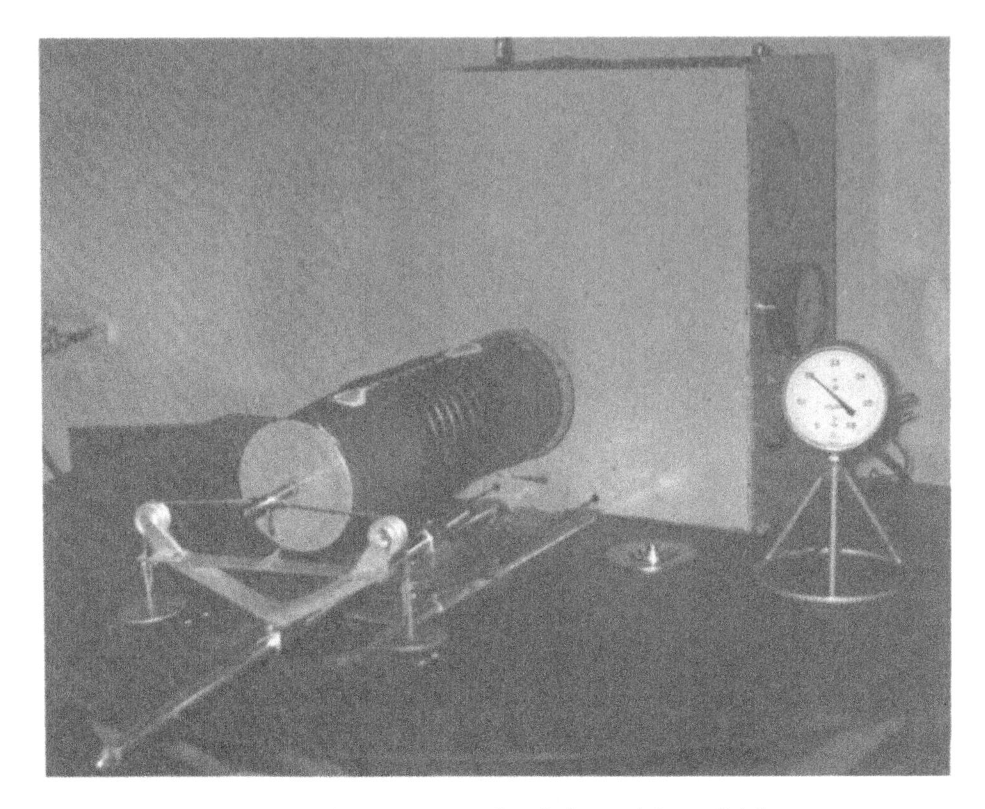

**Figure 6.2.4** Kinemometer test bench for studying soft joints.
From Archive Zvezda.

This new suit differed considerably from its prototype (Figure 6.2.5). Most changes were introduced in the design of the HUT, the hip and shoulder joints. For example, the lower hard body below the waist was replaced by a soft enclosure that was joined to the leg enclosures. It had three length-adjustable restraint cords: two cords ran along the outer sides of the legs and the third one ran along the centre line through the groin (the front adjustment unit), which made it possible both to have a wide degree of body length adjustment (height 164–182 cm) and to expand the anthropometric range significantly.

Height adjustment of the original KRECHET completely hard body torso was made by changing the thickness of spacers on the so-called shoulder rests installed inside the suit. But this adjustment also resulted in changed positions of the head inside the helmet and the crew member's shoulder joints in relation to the armholes. These changes had a negative effect on the field of vision and arm mobility and as a result restricted the anthropometric range of the suit significantly.

The tests had shown that the hard (spherical) shoulder and hip joints used in the KRECHET suit had a very narrow range of deflection angles, though they had zero drag torque. Moreover, their design and production procedures were complicated and they had a large mass. This is why the KRECHET-94 suit was equipped with

**Figure 6.2.5** One of the first KRECHET-94 spacesuit versions (minus the outer garment).
From Archive Zvezda.

shoulder pressure bearings (proven earlier in aviation pressure suits) in combination with a soft single-axis shoulder joint, instead of spherical shoulder joints with two degrees of freedom, and with completely soft convolute joints with two degrees of freedom, instead of spherical hip joints in combination with soft ones, both having one degree of freedom.

A waist load-bearing frame was introduced in the new body design. The frame housed three load-bearing elements (two side catches and a front bracket with a rod) designed to attach the spacesuit to the corresponding elements of the shock-absorbing and attachment system of the lander. The fit checks and tests (Figure 6.2.6) had shown that such a structure enabled the cosmonaut wearing the pressurized suit to detach and reattach in the lunar lander's working station.

Redundant (internal) pressure bladders were introduced in parts of the soft enclosure; the secondary pressurization loops—between the HUT and the backpack—and in the pressure bearings. A fibreglass protective helmet was mounted on the helmet above the light filter. The backpack closing system was updated, the arm's design was improved and some changes were introduced in other elements of the suit (e.g., control panel, upper garment, antenna-feeder arrangement).

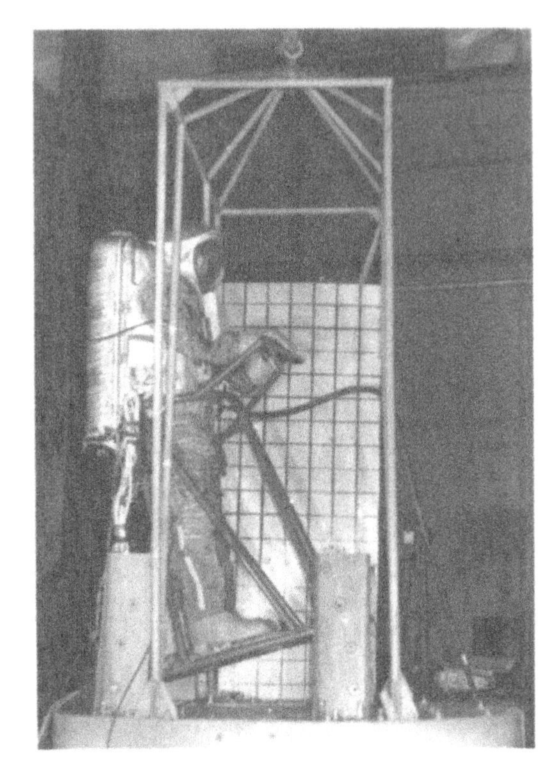

**Figure 6.2.6** Acceleration tests of the KRECHET-94 spacesuit under Moon-landing conditions simulated at a special test facility.
From Archive Zvezda.

In 1967 two experimental KRECHET-94 suits (Figure 6.2.7) were manufactured and the production of three more suits began.

Development testing of the KASPIY SCLSS (Figure 6.2.8) built for the lunar suit on the lines of the LSS of the SKV suit continued. Some new subassemblies and components were introduced. In particular, much design and research went into the development of a two-loop heat exchanger (sublimation type) and a moisture separator with metal–ceramic elements. F.V. Kubar, B.S. Braverman and other engineers made a considerable contribution in the development of these components. Finally, these activities resulted in such sophisticated designs that, with minor modification, are still being used in the ORLAN-type suits today.

The spacesuit's radio and electric equipment consisted of, besides the LSS electric components, the ZARNITSA radio and telemetric system, an antenna-feeder arrangement, an autonomous power supply and a control panel (Figure 6.2.9). The antenna consisted of two cables secured on the foam rubber padding. The padding was attached along the perimeter of the upper portion of the backpack and provided for a 50-mm clearance between the antenna element and the metal enclosure of the suit.

**Figure 6.2.7** General view of the KRECHET-94 spacesuit.
From Archives Skoog and Zvezda.

In the lunar lander, life support for the cosmonaut wearing the KRECHET was provided by an on-board LSS, which included the B-2M interface unit developed by Zvezda. This unit was connected to the suit by an umbilical (Figure 6.2.10).

The Spacesuit Breadboard Review held in 1968 and the following meeting of the Zvezda's Scientific and Technical Council made a decision to consider the KRECHET semi-rigid space suit as the main version for the Moon (lunar lander) spacesuit.

Later the whole experience gained from EVAs from the orbiting stations proved the correctness of the selection of such a suit type.

It should be noted that at that time Zvezda had opponents to this decision (in particular, S.M. Alekseyev). They preferred to keep to the beaten track of "soft" spacesuit application (similar to the US *Apollo* spacesuit). General N.P. Kamanin was also of that opinion. Judging from his memoirs (Kamanin, 1999), the KRECHET-94 spacesuit seemed to him too cumbersome during his visit to Zvezda. He considered it would be right to simplify the first Moon mission programme (reduce the time of cosmonaut stay in the suit and exclude the long-walk capability on the Moon) and to use a BERKUT spacesuit modification as the Moon (lunar lander) suit.

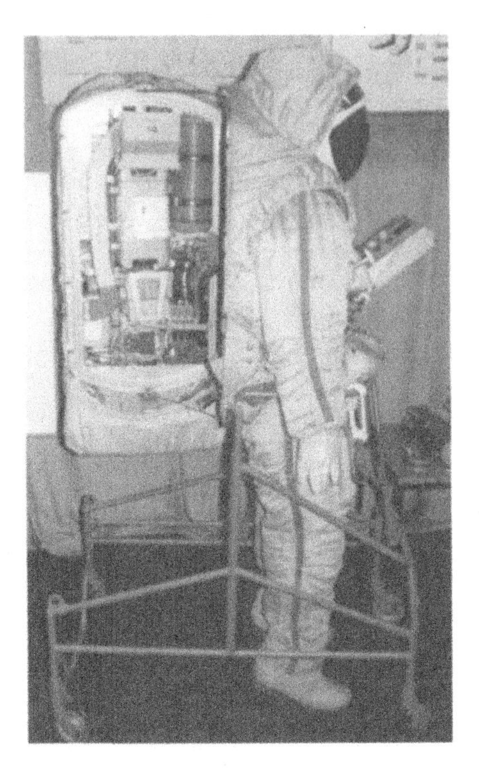

**Figure 6.2.8** KASPIY backpack of the KRECHET-94 spacesuit.
From Archives Skoog and Zvezda.

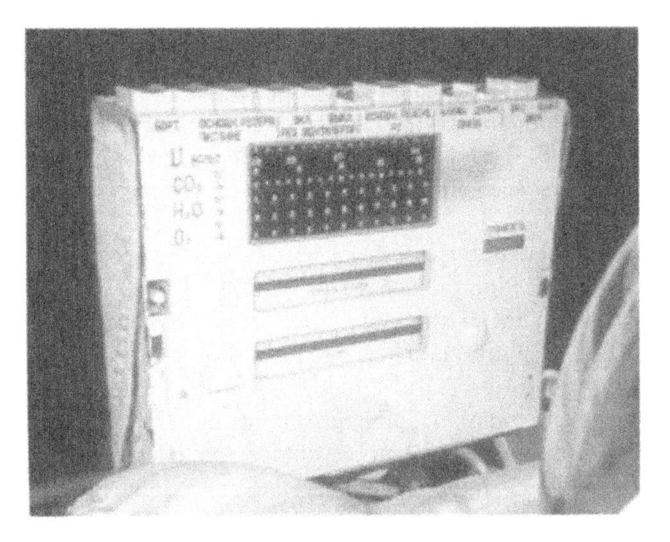

**Figure 6.2.9** KRECHET-94 control panel.
From Archives Skoog and Zvezda.

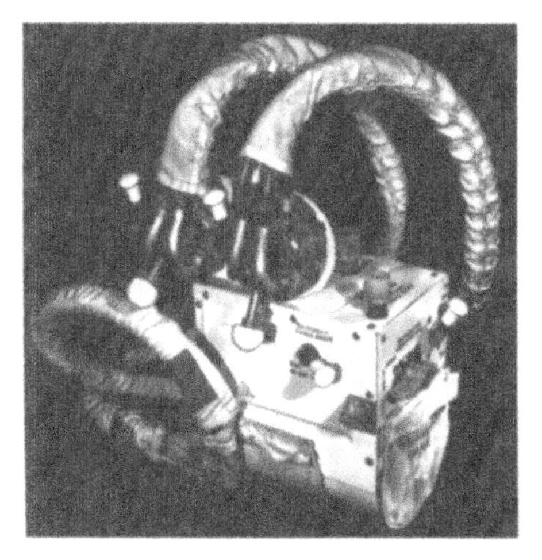

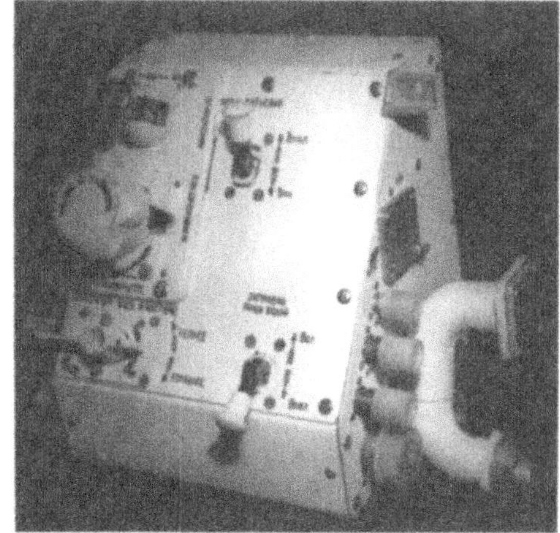

**Figure 6.2.10** General view of the spacesuit's on-board system: (*left*) B-1M (in the Moon-orbiting spacecraft) and (*right*) B-2M (in the lunar lander).
From Archive Zvezda.

But Zvezda was not the body that determined the expedition programme. Moreover, General Kamanin was not fully aware, unfortunately, of each suit's performance characteristics.

In the opinion of most Zvezda specialists, the "soft" type of EVA suit could merit use only if it was used as a rescue suit as well. But this was not required. Moreover, under vacuum conditions the soft suit together with the backpack reached the same or larger dimensions as those of the semi-rigid suit.

The creation of the ORLAN spacesuit and the SELIGER backpack system for the mission commander ran in parallel with development of the lunar lander spacesuit. The ORLAN suit should enable the spacecraft commander to carry out an EVA sortie in order to render assistance to the second crew member, transferring to and from the lunar lander.

The ORLAN suit was developed from the KRECHET-94 modification and had the following advantages typical of a semi-rigid suit: easy, quick and reliable to don and doff, easy operation and speedy readiness to work (no external pneumatic lines and connectors to mate with the SCLSS, optimum arrangement of the controls and monitoring means), high pressure-tightness and reliability (only one operational connector), wide height adjustment range and optimum overall dimensions.

In principle, both spacesuits had the same design solutions for both the suit enclosure and the LSS, but there were some simplifications in the commander's suit (in particular, the ORLAN suit was not completely self-contained). The ORLAN suit operated using an electric umbilical, connecting it with the mother-craft's on-board systems. Therefore it did not have a self-contained electric power supply, radio communication and telemetry units and antenna-feeder arrangement. The electric control panel was also simplified. The ORLAN suit was designed for a shorter operation time (two 2.5-hour cycles); thus the backpack was significantly

**Figure 6.2.11** General view of the ORLAN spacesuit.
From Archive Zvezda.

smaller. The suit enclosure was also simplified: the hip joint had one degree of freedom instead of two in the lunar suit (which was important for the KRECHET suit when walking on the lunar surface), there was no shade visor and the thermal protection enclosure had five layers of shield vacuum thermal insulation instead of the ten layers in the KRECHET. No waste management system or drinking water provisions were included in the ORLAN suit. The ORLAN suit's overall dimensions and mass (59 kg) were correspondingly smaller than those of the KRECHET-94 suit (102 kg), making it more manoeuvrable during EVA.

In 1967 the first sample of the ORLAN suit was manufactured, and its laboratory testing and refinement were carried out in 1968 (Figure 6.2.11).

Because the main type of cosmonaut training given for working outside the

**Figure 6.2.12** ORLAN pressure suit for hydro lab.
From Archive Zvezda.

spacecraft under weightlessness conditions was done in a hydro lab, a special version of the ORLAN pressure suit was developed (Figure 6.2.12). This suit had the following differences from the flight suit: it had special weights providing for cosmonaut neutral buoyancy, a special hoisting system to put the suited cosmonaut in the water pool and take him out, a simplified LSS SELIGER and suit pressurization, ventilation and cooling were provided from the pool-side ("on-board") system.

In 1968 design and refinement work and laboratory testing of five KRECHET-94 suits (Nos 6–11) were carried out, two suits (Nos 12 and 14) were manufactured and transferred for in-house testing and manufacture of the next spacesuit lot was begun.

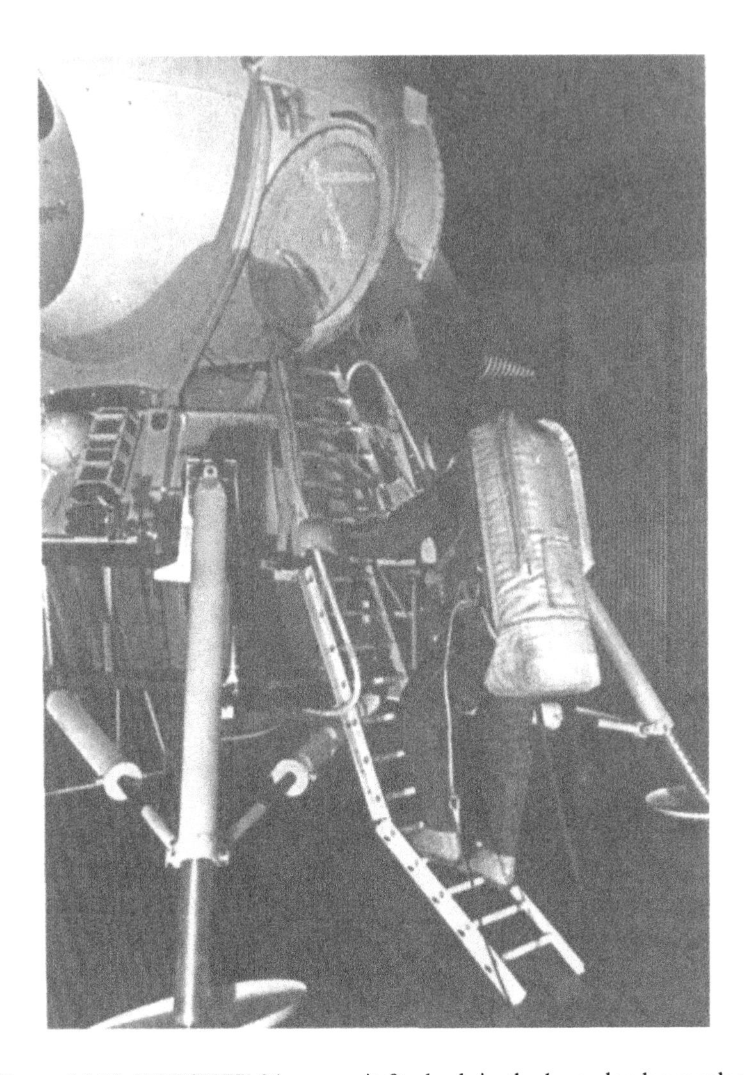

**Figure 6.2.13** KRECHET-94 spacesuit fit check in the lunar lander mock-up.
From Archive Zvezda.

In 1969 nine KRECHET-94 suits (Nos 15–18 and 21–24) were manufactured and transferred for in-house and joint tests. In the same year the following studies, tests and fit checks were either completed or continued:

- fit check work in the lunar lander at OKB-1, including evaluation of the shock-absorbing and attachment system (Figure 6.2.13);
- movement under lunar gravitational conditions was studied on the newly designed test bench, 1/6G (Figure 6.2.14a, b, c)—this work was initiated in 1968;
- movement under short-term weightlessness conditions was studied in a Tu-104 flying laboratory;

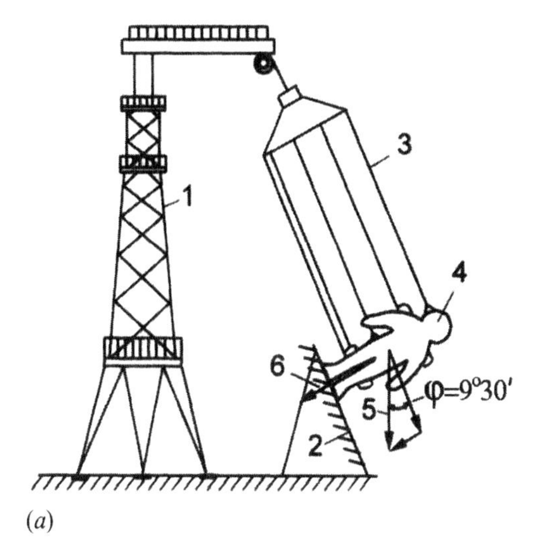


(*a*)

(*b*)

(*c*)

**Figure 6.2.14** Test facility created at Zvezda to study cosmonaut motion under conditions corresponding to lunar gravitation. (*a*) Test facility plan: 1—tower; 2—bearing wall; 3—hanger cable; 4—suited test subject; 5—g-suited test subject mass; 6—$g_M = g \sin \varphi$ mass component, which is 0.165 of a cosmonaut's weight under Earth conditions. (*b*) Suited test subject at the test facility. (*c*) Same as (*b*), but viewed from above.

From Archive Zvezda.

- impact and strength testing of the suit, impact tests of the visor assembly;
- testing of the shield vacuum thermal insulation, heat and vacuum testing of the suit's subassemblies and LSS components (the thermal and vacuum tests of the KRECHET suit were run in the VK 600/300 thermal and vacuum chamber in Zagorsk, near Moscow, which simulated the Moon's conditions. A manikin was used for these tests. Temperatures on the suit enclosure surface itself and of the components attached to its exterior, such as control panels, combined service connectors, etc., were evaluated);

- testing of the main LSS components (useful working life, reliability and other tests);
- in-house tests in the TBK-30 vacuum chamber.

The scope of tests already carried out by 1969 was sufficient to issue (in case of necessity) confirmation of spacesuit readiness for flight tests under the L-3 programme.

But as the date for completion of this programme had been changed, development testing of various items was continued. In 1970 four more suits were manufactured: No. 20 for the tests on compliance with the technical requirements, Nos 25 and 31 for LSS testing and No. 26 for thermal and vacuum chamber tests. In 1971 vacuum and low-temperature tests of the KRECHET-94 and ORLAN suits were continued in the VK600/300 vacuum chamber, which simulated open space and lunar conditions.

The following joint tests in the flying laboratory (Tu-104) were continued:

- development of systems for transfer and airlock operations;
- determination of forces exerted by a cosmonaut on safety tethers and mothercraft handrails; and
- development of tools and welding equipment.

The KRECHET-94 suit was tested in the lunar lander mock-up with simulation of its impact deceleration. Between 1970 and 1971 experimental and research work was carried out on the KRECHET-94 spacesuit (which had a partial pressure glove with a cuff and a suit body with a waist pressure bearing, etc.).

Between 1968 and 1971 a great deal of work was carried out on the ORLAN suit as well. Nine ORLAN suits were manufactured in 1968 for in-house and joint tests, one ORLAN-V suit was manufactured for hydro lab tests and two suits for the Tu-104 flying laboratory (ORLAN-V without weights). Technical tests and tests in the TBK-30 vacuum chamber were carried out in 1970 (Figure 6.2.15), and the first work in the hydro lab for this suit was carried out in 1971.

With the kick-off of the orbiting station programme in 1969, research into the application of the N-1–L-3 ORLAN suit for EVA from an orbiting station was initiated. Further development of the suits for both programmes was carried out in parallel right up to when the work on the lunar suit was suspended in 1972 and finally stopped in 1974 (24 June 1974 on V.P. Glushko's proposal) due to cessation of N-1–L-3 programme (Semeonov, 1996).

It should be noted that in 1972 to 1973 there was an attempt to continue work on the manned lunar programme. Early in 1972 (16 February) a government directive was issued to prepare a technical proposal on the L-3M [Л-3М] programme, which would provide for a three-person spacecraft and two cosmonauts landing on the Moon. In the same year Zvezda prepared a technical proposal on lunar suits for this programme and received draft performance specifications for them from OKB-1. It was planned to modify the KRECHET-94 suit in order to enable six EVAs on the lunar surface over 5 days by refilling and changing expendable LSS elements.

**Figure 6.2.15** ORLAN spacesuit in the TBK-30 thermal vacuum chamber.
From Archive Zvezda.

As subsequent experience of EVA from the orbiting stations has shown, some of the requirements of performance specifications regarding the spacesuit both for the N-1–L-3 and L-3M programmes were difficult to realize. In particular, the planned time periods to make suits operational and their transfer from one spacecraft to another via open space, EVA frequency and the time required for the cosmonaut to stay in the suit all had to be revised to the needs of that particular day.

It should be noted that, simultaneously with the development of the spacesuits for the N-1–L-3 programme, designing and development testing were being carried out of the systems for A.A. Leonov's space walk, of the YASTREB suit and the RVR-1P backpack LSS for *Soyuz* and later for *Almaz*. The experience gained during development of these products and the first EVA sorties was also taken into account for KRECHET-94 and ORLAN systems development.

In parallel with the work at Zvezda, initial work on lunar spacesuit development was also carried out under the leadership of S.M. Gorodinsky in the Institute of Biophysics at the USSR Ministry of Health. But their efforts only resulted in a test sample designed for development of medical and technical requirements.

## 6.3 ORIOL SPACESUIT

The ORIOL spacesuit (soft type) was developed as the second version of the lunar spacesuit mentioned above. It was developed for the same purpose and under the same performance specifications as the KRECHET spacesuit. The ORIOL suit was developed along the classic lines of the soft suit and had a removable backpack that contained the LSS named BAIKAL (Figure 6.3.1).

The suit assembly consisted of an external thermal protection enclosure, a suit enclosure, a removable helmet and a ventilating system. The newly developed water-cooled garment KVO-9 was included in the assembly, as part of the process of suit development.

The suit enclosure had all the typical design elements of soft suits that had been developed at Zvezda. The enclosure consisted of two layers: the restraint layer made of Kapron and the pressure bladder made of a rubberized knitted fabric. The enclosure had a front opening with a lacing (done up by hooks) and a collar with a cable (closed by a special fastener on the neck ring, which served to attach the helmet). The suit body had a restraint system for suit height adjustment and prevention from ballooning. This system was of the same type as used in the Vorkuta aviation suit and the BERKUT spacesuit. The main version of the spacesuit had shoulder bearings and arms similar to those of the SI-5 and S-9 aviation suits. The leg enclosure was similar to that of the ORLAN spacesuit. During suit development, the main attention was paid to improvement of arm and leg enclosure mobility.

Between 1966 and 1969 several fully functional suit enclosure mock-ups were manufactured. Some mock-ups had convolute joints, while others were made of Orthofabric. The mock-up of the BAIKAL removable backpack with the LSS was also developed (Figure 6.3.2). The BAIKAL LSS flow diagram was similar to that of the KASPIY SCLSS. As a result of mock-up development and updates, the ORIOL spacesuit became lighter (mass of the suit proper was 26 kg and of the backpack 60 kg) than the KRECHET-94, but retained all the disadvantages of the soft suit.

As a result of the 1967–1968 decision to use the KRECHET-94 and ORLAN semi-rigid suits for the lunar programme, work of the ORIOL suit programme changed direction.

ORIOL suit development was continued with the purpose of using it as an emergency and rescue suit both for high-altitude aircraft (ORIOL-A) and spacecraft. The suit design was radically changed. The orthofabric was used for the suit's arms and legs, which made it possible to divert the convoluted joints (Figure 6.3.3).

**Figure 6.3.1** General view of the ORIOL pressure suit: (*left*) minus the outer garment; (*right*) with the mock-up of the BAIKAL backpack.
From Archive Zvezda.

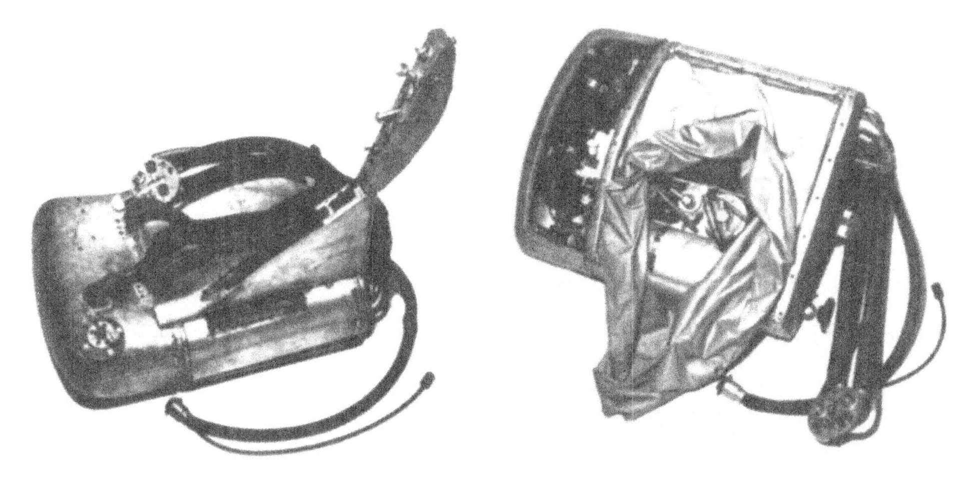

**Figure 6.3.2** Two general views of the BAIKAL backpack mock-up.
From Archive Zvezda.

**Figure 6.3.3** The ORIOL spacesuit, arms and legs of which were made of Orthofabric.
From Archive Zvezda.

Development of lightweight suits for short emergency EVA sorties was also considered.

Later, the experience gained in ORIOL suit development testing was used for further development of the emergency and rescue suits Sokol and Baklan.

# 7

# SOKOL-K and SOKOL-KV-2 rescue suits for *Soyuz*

## 7.1 INTRODUCTION

In the second half of 1969, as a result of development work relating to the *Salyut* orbiting station, the *Soyuz* transport spacecraft was modified. The modified *Soyuz* enabled the spacecrew to transfer from it to the orbiting station (*Salyut*) through an inner passage in the docking system without the need to space-walk (Figure 7.1.1). Thus there was no longer any need for the YASTREB spacesuit and the self-contained life support system (SCLSS) in this spacecraft (now with the index 7K-T, instead of 7K-OK).

The first flight of this new series of spacecraft (*Soyuz-10*) to the *Salyut* orbiter (*Salyut-1*) with a three-member crew took place on 23–25 April 1971. Zvezda supplied the flight garments, shock-absorbing seats, waste management system, drinking water system and survival kits for these new spacecraft. The use of rescue suits in the *Soyuz* spacecraft type 7K-T was, as was the case in the previous 7K-OK spacecraft, according to the OKB-1 (Experimental Design Bureau 1, which became TsKBEM, now RSC Energia) documentation, still not foreseen (see Section 5.2).

The second flight of *Soyuz-11* to *Salyut-1* had a tragic ending. On 30 June 1971 the cosmonauts G.T. Dobrovolsky, V.N. Volkov and V.I. Patsayev perished due to sudden spacecraft depressurization in the descent phase of the mission.

After that, based on the conclusions made in August 1971 of a specially organized governmental commission into the accident (chaired by M.V. Keldysh), a number of actions were taken to improve crew safety. Those actions involved both spacecraft modification and the introduction of protective gear for crew rescue in case of descent vehicle depressurization during the launch, docking and descent phases of the mission. Zvezda was given the task of designing the protective gear, compatible with the personal couch liners of the *Soyuz* spacecraft's shock-absorbing

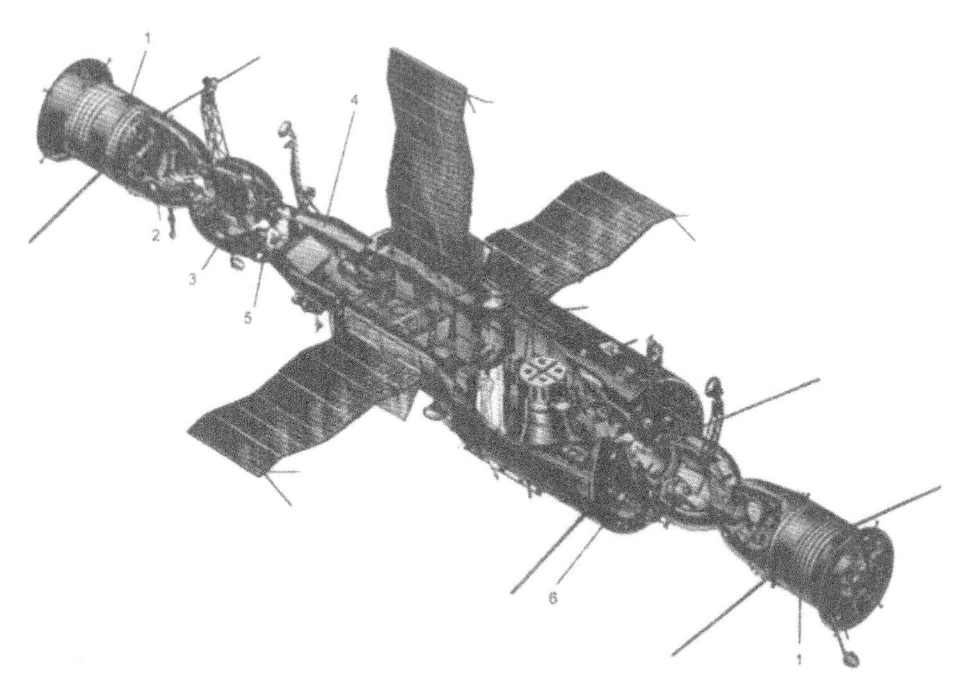


**Figure 7.1.1** Diagram of cosmonaut transfer from the *Soyuz* spacecraft to the *Salyut* orbiting station through the inner hatch: 1—*Soyuz* spacecraft; 2—descent vehicle; 3—utility module; 4—transfer module; 5—inner hatch; 6—intermediate chamber.

seats, in the shortest possible time and with a minimum of modifications to the descent vehicle.

## 7.2 SOKOL-K

None of the earlier developed spacesuits (SK-1, BERKUT, YASTREB) could be used for this purpose, as they were basically designed either for cosmonaut protection outside the spacecraft or were not compatible with the shock-absorbing seat.

Due to weight limitations it was assumed that the maximum possible time period for return of the crew to the Earth in case of an emergency depressurization of the descent vehicle in any of the most critical phases of the mission would be between 105 and 125 minutes.

Several versions of protective equipment were considered: various types of a high-altitude suit with pneumatic and mechanic compensation for the positive pressure in the pressure helmet (partial pressure suit) and a lightweight full pressure suit designed on the lines of the Sokol aviation pressure suit (see Figure 2.2.12). Fit checks of such flight gear mock-ups with the Kazbek seat's couch liners (Figure 7.2.1a, b) and analyses of the flow diagrams for feasible oxygen supply systems and their weight were carried out. As a result of this activity, the

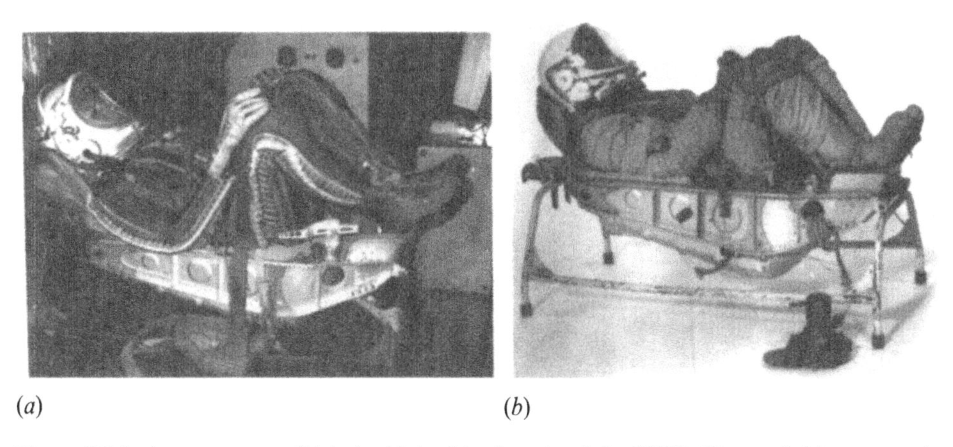

(*a*) (*b*)

**Figure 7.2.1** Arrangement of (*a*) the high-altitude suit of the VKK-47 type (with pneumatic and mechanic compensation) and (*b*) the Sokol aviation suit in the KAZBEK shock-absorbing seat.

From Archive Zvezda.

optimum version was selected: a lightweight full pressure suit with a soft fixed helmet (which had a sliding visor) and an open-type LSS designed for 2 hours of operation.

The Sokol aviation pressure suit adopted as a prototype for the rescue suit required considerable modifications, however. The helmet was of major concern: a lightweight soft helmet was designed and mounted on the suit enclosure, instead of a rigid rotating helmet with a neck pressure bearing. The front portion of the new helmet was a sliding visor shaped as part of a sphere. The nape portion was made as a soft extension of the suit torso enclosure (the restraint layer plus the pressure bladder). The lower edge of the visor was bounded by a half-frame, which formed an interface with another half-frame attached to the torso collar. The half-frames were hinged in the ear zones (Figure 7.2.2). This design concept was developed earlier for the aviation pressure suits of the VSS type (high-altitude full pressure rescue suit) developed in the 1950s (see Figure 2.2.3). The rescue suit's helmet visor was for the first time made of transparent polycarbonate. The "unnecessary" components and back lacing were removed from the suit enclosure, as was the ventilation garment, in order to reduce its weight as much as possible. Thus, the suit enclosure had a minimum of two layers: the restraint layer and the pressure bladder (made of 0.6-mm sheet rubber). The thickness of the enclosure did not differ greatly from that of cosmonaut flight garments used earlier. The pressure suit appendix was made of rubberized fabric. The suit had to be used with the cosmonaut wearing cotton underwear, a communication cap and a harness with biomedical sensors.

The suit was configured in such a way that it helped a cosmonaut to take an "embryo" posture in the shock-absorbing seat, dictated both by the cabin dimensions of the descent vehicle and requirements for withstanding acceleration (Figure 7.2.3). The new rescue suit for the *Soyuz* spacecraft was called SOKOL-K ("K" stands for "space"). The weight of the new suit was 9–10 kg. The suit-donning

**Figure 7.2.2** SOKOL-K suit-donning by G.S. Paradizov at Zvezda.
From Archive Zvezda.

time was 10–12 minutes. A person wearing this suit could survive up to several hours in coldwater, if there was not enough time to don their FOREL suits (see below). A special flotation collar was available (in the survival kit) to provide a stable and comfortable posture while afloat. However, if it was necessary to leave the descent vehicle and stay in coldwater, cosmonauts would doff their rescue suits and put on a special immersion suit (FOREL) with thermal protection (also in the survival kit).

The SOKOL-K suit was developed under time pressure because the launch date of the next *Soyuz* spacecraft depended on it. Another reason for selecting the Sokol aviation suit as a prototype was the fact that Zvezda had a reserve of these suits at that time. Five available Sokol aviation suits were modified in the shortest time possible (by the end of 1971), and laboratory tests and fit checks, including those in the descent vehicle, were carried out. In parallel, a full set of engineering documentation for all suit subsystems and subassemblies (helmet, breathing valve, suit enclosure, in-suit ventilation system, service line feedthrough, etc.) was developed and eight new SOKOL-K rescue suits were manufactured in the first half of 1972 for the in-house and integration tests.

At the same time the spacesuit's LSS flow diagram (Figure 7.2.4) was developed (together with TsKBEM), the main components of the ventilation and oxygen

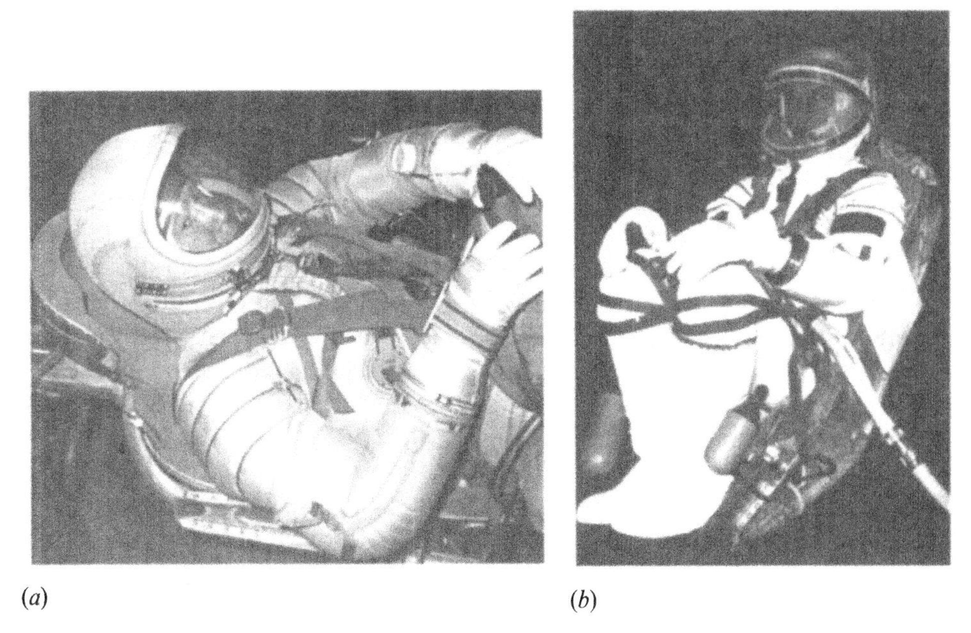

(*a*) (*b*)

**Figure 7.2.3** Cosmonauts A.A. Leonov (*a*) and V.A. Dzhanibekov (*b*) wearing SOKOL-K suits are in the KAZBEK seats (Leonov's suit was pressurized to 400 hPa).

From Archive Zvezda.

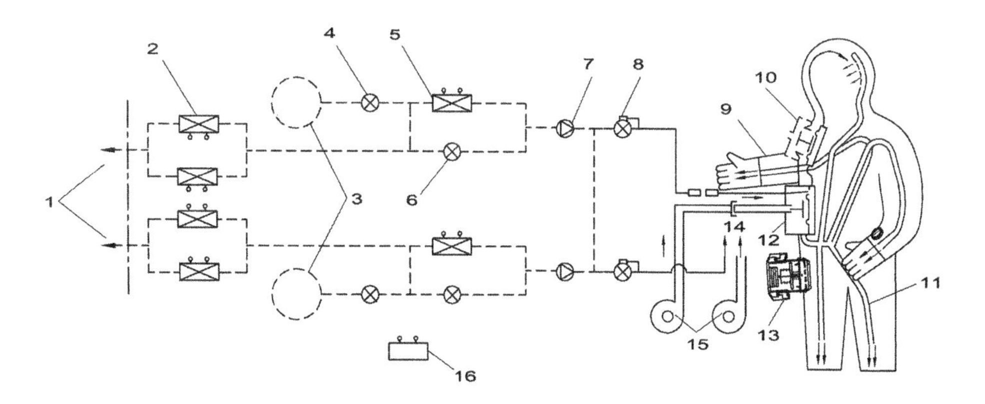


**Figure 7.2.4** Flow diagram of the on-board LSS of the SOKOL-K rescue suit for the *Soyuz* spacecraft: 1—gas discharge to the environment prior to the spacecraft landing; 2, 5—electric valves; 3—gas store in the bottles (at 25 MPa); 4, 6—shut-off valves; 7—check valve; 8—reducer; 9—SOKOL-K rescue suit; 10—breathing valve; 11—suit ventilation system; 12—umbilical/suit interface unit with pneumatic shut-off valve; 13—suit pressure regulator; 14—to the second suit; 15—fans; 16—barometric relay unit.

From Archive Zvezda.

supply systems were manufactured and the test set-up was configured. The SOKOL-K rescue suit was of the ventilation type, as far as its operational principle was concerned. It was ventilated by pressure-tight descent vehicle cabin air supplied to the suit helmet, arms and feet. On cabin depressurization, when the cabin pressure dropped down to ~600 hPa, the gas mixture supply to the suit helmet was activated and the air supply by the fan stopped (the helmet had to be closed manually in all critical phases of the mission). There was 40% oxygen in the gas mixture in order to avoid excessive oxygen concentration in the cabin. TsKBEM was responsible for the delivering system of the gas mixture supply to the suits, which was located in the descent vehicle of the *Soyuz* spacecraft. The gas escaped from the suit through a pressure regulator, maintaining suit pressure of ~400 hPa. The pressure regulator also served as a relief valve. An air inflow valve, which allowed the cosmonaut to breathe environmental air on termination of the gas supply to the helmet (with the visor closed), was located on the helmet.

By autumn 1971 a list of activities calling for a great amount of testing was drawn up and approved by TsKBEM and corresponding Air Force officials (Figure 7.2.5). The following activities were required to be performed at the Air Force Scientific Research Institute (GK NII) in addition to ordinary tests in Zvezda's thermal vacuum chamber:

- evaluation of the capability to stay inside the descent vehicle for 24 hours with the suit on (under conditions of a descent vehicle stay in orbit without a utility module or after splashdown);
- development of a procedure for unassisted suit-donning under weightlessness conditions (flying laboratory).

Moreover, it was planned to perform:

- evaluation of the capability to stay afloat using the flotation means in Zvezda's water tank;
- sea tests of the suit (Black Sea) in order to estimate how long the suit could be worn in coldwater.

The full programme of in-house engineering and physiological tests of the rescue suit was completed in 1972.

Flight operation of the SOKOL-K rescue suit started on 27 September 1973. Vladimir Lazarev and Oleg Makarov, who took off in *Soyuz-12* on that day, were the first to use them (Figure 7.2.6). All subsequent *Soyuz* flights were undertaken using these suits (Figure 7.2.7 and 7.2.8). In 1974 manufacture of six SOKOL-K suits for three crews of the *Soyuz–Apollo* programme began. These suits were labelled "SK-11". *Soyuz-19* (with the crew wearing these suits) got off the pad successfully on 15 July 1975 (Figure 7.2.9).

SOKOL-K improvement work began in parallel with SOKOL-K rescue suit operations in late 1973. At the initial stage, improvement concentrated on suit modification in compliance with the cosmonauts' personal anthropometric

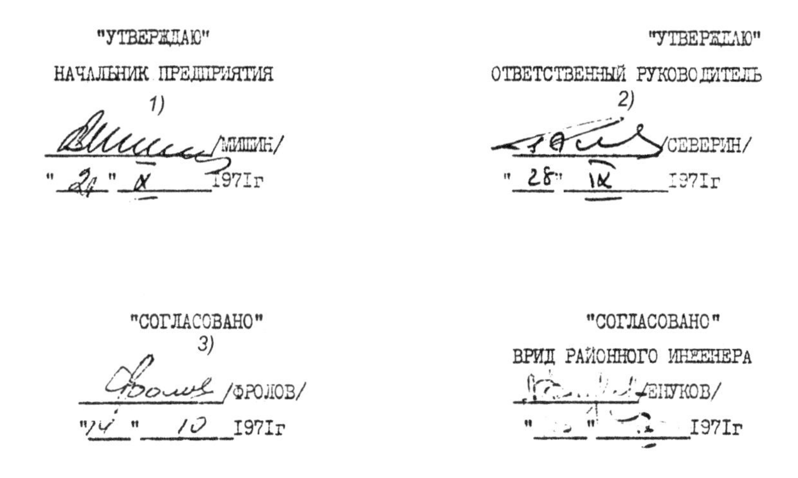

"УТВЕРЖДАЮ"
НАЧАЛЬНИК ПРЕДПРИЯТИЯ
1)
/МИШИН/
"   "      1971г

"УТВЕРЖДАЮ"
ОТВЕТСТВЕННЫЙ РУКОВОДИТЕЛЬ
2)
/СЕВЕРИН/
"28" IX 1971г

"СОГЛАСОВАНО"
3)
/ФРОЛОВ/
"   " 10 1971г

"СОГЛАСОВАНО"
ВРИД РАЙОННОГО ИНЖЕНЕРА
/...ЕНУКОВ/
"   "      1971г

ПЕРЕЧЕНЬ № 147/224-71
испытаний по комплекту защитного снаряжения для
объекта IIФ615.

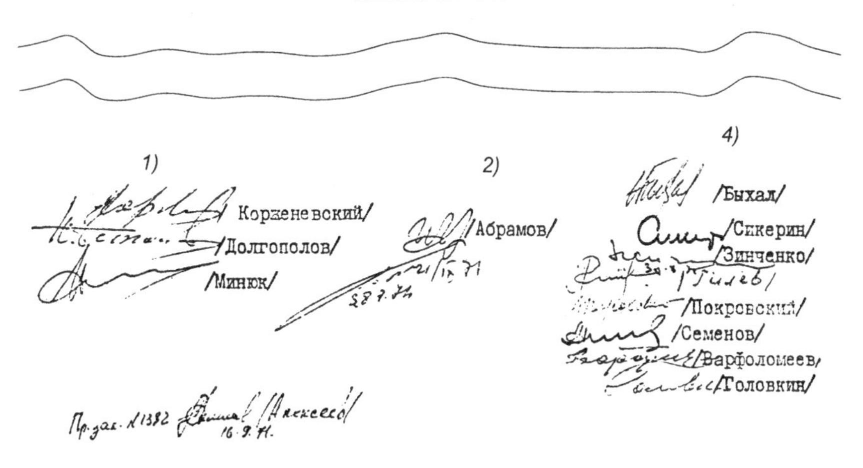

1)
/Корженевский/
/Долгополов/
/Минюк/

2)
/Абрамов/

4)
/Быхал/
/Сикерин/
/Зинченко/
/Гилев/
/Покровский/
/Семенов/
/Варфоломеев/
/Головкин/

**Figure 7.2.5** The title and signature pages of the list of tests for the SOKOL-K rescue suit set. Authors' remarks: (1) OKB-1; (2) Zvezda; (3) Air Force administration; (4) representatives of various Air Force services.

From Archive Zvezda.

parameters (stature, chest circumference, arm and leg length). Special attention was paid to development and manufacture of new standard-size spacesuits and their upgrade in the process of fit checks and training. An extensive study of the Air Force pilot population was carried out in order to develop a dimensional typology (an anthropometric standard).

**Figure 7.2.6** Cosmonauts V.G. Lasarev and O.G. Makarov prior to the first mission in which the SOKOL-K suit was used (Zvezda's laboratory at Baikonur, 1973). *Standing (from left to right)*: N.V. Knyazev, M.M. Ikonnikov, J.P. Abramov, M.M. Balashov, a representative of the Cosmonaut Traning Centre, V.A. Dubrov, V.V. Shuvalov, the two representatives of the Cosmonaut Training Centre, V.S. Agureyev and a representative of the cosmodrome military unit.

From Archive Zvezda.

Rescue suit enclosure development and upgrade was carried out mainly by a team of designers headed by G.S. Paradizov under the management of the leading Zvezda designer A.Yu. Stoklitsky. The LSS development was carried out by a team of designers under the management of the leading Zvezda designer I.P. Abramov and the team head D.V. Kuchevitsky. The SOKOL-K suit and its LSS were tested under the management of B.V. Mihkailov and V.I. Svertshek.

Operation of the new SOKOL-K rescue suits showed that they fully complied with the requirements set forth, but some drawbacks were detected. The work on removal of those drawbacks was carried out in the subsequent years (see Sections 7.3 and 7.4). The following should be mentioned as principal among the suit enclosure drawbacks:

- Variance in the dimensions and shapes between some enclosure parts and posture of the human body in the KAZBEK seat. Such a variance caused discomfort (primarily under the knees) for some cosmonauts, who wore the suit for a long time.

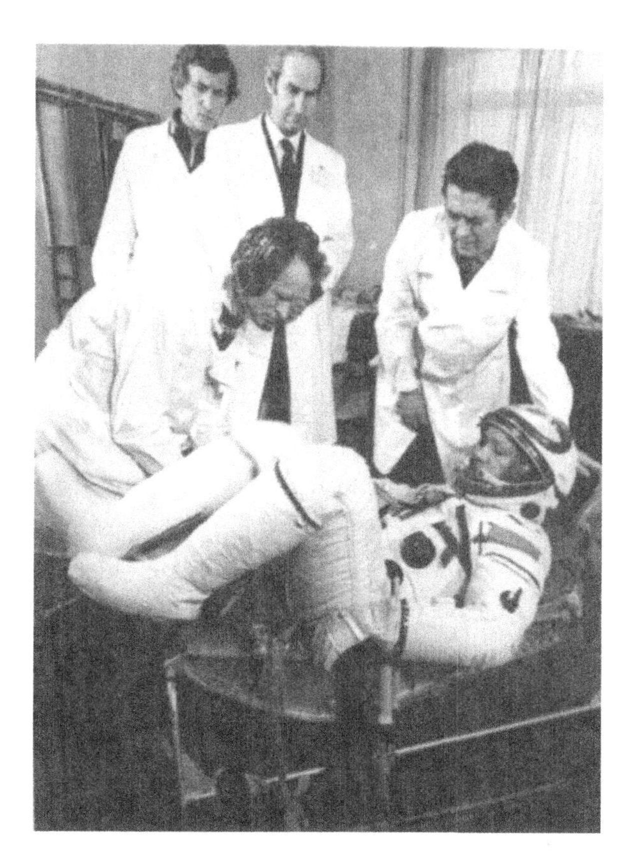

**Figure 7.2.7** Cosmonaut N.N. Rukavishnikov performs final fit checks of his SOKOL-K suit and seat in Baikonur, December 1974. Also shown are V.V. Shuvalov and I.Y. Novachatski, leaning over N.N. Rukavishnikov, and M.M. Balashov and I.P. Abramov, standing behind them.

From Archive Zvezda.

- Suit-donning was both difficult and time-consuming, especially undesirable in an emergency (e.g., spacecraft cabin depressurization), when it was vital to don the suit as quickly as possible. This drawback was caused by a design in the suit enclosure that had been used in all previous types of (aviation and space) soft suits:
  - the front opening that was opened and closed by means of lacing, the so-called "collar tightening" (a device with a cable and a lock), to connect the collar to the lower half-frame of the helmet interface;
  - sealing of the suit by tying the appendix (Figure 7.2.2);
  - complicated usage of the pressure regulator located on the left side of the torso enclosure opening (see Figure 7.2.7).
- Restricted downward view from the helmet caused by the comparatively high lower half-frame of the helmet having a collar connection device.

**Figure 7.2.8** Cosmonauts A.V. Filipchenko and N.N. Rukavishnikov after SOKOL-K donning in Zvezda's laboratory at Baikonur on 2 December 1974, the launch day of *Soyuz-16*. This was part of the *Soyuz–Apollo* programme. *From left to right*: I.Ya. Novokhatsky, I.P. Abramov, A.A. Leonov, V.A. Shatalov, V.F. Bykovsky, G.T. Beregovoy, V.P. Glushko, V.N. Kubasov, M.M. Ikonnikov, Ye.V. Shabarov, E.I. Vorobiev.

From Archive Zvezda.

**Figure 7.2.9** Cosmonauts A.A. Leonov and V.N. Kubasov (wearing the SOKOL-K suits) are about to take the bus to the launch pad surrounded by press photographers and ground support team.

From Archive Zvezda.

The design of the selected LSS also had a significant drawback. Due to weight limitations on the spacecraft and the anticipated short time period of working in the depressurized descent vehicle, the lowest possible gas supply was selected (20 normal litres per minute or 50 volumetric litres under suit pressure). Such a supply provided for $CO_2$ and moisture removal from the helmet, but could not essentially maintain the heat balance for a cosmonaut in an emergency, especially in the case of increased physical activities by the crew. Use of a gas mixture containing 40% oxygen instead of 100%, did not contribute to increasing the crew's capacity for work.

## 7.3 SOKOL-KM

On 13 March 1973 a meeting of the Zvezda Scientific and Engineering Board was held, where proposals for SOKOL-K design improvement were discussed. One proposal was the development of several versions of a mock-up enclosure for a modified SOKOL-KM spacesuit.

The main version was a mock-up having a transverse, pressure-tight waist interface, which consisted of two separate zippers with sealing elements between them. The waist interface divided the enclosure into the "shirt" combined with the helmet and the "pants". Such a design concept provided certain advantages:

- no appendix;
- the lower half-frame of the helmet was fixed to the body enclosure, making it possible to reduce the half-frame height (owing to removal of the collar tightening), and extend the lower edge of the visor downwards and thus increase the field of view from the helmet. At the same time, the dimensions of the whole helmet were increased because the SOKOL-K suit helmet turned out to be too tight for some cosmonauts.

The in-suit ventilation system was moved from the suit enclosure to the underwear overalls. It was assumed that by dividing the suit in this way (shirt and pants) would make it possible to efficiently and easily choose the required standard size of each element of the suit ensuring a better fit to the cosmonaut's body (e.g., a shirt of size 54 and pants of size 52). This mock-up (Figure 7.3.1) was manufactured in 1973 and laboratory tested at Zvezda in 1974.

At the same time other proposals on enclosure improvement were studied and developed:

- reducing the weight by manufacturing the suit body bladder from rubberized Kapron;
- improving mobility by using orthofabric for the suit's arms;
- using rubberized knitted fabric for the pressure bladder of arms and legs;
- updating the Anthropometric Table regarding working posture;
- development of new patterns for the cut-out of the enclosure parts; and
- possibility of using restraint layer parts made of net.

**Figure 7.3.1** General view of the SOKOL-KM rescue suit.
From Archive Zvezda.

The future of the modified suit mainly depended on the directions taken in *Soyuz* spacecraft improvement. As is known, the use of rescue suits in the *Soyuz* spacecraft led to the removal of one of three seats from the cabin of the descent vehicle and installation of the system supplying gas to the suit in its place. This is why TsKBEM ran the activities on development of a modified *Soyuz* spacecraft (*Soyuz-T*, item 11F732), designed for a three-member crew. In this case the gas mixture supplied to the suit was changed to 100% oxygen (availability of an oxygen reserve in the descent vehicle of the *Soyuz-T* became a common feature for the suit and the on-board LSS). This made it possible to implement the second, lower pressure mode of

270 hPa, which was achieved with a new pressure regulator. Moreover, in order to improve heat removal, Zvezda drew up a new design for the rescue suit and the on-board LSS, by using a water-cooling system.

In 1973 Zvezda and TsKBEM made a joint decision to introduce a water-cooled garment into the suit for the *Soyuz-T* spacecraft and agreed on the initial data. Zvezda developed several versions of water-cooled suits and vests as well as an on-board unit for water circulation and temperature control (water cooling was to be the responsibility of the descent vehicle's on-board systems).

Despite the new requirements for heat removal from the suit, the work on the SOKOL-KM suit was stopped. However, SOKOL-KM was used as the basis for development of a new suit modification, SOKOL-KV, with a water-cooled garment as a part of it.

## 7.4 SOKOL-KV

In 1974 a new set of designs were drawn and the manufacture of six SOKOL-KV suits for laboratory tests started. The first three suits were completed in that year.

The SOKOL-KV rescue suit (Figure 7.4.1a, b) had the following principal differences from the SOKOL-K rescue suit:

- new means for suit-donning/doffing and sealing were introduced (namely the waist had an elastic, pressure-tight interface with zippers that separated the enclosure into two parts—a shirt with the helmet and the pants);
- the helmet was larger and had a wider field of vision;
- the pressure regulator was designed to provide for two pressure modes in the suit;
- the pressure regulator was integrated with a breathing valve (in case of splash-down it allowed the cosmonaut to breathe with the helmet closed) and was located in the centre of the chest below the helmet, a place that was easily reached with both hands;
- a water-cooling system was introduced and was integrated with the in-suit ventilation system (it was made as a separate garment, named the KVO-11);
- straight-through flanges for the hoses of the water-cooling system were installed on the suit enclosure (on the pants), while pneumatic feedthroughs for air and oxygen were located on the shirt;
- the restraint layer of the suit's arms and pants as well as the pressure bladder in soft joint areas were respectively made of orthofabric and rubberized knitted fabric.

During testing of the SOKOL-KM mock-up, it was discovered that the waist interface had to have a sufficiently large perimeter to facilitate shirt-donning and thus the transverse diameter of the suit body had to be larger than that of suits with a longitudinal don/doff opening. Therefore, to reduce the transverse diameter of the

**Figure 7.4.1** The SOKOL-KV rescue suit: (*a*) general view; (*b*) lower suit enclosure alone.
From Archive Zvezda.

SOKOL-KV suit, the interface plane was inclined by 30°, which resulted in an increase in the front waist portion of the pants and facilitated shirt-donning.

In 1974 the SOKOL-KV suit underwent laboratory tests, including functional tests, evaluation tests on donning and accommodation in the KAZBEK shock-absorbing seat, and flotation tests in the Zvezda water tank. In 1975 the suit development testing was continued and fit checks in the spacecraft were carried out.

The following new suit elements were developed and integrated in the suit design:

- combined service connector, ORK-21;
- in-suit ventilation and oxygen connectors;
- KVO-12 water-cooled garment designed as pants with built-in chambers of the PPK-S post-flight prophylactic suit; and
- new pressure gloves, GP-7.

A flow diagram and a circuit diagram of the suit and on-board equipment were developed for the three-seat configuration of the *Soyuz-T* spacecraft (see also Figure 7.6.3).

A mock-up approval board was held at Zvezda on 3 December 1975, and an inter-agency board reviewed the SOKOL-KV spacesuit in early 1976. Representatives of the TsKBEM, Institute of Aviation and Space Medicine and GK NII took part. Both boards approved the SOKOL-KV spacesuit for usage in the *Soyuz-T* spacecraft.

## 7.5 SOKOL FOR *ALMAZ*

In parallel with the commencement of flights of the *Salyut* station, the TsKBM enterprise (earlier OKB-52, headed by V.N. Chelomey, today NPO Mashinostroyenia) continued its work to create the *Almaz* station. Due to the fact that no transport spacecraft specially designed for *Almaz* was ready, the first crew missions to this station (popularly named *Salyut-3* in 1974 and *Salyut-5* between 1976 and 1977) used modified *Soyuz* spacecraft that were equipped with the KAZBEK shock-absorbing seats and SOKOL-K rescue suits produced by Zvezda.

At the same time, a branch of the TsKBM continued work on development of its own transport spacecraft. Between 1972 and 1978 Zvezda carried out a great deal of work to adapt the SOKOL-K rescue suit to the systems of the TsKBM re-entry vehicle's design. As far as development of the modified SOKOL-KV rescue suit went, the TsKBM also directed its own efforts to it. Most parts of the re-entry vehicle's oxygen system, as well as the suit ventilation system and the cooling unit, were developed by Zvezda, unlike the suit oxygen supply system developed by NPO Energia that was located in the *Soyuz* and *Soyuz-T* spacecraft's descent vehicle.

It should be noted that, besides the SOKOL-K and SOKOL-KV versions manufactured between 1970 and 1973, Zvezda considered a number of alternative systems for crew protection in case of spacecraft depressurization. Specifically, there were projects to develop various types of partial pressure suits (designed to have a common pressure) and to develop oxygen equipment for the 7K-VI, a military version of the *Soyuz* spacecraft. Experimental samples of the items were manufactured and tested.

The performance specifications for the 7K-VI spacecraft's protective gear, which were issued by TsKBEM as early as 1970, stipulated the use of protective gear together with a self-contained portable LSS for rescue operations in any compartment of the spacecraft for up to 4 hours. Zvezda studied the flow diagrams of the closed-loop LSS and the SOKOL-KR spacesuit (regenerative type), which made it possible to prolong working hours in case of emergency. This study was carried out

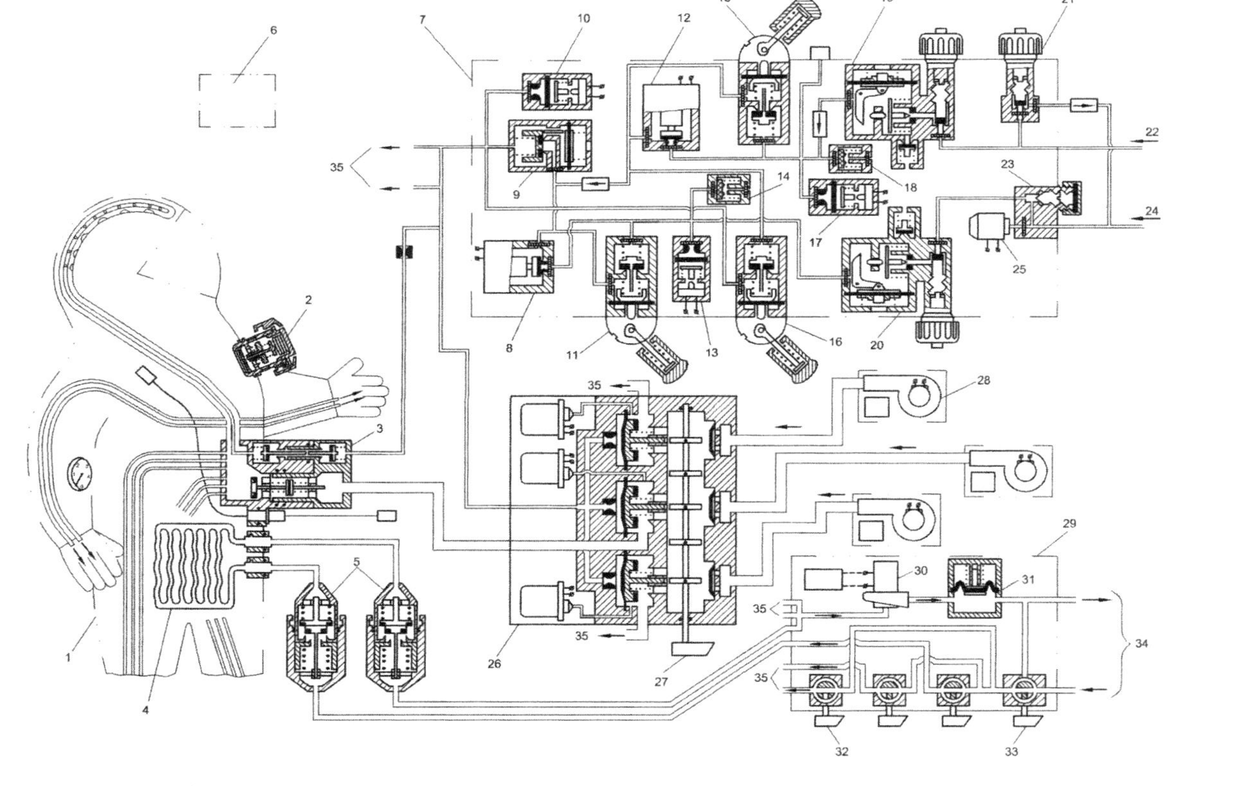


**Figure 7.5.1** Diagram of the on-board LSS for the **SOKOL-KV** suit in the re-entry vehicle of the transportation spacecraft under the *Almaz* programme: 1—**SOKOL-KV** suit; 2—suit pressure regulator; 3—combined service connector; 4—water-cooled garment; 5—connectors of the water-cooling line; 6—barometric relay unit; 7—oxygen supply unit (**BPG**-1); 8, 12—electric and pneumatic valves; 9—low-pressure reducer; 10, 11, 15—annunciators of and valves for actuation of gas supply to the spacesuit; 13, 17—annunciators of actuation of gas supply from the re-entry vehicles gas supply and the orbiting module, respectively; 14, 18—discharge valves; 16—valve for actuation of additional gas supply to the suit line; 19, 20—high-pressure reducers with the shut-off valve; 21—shut-off valve; 22, 24—from the orbiting module and re-entry vehicle's gas supplies, respectively; 23—charging union; 25—pressure sensor; 26—**BR**-1 distribution unit; 27—ventilation mode switch-over handle; 28—fan; 29—**BG**-1M hydraulic unit; 30—water pump; 31—hydraulic accumulator; 32—shut-off valve; 33—water temperature control valve; 34—to the on-board heat exchanger; 35—to the second and third suits.

From Archive Zvezda.

for the purpose of providing rescue equipment for the re-entry vehicle developed by TsKBM. The on-board LSS was to be developed partially by Zvezda and partially by Nauka. However, due to delays in development of this system, the main version for the re-entry vehicle was still the SOKOL-KV spacesuit with its oxygen supply, ventilation and cooling units.

At the same time it was planned to extend the availability of oxygen for the crew, in case of re-entry vehicle depressurization, by 3 hours (105 minutes from the oxygen supply on board the re-entry vehicle and 75 minutes from the orbiting module's oxygen system). The re-entry vehicle oxygen was stored at a pressure of 40 MPa. The adopted diagram of the re-entry vehicle's on-board LSS for the SOKOL-KV spacesuit is given in Figure 7.5.1. In 1975 the following equipment was manufactured and tested: on-board distribution unit, BR-1 (which provided back-up to the suit fans located in the cockpit); cooling unit, BG-1M; and gas supply unit, BPG-1. After cessation of the *Almaz* programme, work on the re-entry vehicle was also terminated.

## 7.6 SOKOL-KV-2

During development of the SOKOL-KV rescue suit the issue about the necessity for water cooling was discussed. The decision was finally made to continue development of the SOKOL-KV suit without the water-cooling system, because this system would complicate operation of the suit as well as the spacecraft and there was no absolute necessity for it (since operational conditions, including the time of flight in the depressurized cabin of the *Soyuz-T*, were the same as for the *Soyuz* spacecraft). A similar decision was made later regarding the water-cooling system for the TsKBM re-entry vehicle.

Suit manufacture, testing and upgrade went on. In 1979 in-house testing of the rescue suit and the on-board part of the system was undertaken, as was joint testing in the mock-up of the re-entry vehicle. These tests showed that the transverse suit waist interface was not reliable enough at operation in spite of its benefits. The main reason was the unreliable separating links of the zippers (provided by general industry), which sometimes failed to close properly. In order to provide the required reliability, maintain all the benefits of the design for the SOKOL-KV spacesuit torso, its subassemblies and units, and to retain experience gained during training and operational programmes for the SOKOL-K suit, the decision was made together with the customer to partially modify the suit and give it the name SOKOL-KV-2.

Instead of the transverse interface, this suit had a front interface in the form of a wedge with its point directed downward and two separating zippers as the sides of the wedge (Figure 7.6.1 and 7.6.2). Again the appendix was used to seal the suit.

Having carried out all the modifications, the following benefits of the SOKOL-KV suit were retained:

- shorter donning time (zippers were used instead of lacing and collar tightening was removed);

**Figure 7.6.1** SOKOL-KV-2 rescue suit.
From Archive Zvezda.

- bigger helmet with increased field of vision;
- the pressure regulator was located in an easy-to-reach place and its functions were combined with the functions of the breathing valve.

The design of the suit's arms, legs and gloves remained the same. As for removal of the water-cooling system, the in-suit ventilation tubes were transferred to the suit enclosure. The ORK-21 connector (a combined service connector for the ventilation, oxygen and electrical power lines) was changed into one that had two hoses for ventilation and oxygen (with bayonet connectors on the loose ends) and a separate electrical connector, similar to those of the SOKOL-K suit. The BG-1M cooling unit was excluded from the on-board equipment. Figure 7.6.3 shows the *Soyuz-T*, *Soyuz-TM* and the new *Soyuz-TMA* on-board LSS flow diagram.

In 1979 documentation for the SOKOL-KV-2 rescue suits was developed, physiological tests and additional joint tests at GK NII were carried out, and production of the flight suits began.

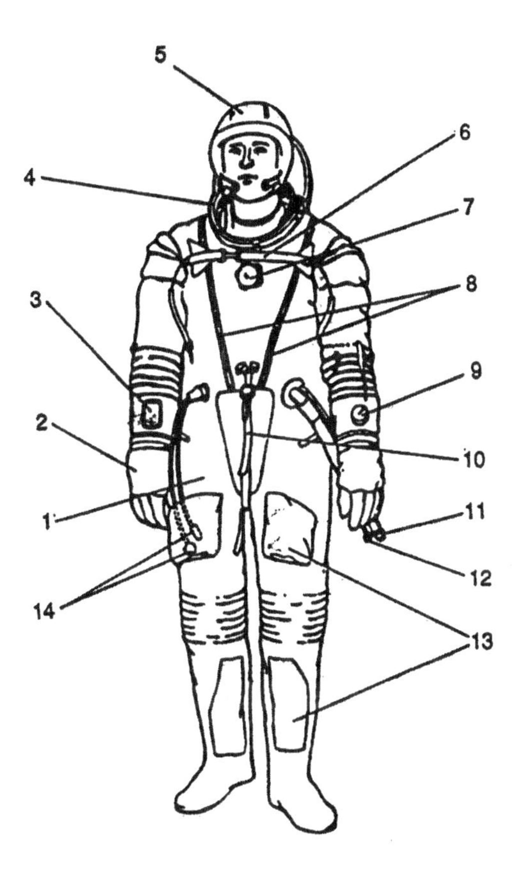


**Figure 7.6.2** SOKOL-KV-2 rescue suit: 1—suit enclosure; 2—gloves; 3—mirror; 4—helmet interface; 5—headset; 6—transverse strap with a snap hook; 7—pressure regulator; 8—front zippers; 9—pressure gauge; 10—front adjustment strap; 11—ventilation hose; 12—oxygen supply hose; 13—pockets; 14—electric connectors for radio communication means and medical sensors.

From Archive Zvezda.

Cosmonauts Yu.V. Malyshev and V.V. Aksenov undertook the first mission using the SOKOL-KV-2 suits in the *Soyuz-T-2* spacecraft on 5 June 1980, and cosmonauts L.D. Kizim, O.G. Makarov and G.M. Strekalov undertook the second mission in the *Soyuz-T-3* spacecraft. SOKOL-KV-2 suits are still being successfully used in the *Soyuz-TM* and *Soyuz-TMA* spacecraft including the *ISS* crews.

Zvezda's experience of the suit fit checks and evaluation by the crews themselves also contributes to the successful operation of the suits. After manufacture of the flight suit (Figure 7.6.4) every crew member evaluates how well it has been adjusted. To do this the suited crew member occupies the KAZBEK seat with its custom-fitted couch liners and sits in the flight posture under positive pressure for 2 hours (under ground conditions). The suit is readjusted or upgraded by the outcome of this. The

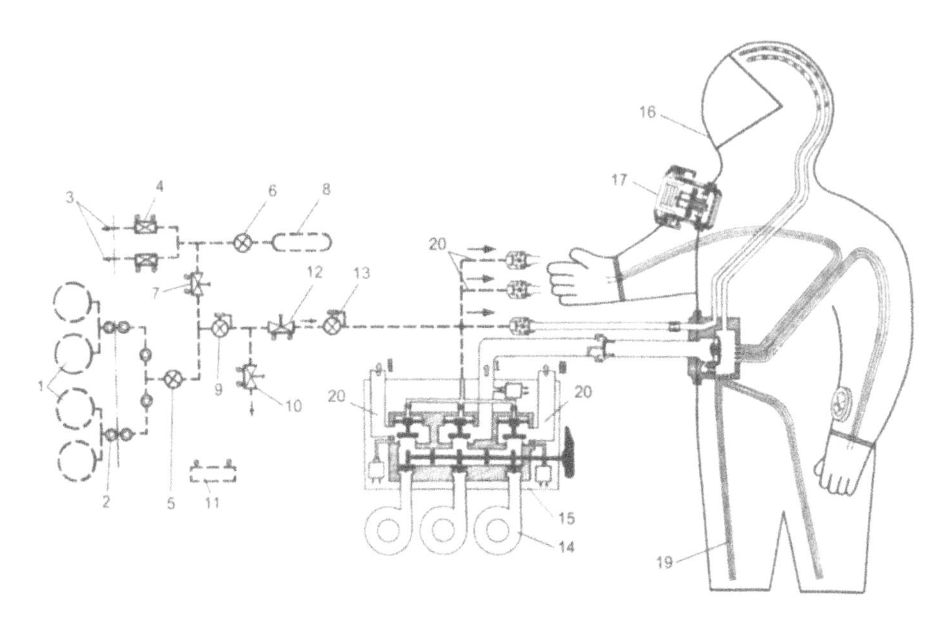


**Figure 7.6.3** Diagram of the on-board LSS for the SOKOL-KV-2 suit for *Soyuz-T*, *Soyuz-TM* and *Soyuz-TMA* spacecraft: 1—oxygen supply in bottles; 2—check valve; 3—oxygen discharge into the environment prior to landing; 4—electric valve; 5, 6—shut-off valves; 7, 10, 12—pilot-operated valves; 8—oxygen bottle; 9, 13—reducers; 11—barometric relay unit; 14—fan; 15—BR-1M unit; 16—SOKOL-KV-2 suit; 17—suit pressure regulator; 18—hose group feedthrough; 19—suit ventilation system; 20—to the second and third suits.
From Archive Zvezda.

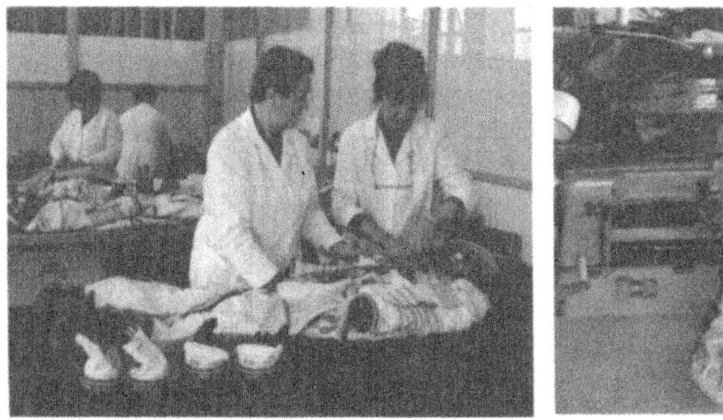
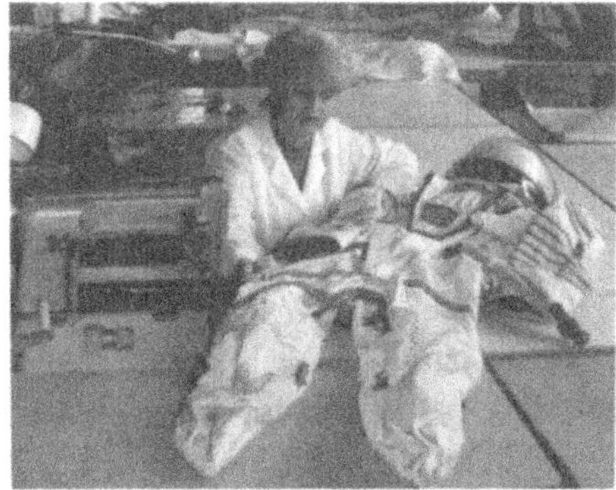

**Figure 7.6.4** Manufacturing SOKOL-KV-2 suits for the crew of the *ISS* expedition (February 2003).
From Archive Zvezda.

next stage of suit evaluation is repetition of the same procedure in the vacuum chamber, with simulation of oxygen supply to the suit helmet. The final stage of suit evaluation is the suit fit check at Baikonur prior to the mission, and spending time in the descent vehicle cabin of the real spacecraft. The final suit check and its

**Figure 7.6.5** Sea training of the Indian pilots Rakesh Sharma and Ravish Malhotra wearing the SOKOL-KV-2 suits (with the flotation collar) as a preparation for the flight to the *Salyut-7* orbiting station.

preflight preparation are carried out at the same time. These activities (at Baikonur) are carried out by a special Zvezda team, which was led by B.V. Mikhailov for a long time and is currently managed by A.V. Alekseyev.

All spacecrews undergo extensive suited training at the Gagarin Cosmonaut Training Centre (Star City), besides the activities at Zvezda. The training includes time in the *Soyuz* spacecraft simulator where a number of unusual situations are simulated: surviving in uninhabited areas after landing, or in case of an unexpected splashdown of the spacecraft (Figure 7.6.5) learning to survive in the open ocean.

Zvezda has developed special ground support test equipment and a set of fixtures for preflight testing. The suits are ventilated with air supplied from cosmodrome ground sources during fit checks at Baikonur. To ventilate the spacesuits while the cosmonauts go to the launch pad, special portable ventilation units, PVU [ПВУ], were developed. They include a fan that takes the ambient air in and supplies it to the suit and a self-contained electric power source. The PVU comes complete with a heat exchanger that cools supplied air with ice (put into the heat exchanger prior to its usage). It should be noted that Zvezda has used these units since the *Voskhod-2* mission.

# 8

# Orbit-based spacesuits of the ORLAN type

## 8.1 HISTORICAL BACKGROUND

The world's first orbit-based spacesuit system was one of the foremost achievements of Zvezda in the development of space technology. The development of spacesuit systems to support the world's first space mission undertaken by Yu.A. Gagarin and the world's first space walk made by A.A. Leonov, as well as work on the orbit-based spacesuit system and its successful utilization for 25 years, demonstrate that Zvezda is a leader in this field of space technology.

As far back as the second half of the 1960s, the TsKBM ([ЧКБМ] the Russian acronym for the Central Design Bureau of Machine Engineering, led by V.N. Chelomey) was involved in the development of the *Almaz* orbiting station. The *Almaz* programme required the use of an extravehicular activity (EVA) spacesuit, primarily to support transfer of crew members from a transportation vehicle to the orbital module.

At the beginning of the work on the *Almaz* programme, Zvezda had already gained considerable experience in the development of EVA spacesuits. Zvezda had already prepared the YASTREB spacesuit with the RVR-1P backpack for flight operations and had developed the ORLAN semi-rigid spacesuit for the commander of the Moon mission, within the L-3 programme. And, at last, it had developed a soft-type spacesuit, ORIOL.

The YASTREB spacesuit with the RVR-1P backpack was the most developed system in 1967. Therefore, the suit and its backpack were selected for use in the *Almaz* programme. The necessary documents were agreed upon and fit checks in an airlock mock-up carried out. Moreover, it was planned to start manufacturing the suit and its backpack.

At the same time, the problem of using rescue suits in the most dangerous phases of the mission was still unresolved. In particular, the use of such suits was vital for the *Soyuz 7K-VI* space vehicle (military application model) and the *Almaz* system transportation vehicle. Thus, when selecting the suit design for the orbiting station, it seemed impossible to get round the problem by using a universal spacesuit. The selection of a universal suit for use both in the EVA and IVA modes would limit considerably the list of potential candidates for the spacesuit design. For instance, it is not practical to use a semi-rigid suit for the rescue mode to handle transportation vehicle cabin depressurization in the take-off and landing phases of the mission, since the shock-absorbing seat cannot accommodate a crew member wearing such a suit.

Moreover, the requirements of the EVA and rescue suits are different. The EVA spacesuit, which operates normally outside the space vehicle under positive pressure, must be very reliable and is equipped with a number of devices not needed for the rescue suit. The latter is used under positive pressure only in an emergency. The crew member normally wears it in the cabin with no positive pressure. The rescue suit must give the crew member as much mobility as possible in a tight cabin and fit the shock-absorbing seat.

Therefore, with this in mind and the fact that EVA from an orbiting station would become routine, Zvezda specialists proposed the use of two different types of spacesuits for these space missions: a lightweight, individually tailored rescue suit and a more sophisticated and reliable EVA spacesuit. Such an approach made it possible to remove certain limitations otherwise imposed on the suit design and to select the more advanced concept of a semi-rigid spacesuit of the ORLAN type (a derivative of the spacesuit for the lunar mission commander in the L-3 programme) to support EVA activities from the orbiting station. The advantages of this type of spacesuit were discussed in Chapter 6 (for further details see also the book by Abramov et al., 1984). Furthermore, it was easy to adjust the size of such suits in orbit and introduce changes (i.e., replace expendable units after each EVA period). Thus, several crews could use the same suits.

After thorough feasibility studies and fit checks of the ORLAN spacesuit in the *Almaz* mock-up, on Zvezda's initiative, in November 1969, the TsKBM jointly with Zvezda made a decision to use the ORLAN spacesuit in the *Almaz* programme.

Items 1 and 2 of the decision document, approved by A.I. Eidis (deputy to Chelomey) of the TsKBM on 20 November 1969 and by G. Severin of Zvezda on 28 November 1969 read:

1 In accordance with results of feasibility studies carried out in line with the government directive dated 3 July 1968, this directive nullifies the earlier approved documents on the application of the YASTREB type spacesuit on the *Almaz* orbiting station and approves application of a new and more advanced spacesuit of the ORLAN type.
2 The ORLAN spacesuit is designed to support extravehicular activity of one or two crew members performed from an airlock with a 785-mm-diameter hatch to maintain on-board equipment on the outer surface of the space vehicle as well as carry out operations with departure from the vehicle and manoeuvring with the use of an individual propulsion system (to be ordered by the TsKBM in accordance with separate performance specifications).

Also under study was the development and application of a separate, self-contained, add-on backpack unit with radio communications, telemetry and power supply equipment for the ORLAN spacesuit.

The same decision document required the spacesuit to support a 5-hour EVA period for the *Almaz* station and from two to four EVAs to be carried out within 2.5 months. It was planned to carry out a feasibility study of refilling the suit's life support system (LSS) with oxygen and water and replacement of the contaminant control cartridge aboard the station.

Because of delays in the work on the *Almaz* programme and the coming launch of the US *Skylab* station, a group of TsKBEM specialists proposed in late 1969 that an orbiting station for scientific and national economy purposes be developed within a short time by combining the available elements of the *Almaz* station orbital module and already proven systems of the *Soyuz* vehicle.

The proposal became the basis of a government resolution on the development of the DOS-7K long-duration orbiting station issued on 9 February 1970. As a result, by 19 April 1971 the first *Salyut* station was placed in orbit.

In line with the above document on the development of the DOS-7K orbiting station, research, development and test activities on the ORLAN spacesuit for the DOS-7K programme began in early 1970. In April 1970, the TsKBEM issued input data for the ORLAN spacesuit for the DOS-7K programme (the document was approved by K.D. Bushuyev on 10 April 1970 and by G.I. Severin on 30 April 1970). The modified spacesuit was called ORLAN-D (D denotes the first letter of the DOS acronym, which means "long-term orbiting station" in Russian). The need to modify the spacesuit resulted mainly from the requirements of long-lasting and multiple utilization of the suit in orbit and maintenance of the suit aboard the station by crew members. Therefore, the suit system was to comprise equipment needed to:

(a) store the spacesuit aboard the station;
(b) undertake the work required to maintain and prepare the suits for operation;
(c) undertake work on the airlock;
(d) dry the spacesuit;
(e) prepare the spacesuits for repeated EVA.

The input data foresaw in-orbit recharging of backpack bottles with oxygen, refilling of water and replacement of the contaminant control cartridge. According to the input data document, spacesuits were to stay in orbit for 3 months, operate for at least 10 hours total time (provided that they are recharged and refilled) and support from three to four EVAs lasting from 2 to 4 hours each. The total weight of two charged and filled spacesuits was not to exceed 216 kg (with on-board components included).

In 1970, detailed drawings of the modified spacesuit and backpack with LSS units were prepared, and studies of long-lasting storage of the spacesuits aboard the orbiting station and their repeated use began. This work continued in 1971. Test models of suit system units were developed and manufactured, including the BSS-1 unit (the spacesuit interface unit designed to connect the spacesuit's pneumatic and hydraulic lines to the airlock on-board equipment), the BVS-2 unit (the on-board ventilation system used to dry the spacesuit) and electric umbilicals designed to receive power supply from the on-board system, radio communication and telemetry data transmission. The length of the electric umbilicals was selected in such a way that a space walk could be made at a distance of up to 15 to 20 metres from the exit hatch. Moreover, during the space walk the electric umbilicals were to serve as safety tethers for the crew members.

Also under development and testing was a set of orbital replacement elements and the SZP-1 [СЗП-1] on-board recharging/refilling and checkout system.

In parallel with the above and in line with the plan approved by the Ministry of Aviation Industry and the Ministry of General Machine Engineering, work continued on the ORLAN-D system such as was applicable to the *Almaz* programme. Zvezda started manufacturing spacesuit models for tests in the hydraulic laboratory of the Cosmonaut Training Centre, for training and for tests in conjunction with the airlock system in the TBK-60 thermal vacuum chamber at the Air Force Scientific Research Institute (GK NII). The B-3 [В-3] on-board docking units were also developed and manufactured for the *Almaz* programme.

Because of delays in the manufacture of hardware for the *Almaz* programme and a decision of the TsKBEM not to use the ORLAN-D spacesuit on the *Salyut-1* orbiting station (DOS-3), work on spacesuits for an orbiting station was limited to design improvements and various tests in 1972 and 1973. In May 1973, Zvezda requested the TsKBEM to find a quicker way of settling the problem of using the ORLAN-D spacesuit on the orbiting station. This would enable Zvezda to gain some experience of operating the spacesuit in the actual environment and use it for further development activities; in particular, for the then planned L-3M programme (see Section 6.2).

An agreement was soon reached with the TsKBEM to start operating the ORLAN-D spacesuit on the DOS-5 station (*Salyut-6*). The agreement was documented in January 1974 by joint TsKBEM/Zvezda Decision No. 2/511-74 and followed by the Inter-agency Committee decision dated 18 September 1974 and the respective directives from the Ministry of Aviation Industry. In June 1974, the final performance specifications for the ORLAN-D spacesuit as applicable to the DOS-5 were agreed upon.

The main cycle of physiological and technical tests of the spacesuit and its systems and life cycle tests (2 years) of the suit's hydraulic system had been completed by 1976. Spacesuit life cycle tests and inter-agency tests (in conjunction with the airlock system) were completed at GK NII in the middle of 1977. Then, ORLAN-D flight units No. 33 and No. 34 and back-up systems No. 35, No. 36 and No. 38 were prepared for installation aboard the *Salyut-6* station (Figure 8.1.1a, b). At the same time, cosmonauts were trained in thermal vacuum chambers (at Zvezda and GK NII), at neutral buoyancy facilities and aboard the Tu-104 flight test bed.

In those years, work on the ORLAN-D spacesuit for the *Almaz* programme did not stop. In particular, the ORLAN-D spacesuit was tested in the TBK-60 thermal vacuum chamber and participated in the inter-agency tests of the *Almaz* station's LSS. On completion of the *Salyut-6* mission and termination of the *Almaz*

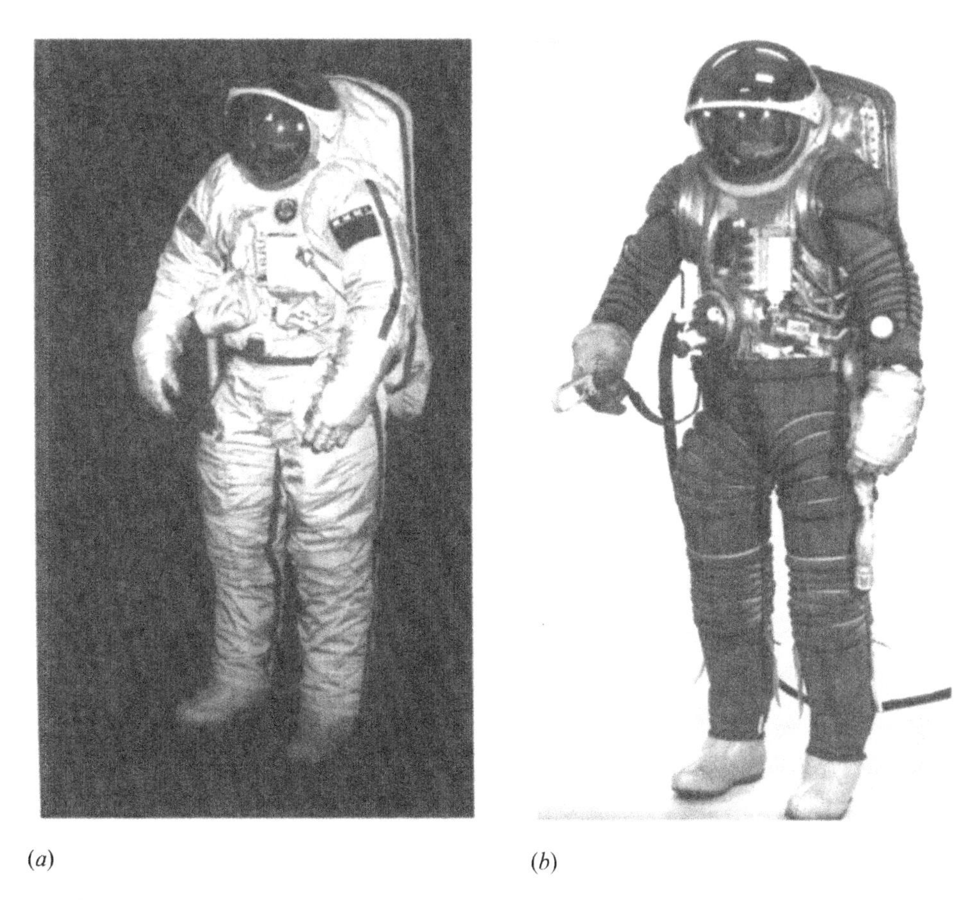

(*a*) (*b*)

**Figure 8.1.1** (*a*) A general view of the ORLAN-D suit; (*b*) without outer garment.
From Archive Zvezda.

programme, further work on the ORLAN-D spacesuit for this programme was stopped.

As is known, the first space walk supported by the ORLAN-D semi-rigid spacesuit was made from the *Salyut-6* orbiting station on 20 December 1977 by cosmonauts Yu.V. Romanenko and G.M. Grechko (Figure 8.1.2 and 8.1.3). Since then, such space suits, continuously improved and produced in several modifications, have become the standard in-orbit-based EVA spacesuits. They have been permanently aboard the *Salyut* and later *Mir* and *ISS* orbiting stations, enabling almost all crew members of these stations to work in free space.

The specific features of each modification of the ORLAN family of spacesuits as well as of cosmonaut training suit models are described in more detail in subsequent sections of this chapter.

Besides the spacesuits and their elements and subsystems, ground test facilities were also developed and manufactured for the testing and verification of these

**Figure 8.1.2** Cosmonaut Yu. Romanenko is preparing the ORLAN-D suit for the first EVA from *Salyut-6*. Photo by G.M. Grechko.

spacesuits and their subsystems. As the spacesuits were modified, ground test facilities were modified, correspondingly.

Starting from the late 1960s, many Zvezda specialists were directly involved in the development of orbit-based spacesuits. It is impossible to present the whole list of design engineers, researchers, test engineers and test subjects, design and process analysts, medical support specialists, workers, technicians and manufacturing specialists in this book.

Other active participants in the development and test work were representatives of the customer (the Energia Rocket Space Corporation), the Gagarin Cosmonaut Training Centre, the Air Force and specialists of various subcontractors.

Permanent project-managers over the whole period of activities were I.P. Abramov (LSS and the suit system in general), A.Yu. Stoklitsky (suit enclosure) and I.I. Chistyakov (electric equipment). Project-managers in separate phases of activities were R.Kh. Sharipov (thermal control system) and E.A. Albatz (LSS). The significant contribution to the development of the ORLAN spacesuit of test engineers and test subjects A.Tz. Elbakyan, V.Ya. Bychkov, G.M. Glazov, among others, should be mentioned.

**Figure 8.1.3** Cosmonauts Yu. Romanenko and G. Grechko are welcomed at Zvezda (1978). *First row (from left to right)*: A.A. Savinov, I.P. Abramov, Yu.V. Romanenko, G.I. Severin, G.M. Grechko, V.A. Dubrov, A.S. Barer. *Second row*: E.A. Ivanov, F.A. Vostokov, A.Yu. Stoklitsky, E.P. Diomin (TsKBEM), an unknown person from the Cosmonaut Training Centre, V.G. Galperin, N.V. Knyazev, and I.A. Sokolovsky (representatives of Cosmonaut Training Centre).

From Archive Zvezda.

## 8.2 DESIGN FEATURES AND DEVELOPMENT OF ORLAN-TYPE SPACESUITS

By 31 December 2002, ORLAN-type spacesuits had accumulated 206 EVAs of a total duration exceeding 800 hours from the *Salyut*, *Mir* and *ISS* orbiting stations. Members of 51 crews (including French and ESA cosmonauts and NASA astronauts) participated in these EVAs.

The total number of ORLAN-type spacesuits (four modifications) used aboard the stations over the given period is currently (March 2003) 25 and some of them were in orbit for up to 3 years.

The principle of keeping the suit in orbit, quick unassisted donning/doffing, one standard size for crew members with different anthropometric characteristics and multiple use with the possible replacement of failed elements were invariably the principal considerations for all improvements and changes made to the spacesuit.

The design concept and LSS flow diagram of the ORLAN-M spacesuit, the most recent member of the ORLAN family, are given in Figures 8.2.1 and 8.2.2.

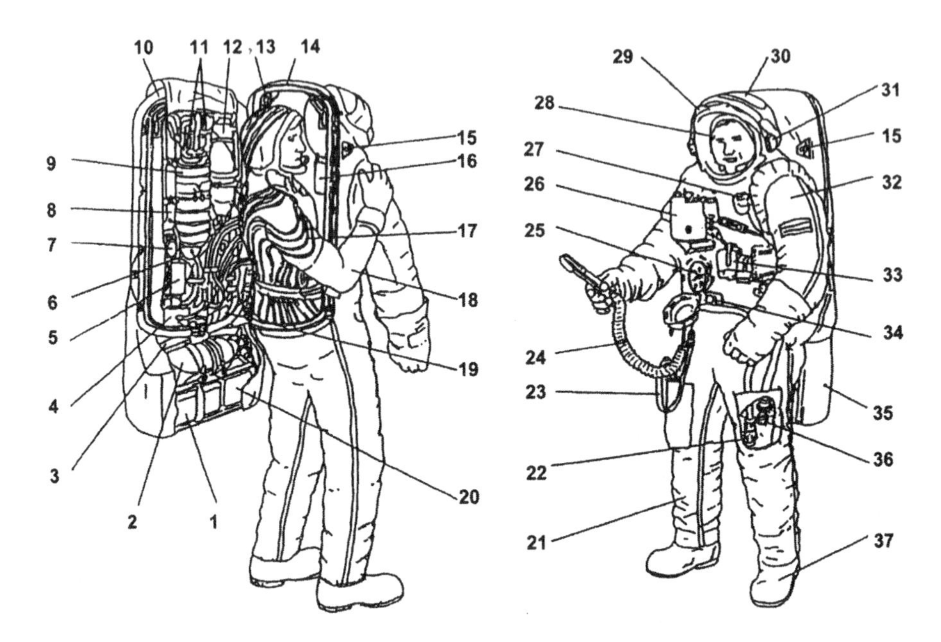


**Figure 8.2.1** Design concept of the ORLAN-M spacesuit: 1—battery; 2—primary $O_2$ bottle; 3—telemetry unit; 4—moisture collector; 5—measuring unit; 6—heat exchanger; 7—filter; 8—feedwater tank; 9—$CO_2$ removal cartridge; 10—backpack; 11—primary and redundant fans; 12—redundant $O_2$ bottles; 13—suit pressure regulators; 14—hard upper torso; 15—SAFER (simplified aid for EVA rescue) attachment points; 16—drink bag; 17—water-cooled garment; 18—underwear; 19—primary and redundant pumps; 20—radio station; 21—soft lower torso; 22—electrical connector; 23—safety tether; 24—adjustable length safety tether; 25—combined connector; 26—electrical control panel; 27—suit pressure gauge; 28—communication cap; 29—light visor; 30—top window; 31—lights; 32—suit arms; 33—pneumatic control panel; 34—suit attachment point; 35—removable BRTA (battery, radio and telemetry assembly); 36—connector for emergency oxygen hose; 37—boots.

From Archive Zvezda.

Besides the foregoing, there were initial problems that needed to be solved concerning the ability of suited crew members to work effectively and their safety during EVAs as part of the development of the spacesuit and introduction of improvements and changes.

Measures that ensured settlement of these problems were conventionally divided into several groups. Their aims were to:

- select those designs of the spacesuit and its systems that meet the specifications to the maximum degree;
- develop procedures and techniques of spacesuit preparation for EVA, along with preventive and scheduled maintenance work;
- demonstrate the proper operation of the spacesuit and its components in a ground development test programme.

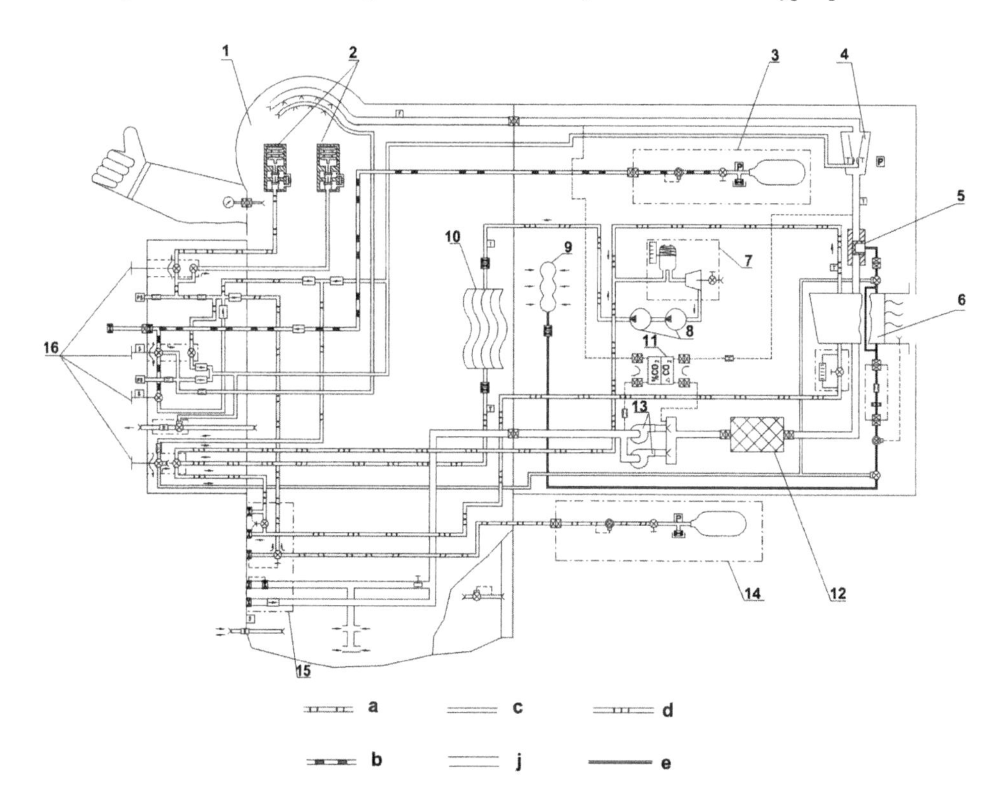


**Figure 8.2.2** Principal flow diagram of the ORLAN-M self-contained LSS (SCLSS): Primary (*a*) and redundant (*b*) oxygen system; tubes for oxygen (*c*), cooling water (*d*) and feedwater (*e*); ventilation loop (*j*): 1—inner cavity of the suit enclosure; 2—pressure regulators; 3, 14—redundant and primary oxygen supply units; 4—injector; 5—moisture collector/separator; 6—sublimation heat exchanger; 7—hydraulic accumulator with separator; 8—primary and redundant water pumps; 9—tank with water to be supplied to the heat exchanger; 10—water-cooled garment; 11—measuring unit; 12—$CO_2$ and contamination control cartridge; 13—primary and redundant fans; 15—combined connector to connect the suit to the on-board systems by means of the umbilical during airlock; 16—SCLSS controls.
From Archive Zvezda.

In the design selection process for the spacesuit and its components as well as during subsequent modifications/improvements, special attention was given, above all, to their reliability, suit mobility and effective operation of the SCLSS. These were the principal factors that assured both crew member safety and effective execution of EVA for the entire service life of the suit.

ORLAN-type spacesuit design concepts and flow diagrams of its systems were selected with the following considerations in mind. No single failure of any spacesuit elements should result in the failure to carry out an EVA programme. Failures resulting in catastrophic consequences should be excluded completely. The reasoning behind these concepts was to make the principal, vitally important functions redundant either at the element level or at the subsystem level. Those spacesuit elements whose damage could result in the loss of pressure tightness were made redundant. Thus, the spacesuit uses a dual visor system and there are dual pressure bladder enclosures of soft pressure parts. Moreover, the back-up pressure bladder becomes operational automatically in case the main bladder fails. The disconnects and movable joints of the spacesuit (suit entrance interface and pressure bearings) feature dual barrier sealing. Butt joints of suit elements, welds and other important places of the hard metal torso are additionally sealed with fabric or rubber material glued to the internal surfaces.

Selection of a semi-rigid design concept for the spacesuit with the application of joints and pressure bearings on the soft parts of the suit made it possible to use a high operating pressure (400 hPa) and, thus, minimize pre-breathing time for crew members prior to the EVA period, with almost no risk of decompression sickness.

Materials for the elements of the orbit-based spacesuit were selected on the basis of their wear and corrosion resistance, zero release of harmful substances and fire safety (the latter is of particular importance for electric units and cables). One of the most important and difficult problems to solve was the long-lasting storage of the water that circulates in the spacesuit's hydraulic system. System development was preceded by selection of a water-preserving agent that would assure both water quality and zero corrosion of the materials over the whole operational life of the spacesuit. Studies were carried out both at the material sample level and at the assembled system level. The best results were obtained for water with silver ions added.

Studies resulted in the development of the following procedures for the in-house preparation and filling of the suit's water-cooled loop with water:

- cleaning of the system elements and washing the system with distilled water;
- disinfection of the hydraulic system loop by filling it with water containing silver ions ($2\,mg\,l^{-1}$) and keeping the water in the system for 1 or 2 days;
- filling the system with chemically demineralized water with an addition of silver ions ($0.2\,mg\,l^{-1}$).

The hydraulic system mainly uses stainless steel parts, polyvinyl chloride (PVC) tubes and a few rubber elements (flexible membranes and sealing rings). The heat exchanger housing is made of aluminium alloy with a special coating. The cooling system's water loop elements are designed in such a way that water could circulate within the required limits even if severely contaminated. The suit's feedwater supply subsystem is isolated from the cooling system loop. The water tank is made of a fluoroplastic film and the outlet is made of a PVC tube. The preparation procedure for this system is close to that for the cooling system loop.

In the operational period, a special filter regularly cleans the water in the hydraulic system loop and a certain quantity of silver is added to it. The subsequent, long-lasting operation of the spacesuit has confirmed the effectiveness of the selected hydraulic system maintenance procedure.

To decrease the time needed to maintain the in-orbit-based spacesuit, the oxygen bottles, contaminant control cartridge, feedwater tank and other orbital replacement elements are made easy to disconnect. The main and back-up oxygen bottles are identical to those used in the on-board system in the airlock process.

To increase the sublimator's operational life and exclude contamination of its pores, an expendable quick-disconnect filter is installed upstream of the sublimator, and the separator, which is used to remove condensate to the sublimator, is made replaceable.

The SCLSS (closed regenerative type) (Figure 8.2.2) maintains the needed microclimate inside the spacesuit. The system incorporates a number of technical features: the oxygen supply system has devices that maintain the pressure in the suit, provide ventilation and control of the gas mix system; thermal control system; electrical equipment; and control and monitoring units.

In the nominal operating mode, the oxygen supply system feeds the spacesuit with oxygen via two redundant pressure regulators that automatically supply oxygen as necessary, in proportion to the suit pressure drop rate. Gas supply, with a flow rate of about $1.0\,kg\,h^{-1}$, to the spacesuit can also be switched on manually using the injector, which is built into the ventilation system and arranged in the line supplying gas to the helmet.

Furthermore, the suit's SCLSS is fitted with a separate back-up oxygen system, which can be used both in the nominal operating mode (described above) and in an emergency when oxygen enters the spacesuit in a continuous flow at a rate of about $2\,kg\,h^{-1}$ (with approximately half the flow going directly to the helmet and the remaining flow going through the injector). The emergency oxygen supply from the back-up system is switched on manually by the crew member. When the pressure in the suit drops below 220 hPa (for the ORLAN-D, ORLAN-DM and ORLAN-DMA suits) or below 270 hPa (for the ORLAN-M model), the emergency gas supply to the suit is automatically switched on by the pressure regulators. The maximum possible emergency flow rate amounts to about $3\,kg\,h^{-1}$ (including the emergency supply switched on manually).

Ventilation of the suit's internal cavity is an essential prerequisite to keeping the gas mixture parameters within the specified limits. The ventilation gas removes metabolic process products (carbon dioxide, harmful impurities, moisture) and, partially, heat from the suit's internal space. A centrifugal fan driven by a brushless electric motor provides gas circulation along the closed loop. The system is equipped with two fans—main and back-up. The latter is switched on automatically in case the main fan fails. If both fans fail or the electric power supply is lost, ventilation is provided by switching on the injector.

Ventilation tubes attached to the interior of the suit enclosure supply gas to the suit helmet and remove gas from the suit's arms and legs, returning it to the system. The suit ventilation rate is between 150 and $200\,l\,min^{-1}$. Such a rate provides removal of moisture and carbon dioxide from the suit when the crew member is engaged in the hardest workload.

Continuous supply of the suit with oxygen and ventilation gas cleaning maintains the gas mix in the suit within the given limits. Removal of carbon dioxide and harmful impurities from the circulating gas occurs in the contaminant control cartridge mainly filled with a LiOH-based substance. An initially high percentage of oxygen in the suit atmosphere (exceeding 90%) results from purging the suit with pure oxygen from the on-board system prior to the pre-breathing session.

The thermal control system uses a very efficient method of removing man-generated heat by a water-cooled garment. The cooling rate is controlled by the crew member manually with a multi-position "heat–cool" valve that can direct water into the water-cooled garment through the heat exchanger or bypass it.

Man-generated moisture is transferred by the ventilating gas to the sublimation heat exchanger where it is condensed. Condensate is collected in the moisture collector/separator and then removed to the sublimation cavity of the heat exchanger for evaporation in the ambient vacuum.

Application of the water-cooled garment in combination with effective operation of the sublimation heat exchanger keeps the crew member's thermal balance stable for workloads up to 600 W.

The suit's electrical equipment comprises a control and monitoring panel, devices for the operation of fans and pumps (electric motors and electric units), radio communications and telemetry means and control and measurement means, and the electric cable connecting the suit to the on-board systems.

The radio communications system of the ORLAN-DMA and ORLAN-M (see Section 8.3) models comprises a transceiver arranged in the removable, autonomous suit operation unit, a headset and an antenna-feeder device. The system provides wireless radio communication between two crew members as well as with the orbiting station's radio set with its subsequent downlink relay. The transmitters are switched on by automatic voice recognition or by a push-button. The self-contained power supply and telemetry units are also arranged on the suit's removable unit.

Development testing was one of the most important phases of the ORLAN-type spacesuit development and modification work. Conventionally, it was a standard programme, which included:

- experimental development of newly introduced or modified elements;
- technical tests of the spacesuit system and its separate elements, which included verification of the system for proper operation after exposure to mechanical impacts under changing environmental conditions, and equipment life expectancy tests;
- ergonomic evaluation of the spacesuit including tests performed in a hydro lab;
- man-evaluated tests of the spacesuit in a vacuum chamber, with simulation of crew activity in the airlock process and during autonomous operation.

If the necessity arose, the standard spacesuit development test programme could be expanded to include experimental development tests of LSS subsystems and the components that were involved in the expansion/improvement of suit functions.

On the one hand, the experimental development test programme facilitated the selection of optimal design concepts and flow diagrams and, on the other, it made it possible to demonstrate the performance of manufactured systems to meet given requirements.

The task set for the in-house development test programme was to simulate, as close as possible, the actual operation of systems aboard the orbiting station.

Feasibility of operation for long periods and multiple use of the orbit-based spacesuit were demonstrated by multiple repetition at sea level of its duty cycle (simulating an EVA period) in certain tests:

- life tests:
  - multiple donning/doffing of the spacesuit by a test subject;
  - pressurization and depressurization of the pressure suit;
  - execution by a suited test subject (operating under positive pressure) of various movements by his arms and legs using all the suit joints and pressure bearings;
  - lengthy operation of LSS elements (fans, pumps, etc.) (the number of operating cycles and the accumulated operation time during equipment life expectancy tests were three times larger than those expected in an actual EVA or specified in the requirements of the suit);
- independent tests of separate elements and units of the spacesuit;
- man-evaluated simulation of spacesuit operation in the airlock process and during an EVA in a vacuum chamber;
- fulfilment by a suited test subject of real EVA tasks in a hydraulic weightlessness simulator (in the hydro lab).

The operation time of the spacesuit and its elements was recorded over the whole suit development programme and during training sessions of cosmonauts in suited mode. This made it possible to additionally evaluate and demonstrate suit performance capabilities in the process of lengthy operations.

Since each new, modified model of the spacesuit was a derivative of earlier models, which had already undergone the whole development and demonstration test cycle, results of earlier test activities were partially used to evaluate new models. If the necessity arose, computation of analytic data was done to take into account differences in the design of the modified and foregoing versions (strength and thermal computations predominated).

Development and demonstration tests were mainly carried out at Zvezda's test facilities, including the very important man-evaluated tests in a vacuum chamber. To support the test programme, Zvezda developed the TBK-50 [TBK-50] vacuum chamber, which enabled Zvezda to simultaneously run man-evaluated tests and train cosmonauts with two ORLAN-type spacesuits in the same test/training session (Figures 8.2.3).

Inter-agency tests were carried out in conjunction with the station airlocks in a vacuum chamber at the GK NII.

Development and demonstration tests of the spacesuit under neutral buoyancy conditions and in a flying test bed were carried out at the Gagarin Cosmonaut Training Centre.

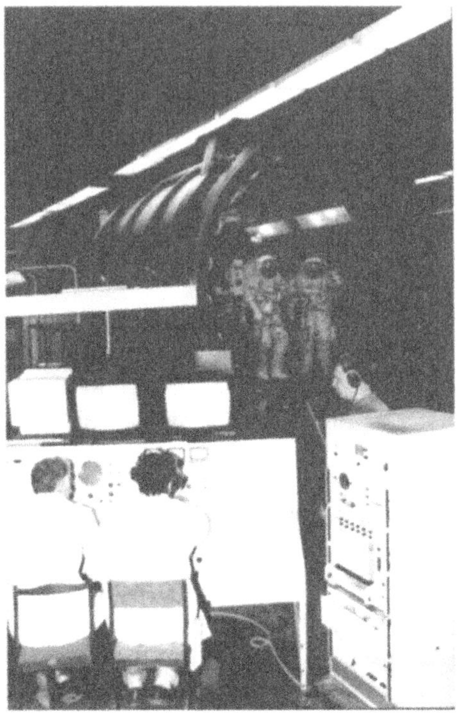

**Figure 8.2.3** ORLAN-type suit-testing in the TBK-50 thermal and vacuum chamber at Zvezda.

From Archive Zvezda.

## 8.3 MODIFICATION OF SPACESUITS WHILE OPERATING ABOARD THE *SALYUT*, *MIR* AND *ISS* ORBITING STATIONS

Four main modifications of the ORLAN-type suit have taken place. Changes in the design of the spacesuit and its systems resulted from new operating conditions, new tasks set for crew members involved in EVA and experience gained in suit operations, known drawbacks, the recommendations of cosmonauts, ideas for new design and improvement of reliability and operating life.

The ORLAN-D spacesuit (Figure 8.1.1), the first modified member of the ORLAN family, was used in the *Salyut-6* and *Salyut-7* missions. The next modification, the ORLAN-DM spacesuit, was an intermediate stage before the ORLAN-DMA came into being.

EVA suits used in the *Mir* programme were the ORLAN-DM (the first two years), the ORLAN-DMA and ORLAN-M, while those used within the *ISS* programme are the ORLAN-M spacesuits.

The ORLAN-D spacesuit modification (a derivative of the spacesuit for a commander of the lunar mission) was mainly developed between 1969 and 1974

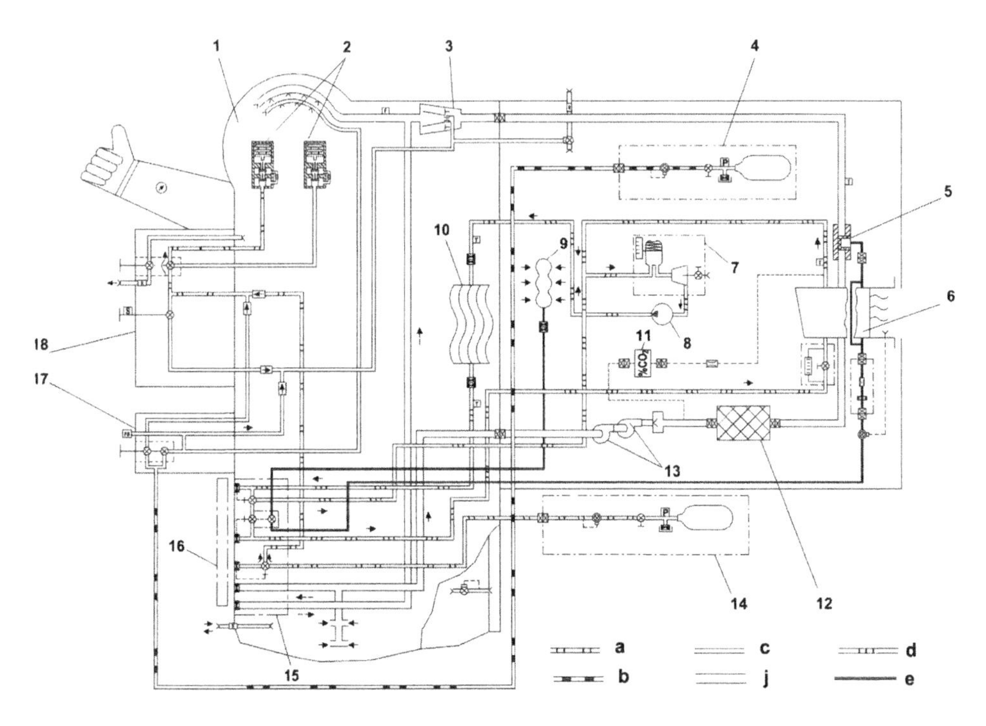


**Figure 8.3.1** Principal flow diagram of the ORLAN-D self-contained LSS (*Salyut-7* version). Primary (*a*) and redundant (*b*) oxygen system; tubes for oxygen (*c*), cooling water (*d*) and feedwater (*e*); ventilation loop (*j*): 1—inner cavity of the suit enclosure; 2—pressure regulators; 3—injector; 4, 14—redundant and primary oxygen supply units; 5—moisture collector/separator; 6—sublimation heat exchanger; 7—hydraulic accumulator with separator; 8—water pump; 9—tank with water reserve to be supplied to the heat exchanger; 10—water-cooled garment; 11—measuring unit; 12—$CO_2$ and contamination control cartridge; 13—primary and redundant fans; 15—combined service connector to connect the suit to the on-board systems with the umbilical during airlock; 16—removable portion of the combined service connector (with handles to switch on the sublimator and to control the cooling water temperature); 17—cock to actuate the redundant oxygen supply and emergency oxygen supply; 18—remote control (for suit pressure mode switch-over and injector switch-on).

From Archive Zvezda.

with the requirements and specific features of operations aboard the orbiting station in mind (namely: multiple use with the suit remaining long term aboard the station without return to the Earth, change of crews with different anthropometric characteristics, necessity to check and prepare suits for repeated EVA aboard the station by crew members, adaptation to on-board interfaces of the space stations airlock, increased duration of EVAs, etc.).

The units in the spacesuit's backpack were completely rearranged and the suit's LSS flow diagram was considerably changed (Figure 8.3.1). To begin with, the oxygen bottles and contaminant control cartridge were made easy to access and the facility to refill the LSS cooling system with water was provided. The evaporating heat exchanger was replaced by a sublimator. The main and back-up oxygen bottles were made similar. The capacity of each bottle was 1 litre at 42 MPa operating pressure (the ORLAN spacesuit for the L-3 programme was fitted with an emergency oxygen bottle of 0.4 litre capacity). The main bottle was transferred from the backpack's internal cavity to its external surface (in the lower part of the backpack) to facilitate its replacement. And, finally, the flow diagram of the water cooling system was changed to provide circulation of water in the suit's water-cooled garment during airlocking with no involvement of the on-board pump (as was the case in the ORLAN spacesuit for the L-3 programme). Some additional changes, compared with the suit used in the *Salyut-6* orbiting station, were introduced in the ORLAN-D suit design during suit manufacture for the *Salyut-7* orbiting station. In particular, the injector was moved from the backpack to the suit enclosure, a separator was introduced to remove gas bubbles from the cooling water and the design of the suit's on-board unit was modified. Between 1969 and 1984 a total of 34 ORLAN-D suits, including 7 flight suits, were manufactured.

Between 1977 and 1984, crew members wearing the ORLAN-D spacesuits carried out three EVA sessions from the *Salyut-6* orbiting station and ten EVA sessions from the *Salyut-7* station.

Utilization of the ORLAN-D spacesuit was also intended for the *Buran* reusable space transportation system (as agreed with the TsKBEM in the protocol dated 16 October 1980). The plan was to undertake three 5-hour EVA sessions during a 7-day *Buran* system mission and from six to eight EVA sessions during a 30-day mission. Airlocking could be carried out either in the vehicle's airlock ShKK [ШКК] or in the vehicle's docking module.

The ORLAN-D spacesuit was connected to the station's on-board systems by a 20-m multi-wire cable, which was used to supply the spacesuits with electric power, support radio communications and transmit telemetry data about the operation of the suits and the state of the crew members. Cable communication of the spacesuit with the station's on-board systems was acceptable when crew members undertook activities on the station surface near the airlock. Because of the successful operation of the EVA spacesuit in the *Salyut-6* mission, the go-ahead was given for the ORLAN-DM modification, a new member of the ORLAN family, in which experience gained from operation of the ORLAN-D spacesuit would be taken into account. This suit could be transformed, in the future, to the completely self-contained ORLAN-DMA spacesuit (mainly by adding a removable unit with additional equipment). In 1983, documentation for the spacesuit was developed and 16 ORLAN-DM spacesuits were ordered.

A new panel designed to control the electrical and radio communications systems was developed for the ORLAN-DM spacesuit and the suit's electric circuit and cabling was changed accordingly. Certain changes were also introduced in the suit design to improve operating characteristics and the reliability of the spacesuit. The units in the backpack were rearranged and the suit body was improved (in particular, the injector was transferred from the backpack to the front part of the enclosure body). Mobility of the suit's arms and legs was

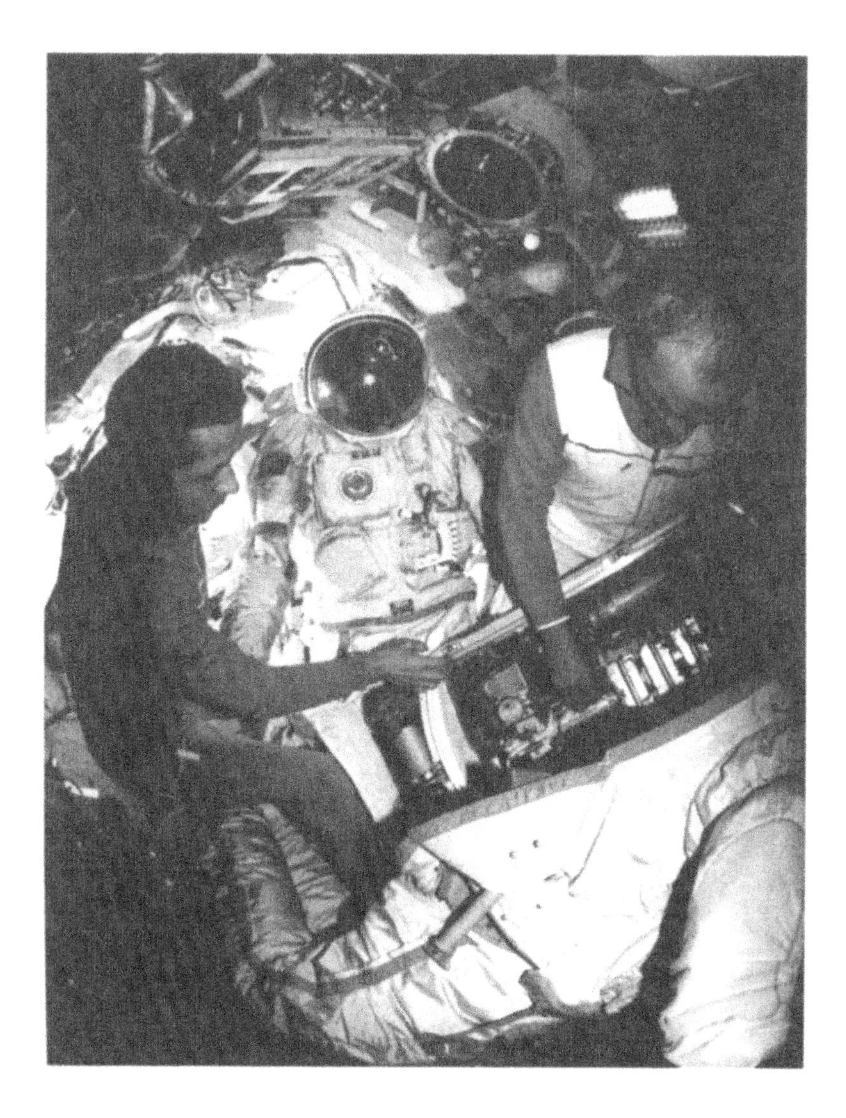

**Figure 8.3.2** Cosmonauts V.A. Dzanibekov and V.P. Savinykh preparing the ORLAN-DM suits for the EVA from *Salyut-7*.

improved, a protective casque introduced in the helmet and a combined control panel for the pneumatic and hydraulic systems developed. The main valves of the combined connector were made redundant. The measurement system, the water pump and the water tank (replacement of the latter was made easier) were all modified. A new reinforced material was used for one of the suit's pressure bladders. An emergency oxygen hose was introduced and the cooling areas of the water-cooled garment decreased. Lamps were installed on the suit's helmet.

ORLAN-DM spacesuits were delivered to the *Salyut-7* station in the middle of 1985 (Figure 8.3.2) and to the *Mir* station in March 1986.

**Figure 8.3.3** ORLAN-DMA suit with the 21KS unit.
From Archive Zvezda.

Within the *Mir* programme, the ORLAN-DM spacesuits carried out five EVA sessions each (the date of the last EVA was 30 June 1988). Then they were replaced by the two ORLAN-DMA spacesuits delivered to the *Mir* orbiting station by two *Progress* spacecraft.

The ORLAN-DMA spacesuit (Figures 8.3.3, 8.3.4 and 8.3.6) could be used without an electric umbilical connecting the suit to the on-board station systems. It was equipped with a special removable unit (Figure 8.3.5) that contained the power supply, radio communications and telemetry system units and antenna-feeder device. The radio antenna element was built into the modified suit's outer garment.

Work on the development of a completely self-contained spacesuit system was carried out within the "Suit–UPMK System" project (UPMK stands for "Cosmonaut Transference and Manoeuvring System") in line with the government resolution dated 25 September 1985 and government directive dated 31 October 1985

*Скафандр для работы в открытом космосе "Орлан–ДМА"*

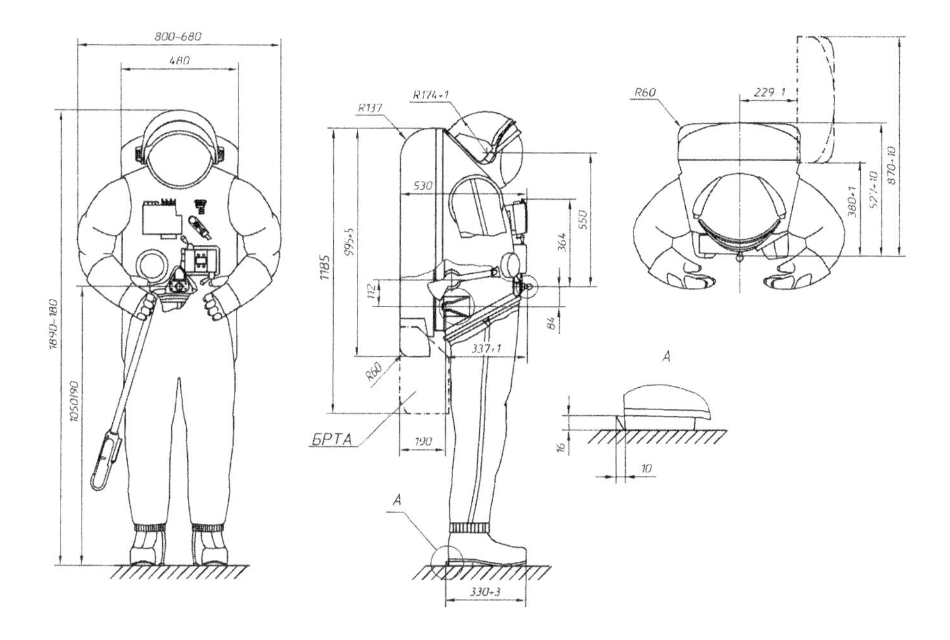


**Figure 8.3.4** ORLAN-DMA main dimensions.

Original Zvezda.

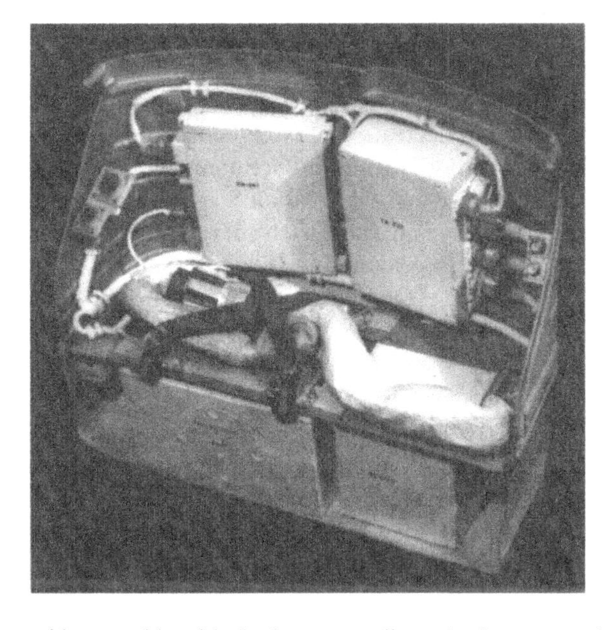

**Figure 8.3.5** Removable assembly with the battery, radio and telemetry equipment (BRTA).

From Archive Zvezda.

(see Chapter 9). Therefore, the ORLAN-DMA suit was fitted with elements that were compatible with the UPMK (identified as the 21KS system). Flight tests of the 21KS system were successfully carried out in 1990.

Furthermore, some major changes were introduced in the ORLAN-DMA suit design. To begin with, the rigid part of the suit enclosure was again improved. Its lower part provided for attachment of leg enclosures via a special removable flange, which made it possible to replace soft parts of the suit in case they were damaged or worn out. At the same time, the internal volume of the body was slightly decreased and the design of the front lock (made in the form of a "pin") for spacesuit restraint changed. The suit's pressure gloves were improved and the utilization of pressure cuffs, which retain the suit pressure for some time in case of damage to the gloves, was provided for. Other improvements and changes included installation of the new LP-6 contaminant control cartridge (absorbing $CO_2$ and harmful impurities) of increased capacity (with an increase in duration of the suit's autonomous operation mode by up to 6 hours), modification of the fan (a new electric motor was used) and headset (with improved electrical acoustic means) and introduction of an additional safety tether.

After development of the ORLAN-DMA spacesuit, in 1987 the decision was made to also use it in the *Buran* programme instead of the ORLAN-D spacesuit. At the same time, the go-ahead was given to work on further modification of the spacesuit, as applied to the planned *Mir-2* orbiting station.

Early in the 1990s, Zvezda started to actively cooperate with a number of Western companies and organizations with the aim to jointly develop or modify spacesuits for advanced space projects.

Since the requirements of the spacesuit developed in Europe for the *Hermes* vehicle were close to those used in Russia for development of its suits, the idea came up to combine the efforts of European and Russian companies in the development of technologies for EVA spacesuit systems. This would make it possible to share development costs and decrease the overall costs of the programmes.

In 1992, on the initiative of Dornier in Germany and Zvezda supported by the European Space Agency (ESA) and the Russian Space Agency (RSA), the go-ahead was given for a feasibility study and, then, to the development of a design concept for a joint Russian/European new generation EVA spacesuit to be used both for the *Hermes* and *Buran*-type vehicles and for the then planned Russian *Mir-2* station. Other participants in this work were Laben in Italy and Dassault in France.

The idea was to combine the available experience of Russian industry and novel European technologies, during work on the joint spacesuit called the EVA SUIT 2000. This supplied the prerequisites for successful materialization of the joint plans. Utilization of already available Zvezda test facilities and experience would enable ESA to decrease the cost of development of the spacesuit.

Russian specialists found many similarities between the European *Hermes* spacesuit concept's ESSS design and that used in the ORLAN-type spacesuits. As a result, development of the EVA SUIT 2000 was based in many respects on the same concepts. Joint work on the EVA SUIT 2000 project are described in more detail in Chapter 11.

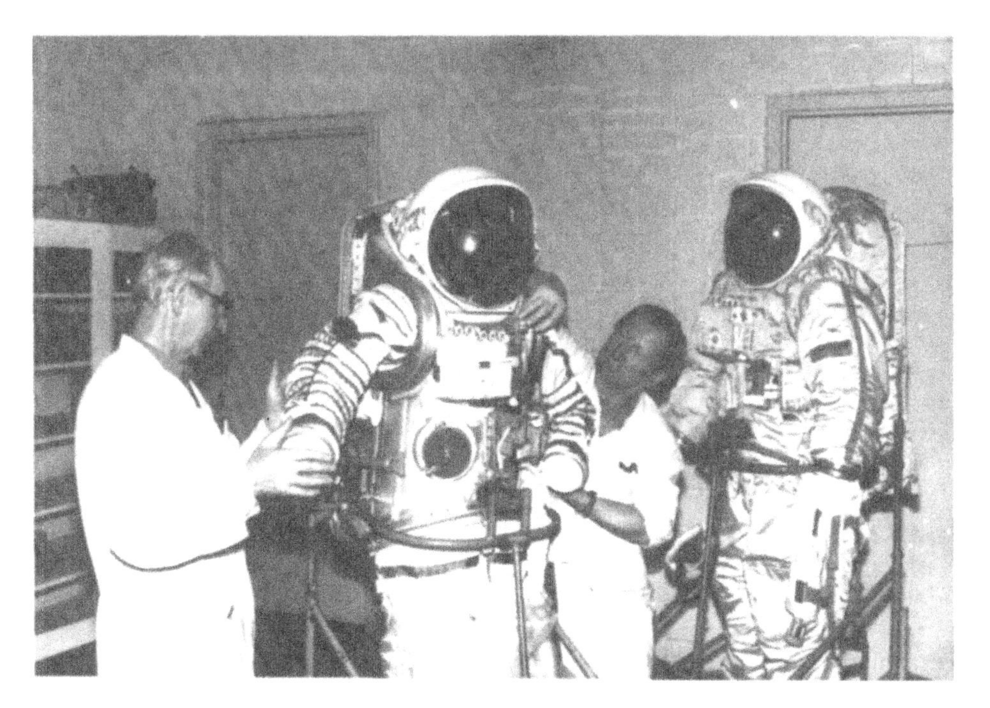

**Figure 8.3.6** The ORLAN-DMA assembly shop.
From Archive Zvezda.

When the joint US/Russian Statement on Space Cooperation (Gore–Chernomyrdin Commission) about the participation of Russia in the *ISS* programme was signed in 1993 and when its work on the *Hermes*, *Buran* and *Mir-2* programmes terminated, the EVA SUIT 2000 was planned to be used in the Russian segment of the *ISS*. However, late in 1994, when work on the *ISS* intensified, ESA initiated the termination of the EVA SUIT 2000 project because of financial constraints. So, Zvezda renewed work on the earlier started modification of the ORLAN-type suit, but with the *ISS* programme in mind. Certain ideas conceived by Zvezda in the EVA SUIT 2000 project became part of the new spacesuit's modification. When work started on the *Mir–Shuttle* programme and, subsequently, on the *ISS* programme with the planned utilization of Russian EVA suits by international crews, Zvezda made the decision in 1995 to significantly modify the ORLAN-DMA suits of the next batch to be manufactured for the *Mir* station and add the letter M to the suit name.

The tasks of this modification work were to:

- improve the general performance of the spacesuit (primarily the mobility and accommodation of cosmonauts/astronauts of all shapes and sizes, as well as increasing the amount of time to be spent in autonomous operation mode);
- improve spacesuit reliability and crew member safety.

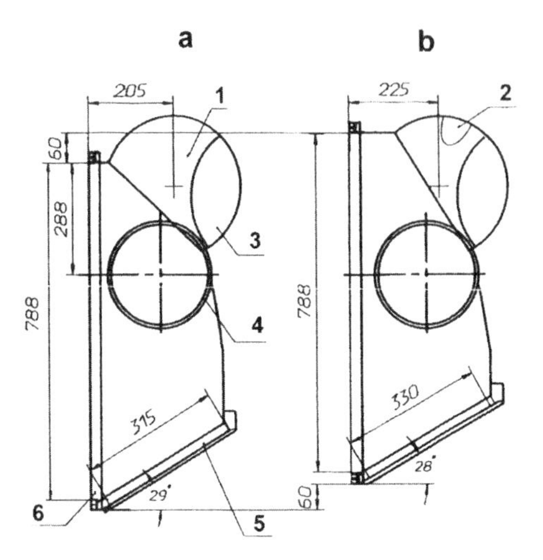


**Figure 8.3.7** Comparison of the ORLAN-DMA (*a*) and ORLAN-M (*b*) HUTs: 1—helmet; 2—additional visor; 3—visor; 4—shoulder opening; 5—waist flange; 6—HUT frame.
From Archive Zvezda.

By taking the new requirements and operating experience on the *Mir* station into account, many new features had been introduced in the ORLAN-M spacesuit design by the beginning of operations aboard the *ISS*: increased spacesuit body dimensions and an enclosure height adjustment envelope; additional visor improving the upper field of view (Figures 8.3.7 and 8.3.8a, b, c); protective visor to prevent helmet visor misting up; elbow and ankle pressure bearings (Figure 8.3.9 and 8.3.10); pressure glove of improved mobility and strength; wrist pressure disconnect of improved reliability; water-cooled garment of improved performance; improved snap hook for safety tethers; safety tether of variable length widening the cosmonaut operating envelope; back-up pump; modified fan; modified radio set; $CO_2$ absorption cartridge of increased capacity; etc.

It is important to indicate that some earlier introduced changes proved to be ineffective and, thus, were excluded: in particular, the application of pressure cuffs was excluded because of improved reliability of the pressure gloves and their interfaces; the injector was again transferred to the backpack to widen the suit torso; simultaneously, the backpack/HUT interface of ventilation tubes was transferred from the "waist" area to the helmet area and a decreased suit pressure mode corresponding to changes in the operational modes of emergency systems was also excluded.

Changes introduced in the hard torso by moving the exit hatch up, enlarging the rear part of the helmet and increasing the length of the leg enclosures made it possible to significantly facilitate suit-donning/doffing and accommodation and increase crew member anthropometric dimensions—112 cm for the chest (instead of 108 cm in the ORLAN-DMA case) and 190 cm for the height (instead of 185 cm).

Modification of the new suit allowed hygienic underpants to be worn (Pampers-type) and installation of a potable water tank like that of the US EMU spacesuit.

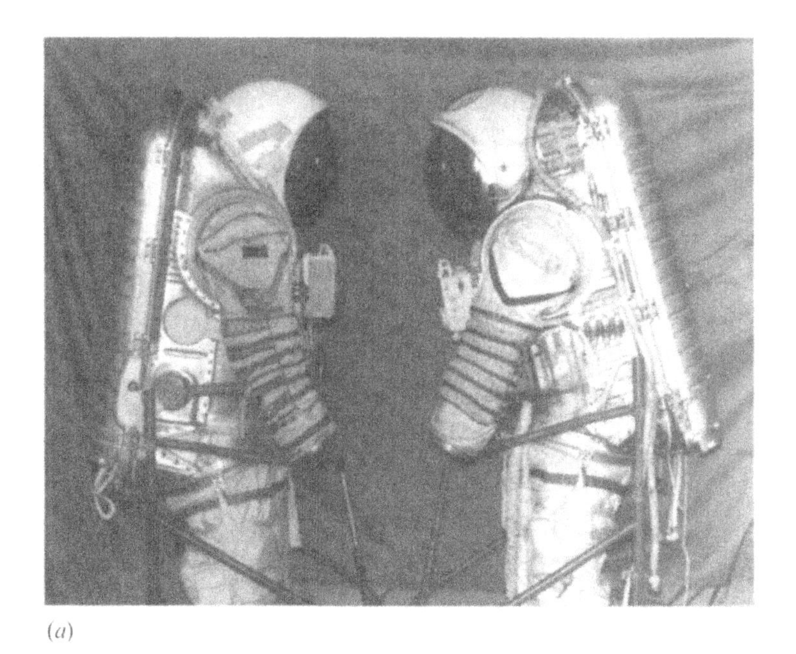

(*a*)

(*b*)

(*c*)

**Figure 8.3.8** (*a*) External configuration differences between ORLAN-DMA (*left*) and ORLAN-M (*right*). Backpack configuration differences: (*b*) ORLAN-DMA; (*c*) ORLAN-M. From Archive Zvezda.

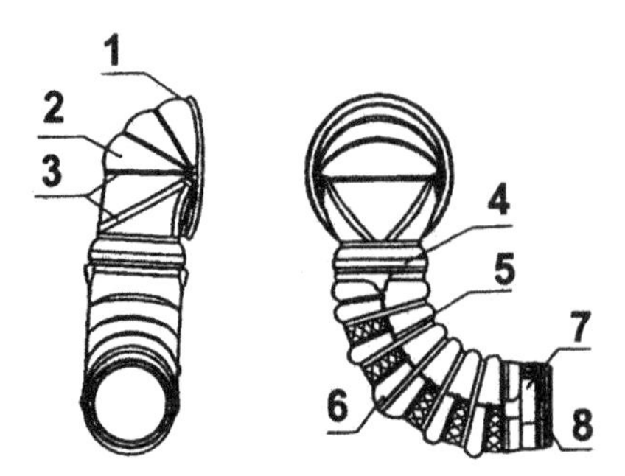


**Figure 8.3.9** ORLAN-M suit arm (without thermal protection garment): 1—shoulder bearing; 2—soft shoulder joint; 3—restraint elements of the shoulder joint; 4—arm bearing; 5—axial restraint cord of the elbow joint; 6—soft elbow joint; 7—length adjustment device for the suit's arm; 8—wrist bearing.

From Archive Zvezda.

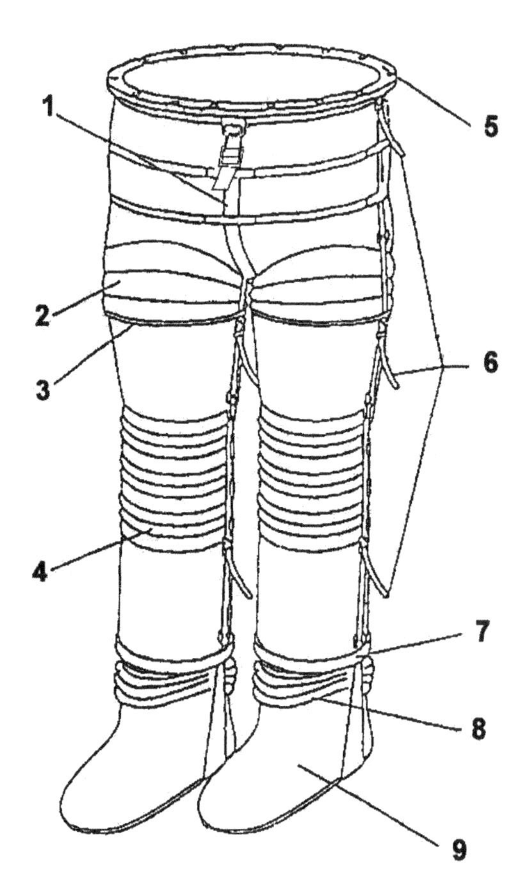


**Figure 8.3.10** Lower torso (legs) without the thermal micrometeorite (protection) garment: 1—central adjuster; 2—hip joint; 3—metal ring; 4—knee joint; 5—waist interface; 6—length adjustment straps; 7—calf bearing; 8—ankle joint; 9—soft boots.

From Archive Zvezda.

It goes without saying that the above changes improved the efficiency of EVA sorties.

*Main technical characteristics of the ORLAN-M spacesuit*

| | |
|---|---|
| Nominal duration of the autonomous mode | 7 hours |
| $CO_2$ absorption cartridge operating time (with airlock time included) | 9 hours |
| Suit positive pressure: | |
| —nominal mode | 392 hPa |
| —emergency mode | 270 hPa |
| Oxygen available (main and back-up) | 1 kg each |
| Cooling water available | 3.6 kg |
| Assured heat removal: | |
| —average | 350 W |
| —maximum | Up to 600 W |
| Total consumed power by the suit systems | Up to 54 W |
| Quantity of telemetry measured parameters | 29 |
| Spacesuit weight (wet) | ~112 kg |
| Service life | Up to 15 EVAs over 4 years (no return to the Earth) |

When all the required tests were completed, ORLAN-M spacesuits Nos 4, 5 and 6 (Figure 8.3.11) were delivered to the *Mir* orbiting station in 1997. They accumulated 36 EVAs, and the date of the last one was 12 May 2000 (Figure 8.3.12).

Two ORLAN-M spacesuits (Nos 12 and 23) were delivered to the *ISS* in the service module in July 2000 and one ORLAN-M suit (No. 14) was delivered to the *ISS* in the SO-1 docking module in September 2001.

By 31 December 2002, 11 Russian and US crew members had undertaken 18 EVAs using ORLAN-M spacesuits.

It is worth saying a few words about the cooperation with NASA. After the US/Russia Agreement on Cooperation in Space Exploration was signed (by G. Bush and B. Yeltsin) in 1992, specialists of NASA and Hamilton Standard showed interest in Zvezda's experience in the development and operation of orbit-based EVA suits. Hamilton Standard (now Hamilton Sundstrand) developed the US EMU suit for the *Shuttle* and *ISS* programmes.

Russia already had 15 years of experience of such operations aboard the *Salyut* and *Mir* stations by that time.

Subsequent years have brought about several contracts (including those signed within the framework of the main contract between the RSA and NASA on the *Mir–Shuttle* programme): comparative analysis of US and Russian EVA suits; feasibility of bringing them together; provision for EVA in the Russian suit undertaken from the US airlock; development of means for unassisted rescue of *ISS* crew members

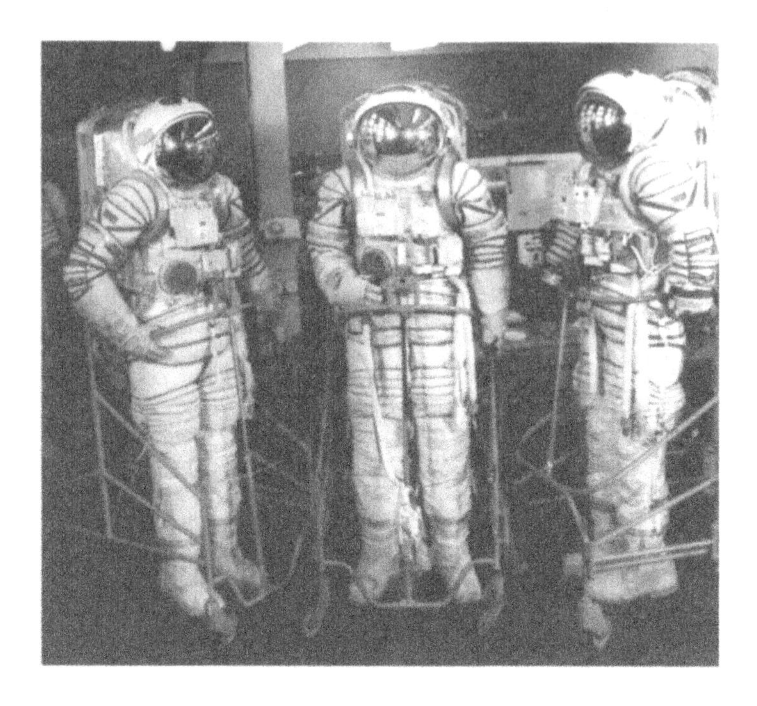

**Figure 8.3.11** Manufacture of the first three flight units of ORLAN-M for *Mir*.
From Archive Zvezda.

**Figure 8.3.12** ORLAN-M suit during EVA on *Mir*.
From Archive Zvezda.

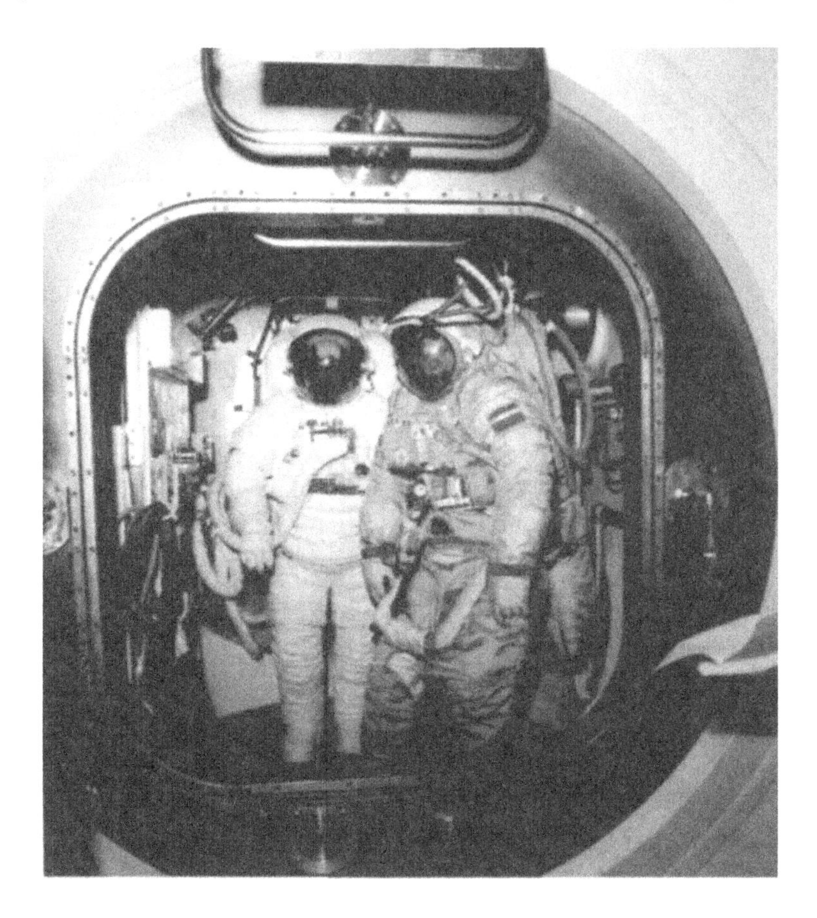

**Figure 8.3.13** G.M. Glazov, Zvezda's test engineer (suited in ORLAN-M), and J. Marmolejo, NASA's test subject (suited in EMU), are in the *ISS*'s common airlock located in the NASA thermal and vacuum chamber.

Photo by courtesy of J. Marmolejo (NASA).

during EVA (SAFER); training US specialists in operating ORLAN suits aboard the *ISS*; training US astronauts in wearing Russian EVA suits, etc.

Zvezda delivered two special spacesuits as well as the technical documentation needed to support the training of *ISS* crews in NASA's hydro lab. Moreover, an ergonomic model of the ORLAN spacesuit was delivered to Zvezda's US partners for comparative analysis.

A considerable amount of work was done to integrate the ORLAN spacesuit with the *ISS*'s common airlock. Zvezda specialists managed to make a new, small, on-board system for the spacesuit that provided interoperability of the ORLAN spacesuit both with the Russian segment's airlocks and the station's common airlocks. The spacesuit was modified correspondingly. The developed system was successfully tested with common airlock interfaces both at Zvezda and NASA (in a vacuum chamber and in conjunction with an airlock flight model) (Figure 8.3.13).

**Figure 8.3.14** NASA astronaut Susan Helms during ORLAN-M suit fit check at Zvezda. From Archive Zvezda.

Furthermore, joint work also included studies on adaptation of the ORLAN spacesuit to US requirements and feasibility of using EMU suit elements in the ORLAN suit, etc.

Over the whole period of the joint work, Zvezda specialists were in constant contact with NASA specialists (mainly with those working for the NASA EVA design office and Hamilton Sundstrand). Zvezda specialists constantly participate in work associated with routine operations of the systems used to support EVA sessions undertaken by both Russian cosmonauts and US astronauts. Preparations for these EVA sessions are partially carried out at Zvezda (crew member fit checks—Figure 8.3.14—and training in a vacuum chamber).

It is worth discussing the work on modification of the spacesuit's on-board systems. In the *Salyut-6* mission, the ORLAN-D spacesuit operated in conjunction with the BSS-1 [БСС-1] on-board docking system, which was used during preparation for the EVA sortie and in the airlock process to connect the spacesuits with the on-board oxygen bottles and heat exchanger. The latter cooled water circulating in the water-cooled garment. The valve controlling the temperature of water entering the water-cooled garment was installed in the BSS-1 on-board docking system.

The *Salyut-7* and *Mir* orbiting stations used the modified BSS-2 on-board docking system (and then the BSS-2M system—Figure 8.3.15), which comprised

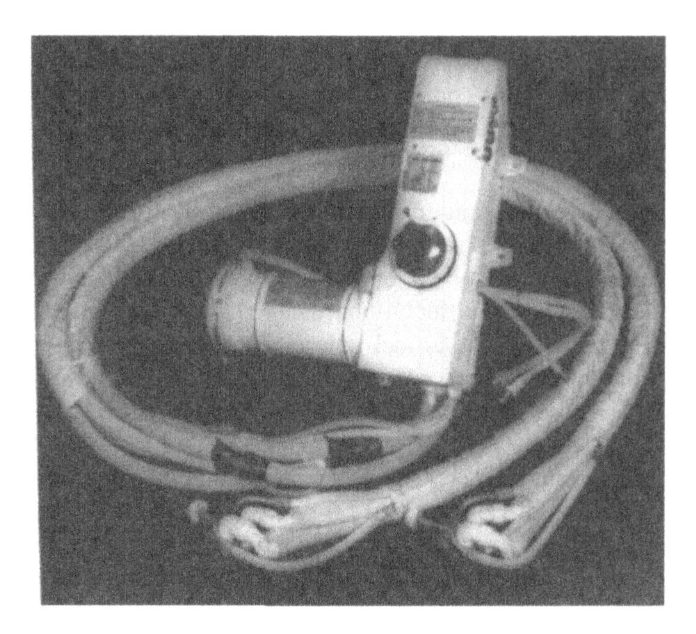

**Figure 8.3.15** General view of the BSS-2M unit.
From Archive Zvezda.

fans, moisture collectors and $CO_2$ absorption cartridges. With the spacesuit operating from the station's on-board system, the suit no longer needed to use its own consumables and, thus, duration of the suit's autonomous operation mode increased. The on-board systems of these orbiting stations also comprised the hose bunch, electric umbilicals and oxygen bottles. Station on-board systems provided the suit with power supply and cooling of the circulating water during the airlock process.

As already mentioned, with the advent of the *ISS* the task was set for operation of the ORLAN suits not only in the airlock of the Russian segment, but also in the stations's common airlock, designed for operation with the US-made EMU spacesuits. To simplify interoperability of the ORLAN suit and the *ISS*'s common airlock systems, a new on-board system was developed. It simply comprised a suit control panel (BUS) and a bundle of hoses with electric cables. The system is now used to test two ORLAN-M spacesuits and support operation of the suits in the process of direct and return airlocking[1] with EVA undertaken both from the Russian segment's airlock and the common airlock. The system is designed to:

- feed the suit with oxygen from the on-board storage system;
- pressurize the suits to check their pressure-tightness;
- purge the suits with oxygen to change the gas mix in them;
- depressurize the suits in the process of airlocking;

[1] By "airlocking" we mean the process of airlock pressurization and depressurization.

- connect the suits to on-board systems via hoses and electric cables, providing, among other things, connection of the suit's water circulation loop to the on-board heat exchanger and connection of suit systems to the on-board power supply as well as to the radio communications and telemetry systems (for the Russian segment).

Figure 8.3.16 shows pneumatic and hydraulic flow diagrams for the BSS-2M and the on-board system, with the latter arranged in the *ISS*'s Russian segment (identified as BSS-4). The common airlock uses the on-board system in a similar way.

The only differences are that the on-board suit control assembly (OSCA) in the common airlock is a part of the airlock interface panel and the water tubes and electric cables of the bunch of hoses are terminated by a unified combined connector, similar to the EMU hose bunch connector and interfaces connected to the same panel.

From the design point of view, OSCA is a device that incorporates a controlled cam mechanism and a set of valves that are sequentially opened or closed, according to the given programme, depending on the position of the control handle. The assembly's front panel with control handle position marks is shown in Figure 8.3.17. The control handle is set in this or that position, depending on the selected operating mode, by crew members manually. Oxygen supply, pressurization and purging modes are monitored by pneumatic indicators (blinkers).

Removal of the absorption cartridge and moisture collector/separator from the simplified on-board system called for modification of the spacesuit's LSS, resulting in increased capacity of the $CO_2$ absorption cartridge and moisture separator installed in the spacesuit. This was brought about because the given components of the modified spacesuit can then operate for the entire suited mode, including the airlock period (up to a total of 9 hours). As there is no fan in the on-board system, post-EVA drying of the spacesuit has to be done by the suit's LSS fans. In its turn, this resulted in changes in the suit ventilation system flow diagram, installation of a special valve in the suit (see Figure 8.2.2) and development of an additional set of accessories.

To meet requirements for EVA from the station's common airlock with the minimum amount of additional equipment, a proposal was made to transfer equipment, normally arranged in the Russian segment, to the common airlock for the time needed to carry out the EVA and use it to support the EVA session from the common airlock, only equipping the common airlock with the on-board system that is most compatible with Russian segment equipment and features for both the EMU and the ORLAN-M suits.

## 8.4 SPACESUITS FOR CREW-TRAINING

Besides these modifications to the orbit-based spacesuit, Zvezda developed several special test and cosmonaut training models: ORLAN-V, ORLAN-GN and ORLAN-T.

The letters accompanying the names denote the following. The V [B] stands for

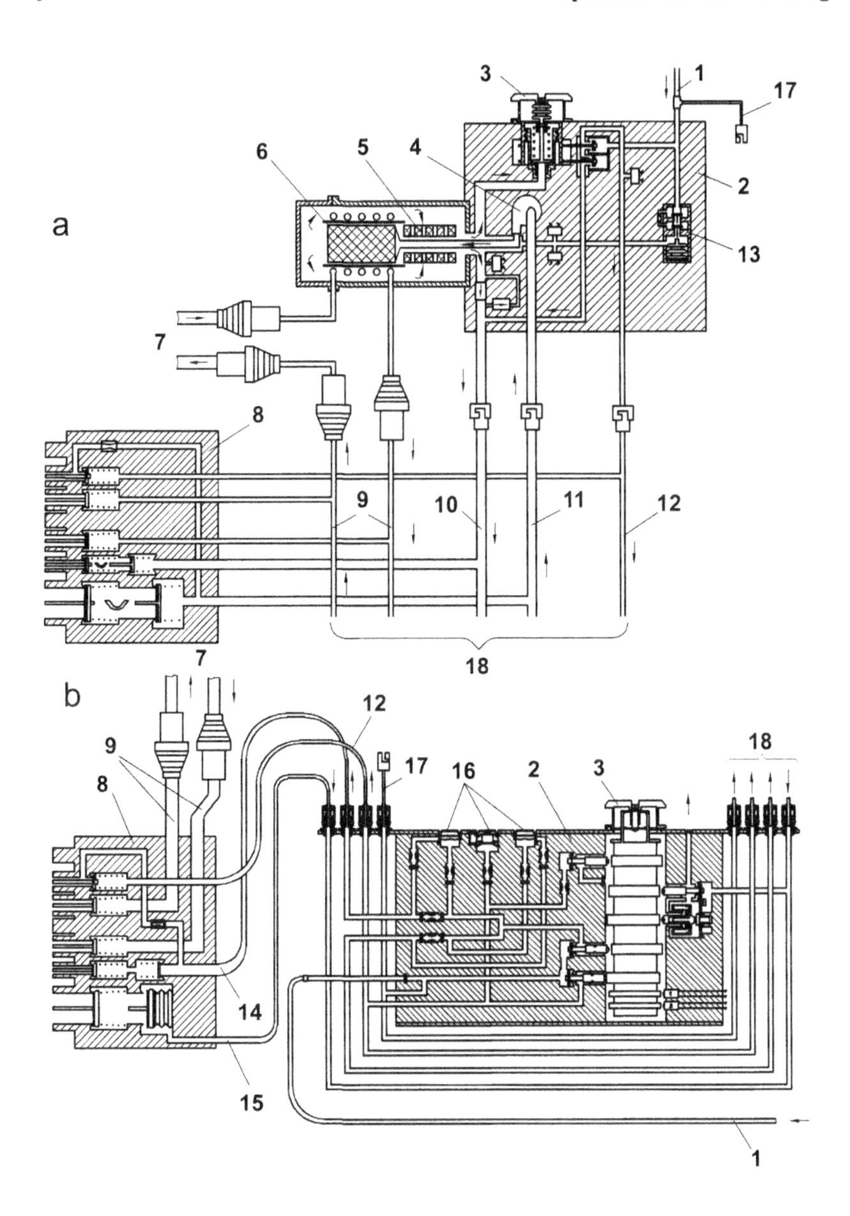


**Figure 8.3.16** Principal flow diagrams of the BSS-2M unit (*a*) for the *Mir* and the BSS-4 unit (*b*) for the *ISS*: 1—oxygen from the on-board bottles; 2—on-board suit control assembly; 3—control handle for spacesuit operation modes; 4—fan; 5—moisture collector; 6—$CO_2$ control cartridge; 7—to the on-board heat exchanger; 8—combined connector; 9—cooling water tubes; 10—to the spacesuit's ventilation line; 11—from the spacesuit's ventilation line; 12—line of oxygen supply to the spacesuit; 13—spacesuit pressure regulator; 14—line of oxygen supply for spacesuit-purging; 15—spacesuit's purge gas discharge line; 16—spacesuit operation signalling elements; 17—emergency hose; 18—to the second suit.

From Archive Zvezda.

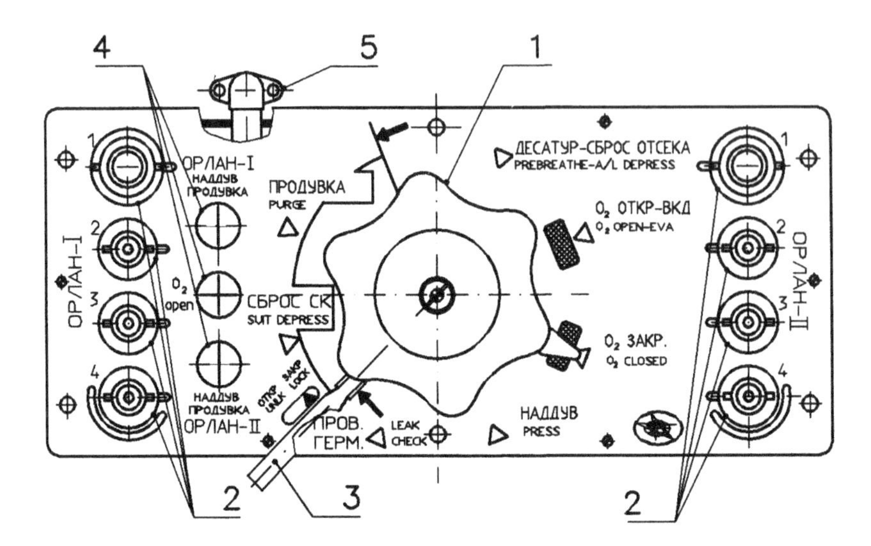


**Figure 8.3.17** Front panel of the BUS BSS-4: 1—control; 2—couplings to connect the umbilical assembly; 3—locking device; 4—pneumoindicators (ORLAN-I press, depress; ORLAN-II press, depress); 5—coupling to supply oxygen from the on-board gas capacity. From Archive Zvezda.

"ventilation" in Russian. The GN [ГН] is the Russian acronym denoting "hydraulic weightlessness" and the T [T] denotes that the model is used for "training".

The ORLAN-V models (for each of the modifications) were designed for use aboard a flying laboratory that provided weightlessness conditions for several dozen seconds. It featured a suit enclosure, including the backpack housing, that was identical to the normal one. The LSS units were missing inside the backpack. Suit ventilation and positive pressure were provided by supplying the suit with air from on-board sources via a hose (with a flow rate up to $250\,\mathrm{l\,min^{-1}}$). Air was removed from the suit through a pressure regulator arranged on the combined connector. With the hose disconnected, bottles of compressed air could be installed in the backpack to pressurize the suit. The ORLAN-V models were and still are used to carry out various fit check work under ground conditions.

The ORLAN-GN models (Figures 8.4.1, 8.4.2 and 8.4.3) were designed for work in the hydraulic laboratory of the Cosmonaut Training Centre. The enclosure of these models was the same as those of normal suits (the outer garment was however missing on the spacesuit body). Also missing was the normal LSS. Supplying air and cooled water from ground sources supported the vital functions of the suited person. Water entering the spacesuit system cooled the air circulating in the suit by means of a special water/air heat exchanger located in the suit's backpack. The backpack also housed an emergency air bottle, designed to operate for 15 minutes and switched on manually in case of inadvertent disconnection of the ground hose. Air was removed through a pressure regulator arranged on the front part of the body. The pressure regulator made it possible to maintain any positive pressure in the suit in the range 0

**Figure 8.4.1** Neutral buoyancy training with the ORLAN-D suit at the Gagarin Cosmonaut Training Centre (1983–1984). *From left to right*: G.I. Severin, I.P. Abramov, L.D. Kisim, B.V. Michailov.

From Archive Zvezda.

to 400 hPa. An electric umbilical was connected to the spacesuit. It was used to supply electric power and pick up data from measuring instruments and biomedical sensors.

This model featured easy-to-remove weights (located on the chest, back and arm and leg enclosures) used to control the neutral buoyancy of the suited person, changes in the design of certain suit components (outer garment, pressure bearings and suit handling system), introduced to adapt them for operation in water, and a lack of normal control panels on the front part of the spacesuit body.

By the beginning of joint NASA/RSA work on the *ISS* programme, Zvezda had manufactured two ORLAN-M-GN models (Figure 8.4.4) and delivered them to

**Figure 8.4.2** General view of the ORLAN-DMA-GN suit (prior to suit-donning for training at the Gagarin Cosmonaut Training Centre).

NASA's hydro lab to support the training programme of US astronauts. The models were specially adapted to NASA ground support equipment interfaces.

The ORLAN-T model is used by cosmonauts for training at the Vykhod ["ВБЫХОД", Russian for "EVA sortie"] training facility at the Cosmonaut Training Centre. The suit model makes it possible to train cosmonauts in airlock procedures at sea level without a decrease in ambient pressure.

Development of the ORLAN-T suit was stipulated by the same government resolution dated 29 September 1985, and respective government directives, which gave the go-ahead to development of the self-contained spacesuit system and the 21KS system. The ORLAN-T suit is a normal spacesuit with a modified ventilation and oxygen supply system, pickup means and electrical equipment control system. The changes introduced make it possible to pressurize the suit from external sources of air and to simulate, under ground conditions, various emergency situations (failure of units, leakage from the spacesuit, actuation of the emergency warning system, etc.).

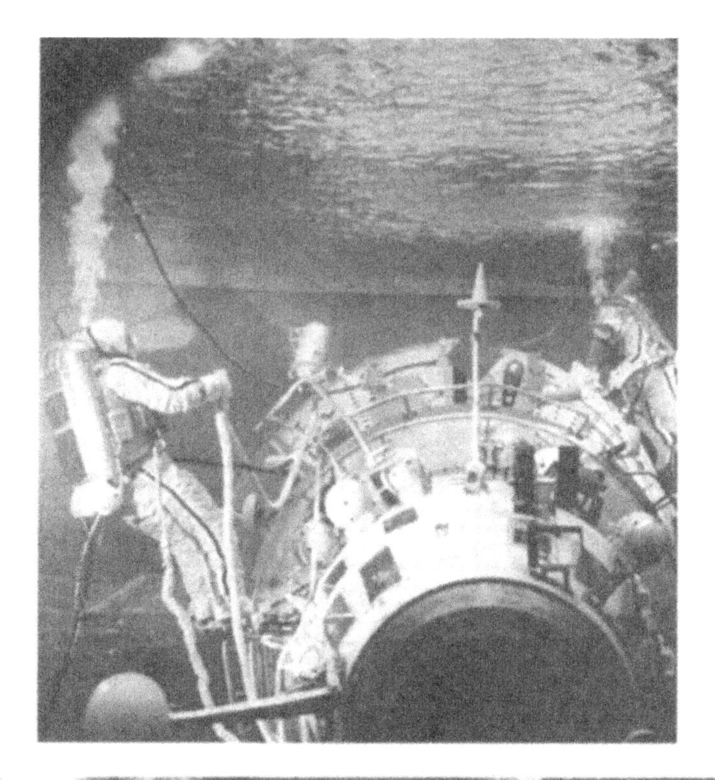

**Figure 8.4.3** Crew-training in the hydro lab at the Gagarin Cosmonaut Training Centre.

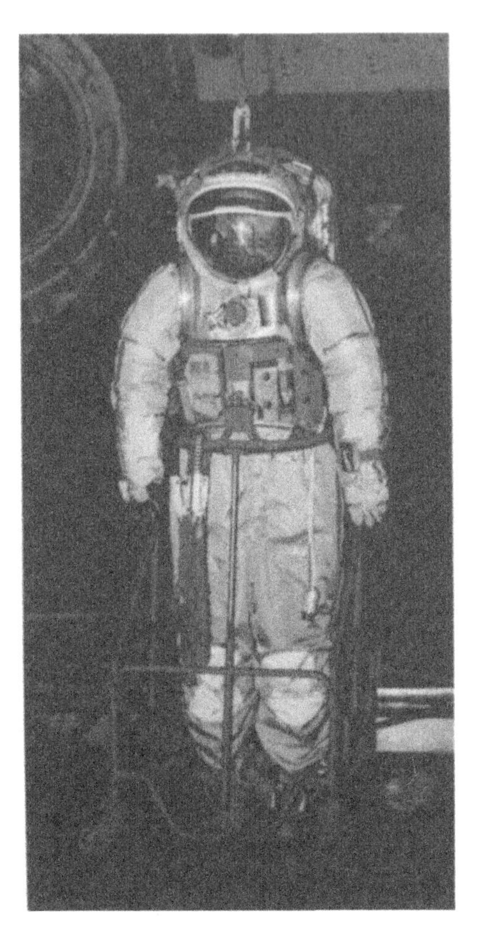

**Figure 8.4.4** ORLAN-M-GN.
From Archive Zvezda.

## 8.5 SOME RESULTS OF OPERATIONS OF THE ORLAN-TYPE SPACESUITS

The list of all EVA sorties undertaken from the *Salyut-6*, *Salyut-7*, *Mir* and *ISS* orbiting stations and supported by ORLAN-type spacesuits is given in Appendix 1.

Considering all the EVA sorties collectively, the spacesuits and their systems operated properly on the whole and, thus, made it possible to complete practically all the scheduled and unscheduled work in free space or in depressurized station modules.

There was not a single case of decompression sickness in the crew members involved in EVAs. Experience has proven that a semi-rigid-type spacesuit was the most suitable for EVAs carried out to support the operations of a long-duration orbiting station. Suit operation results have also demonstrated the advantages of using this type of spacesuit in EVAs. It is reasonable to point out that, for the first time in space history, in-orbit operations of spacesuits aboard orbiting stations, over many years and without returning to the Earth, and their utilization by many crews have been accomplished.

A lot of important and complicated repairs, assembly and other work were done during EVA sorties from the *Salyut*, *Mir* and *ISS* orbiting stations, providing for both extended operation of the stations themselves and their further development. They, in particular, include the uncoupling of the KRT [KPT] antenna (which failed to separate) from the *Salyut-6* station, repair of the hydraulic system on the *Salyut-7* station's surface, installation of additional solar arrays, inspection and repair of the *Spectr* module on the *Mir* station, installation of additional stand-away engines on the *Mir* station (Figure 8.5.1), inspection of the station system for docking with transport/cargo vehicles and removal of foreign objects from docking system units.

The total number of EVA sorties supported by ORLAN-family spacesuits between December 1977 and 31 December 2002 is given in Tables 8.1 and 8.2.

The number of EVA periods supported by each spacesuit mainly depended on the amount of scheduled activities on the station, scheduled time, station life time, warranted life, hardware condition and some other things. Over the 26-year operational period of the *Salyut*, *Mir* and *ISS* programmes, 25 ORLAN-type spacesuits have been used for EVAs and each of them took part in from 3 to 15 EVA sessions. Members of 45 crews carried out 104 EVA sorties of total duration exceeding 850 hours. Besides Soviet and Russian cosmonauts, there were participants in EVA sessions from France (Figure 8.5.2), the ESA and NASA.

The long-lasting operation of the spacesuit in many ways depended on the quality of test carried out and maintenance work. The main tasks of this work are to make sure that the spacesuit and its systems operate properly prior to the EVA period and to preserve the suit's system operability for subsequent EVA periods. Most crews cope with these tasks very well.

The amount of maintenance and pre-EVA preparatory work varied, depending on the condition of the spacesuit and the time intervals between EVA sessions. The experience gained resulted in selection of the following kinds of suit maintenance:

- maintenance of the spacesuit prior to its first EVA sortie on delivery to the station or after a long break between EVA cycles;
- maintenance related to changing of the crew;
- maintenance between repeated EVAs carried out by the same crew (every 3–15 days);
- post-EVA maintenance.

Suit maintenance and preparatory work were carried out, as a rule, several days or the day before the EVA period, and final checks were made on the EVA day. Most time was required for the first kind of maintenance (about 6.5 hours). Maintenance between repeated EVAs took about 1.5–2 hours. The pre-EVA preparations on the EVA day lasted about 1.5 hours (for all kinds of maintenance). Post-EVA

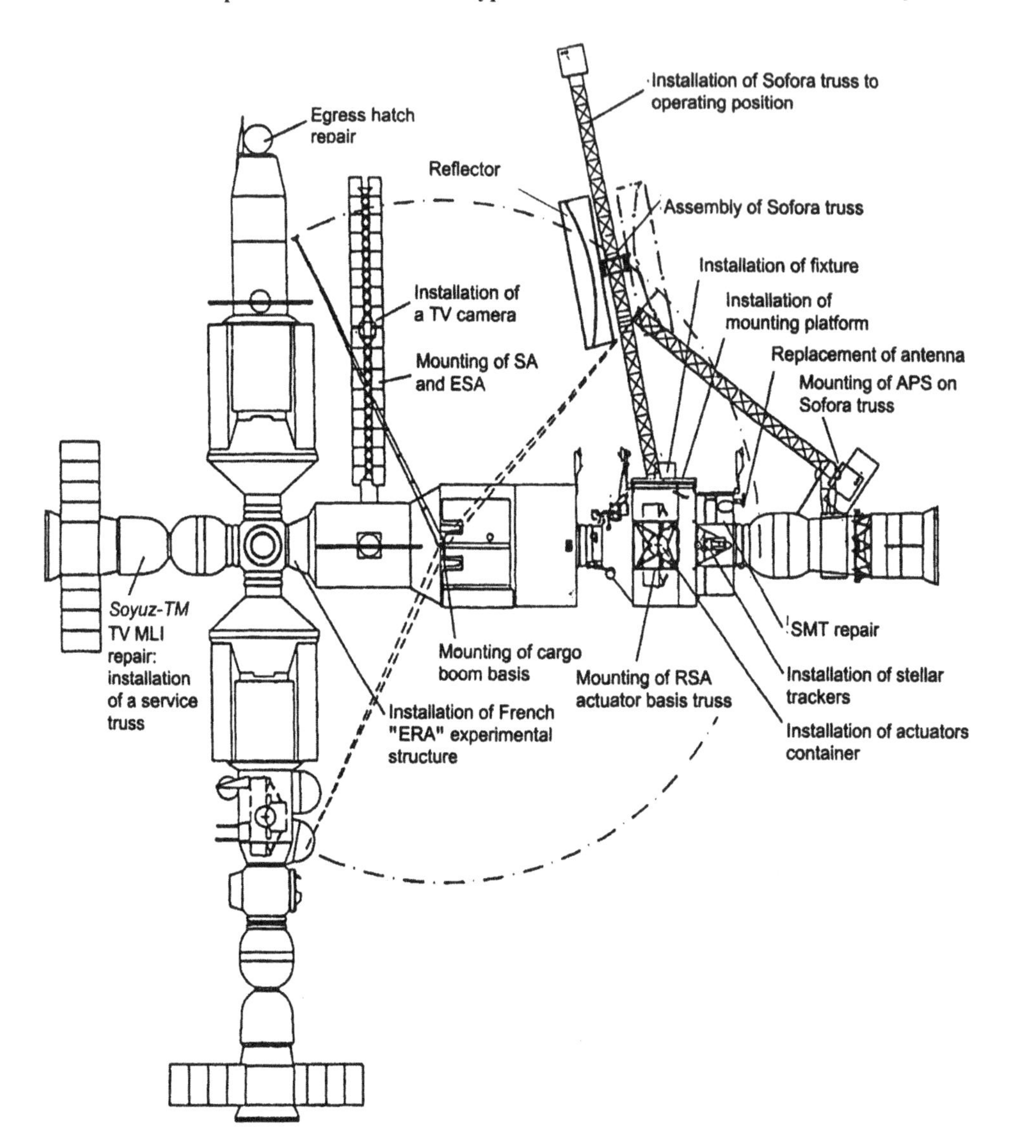


**Figure 8.5.1** *Mir* station maintenance and repairs carried out during EVAs. The principal work carried out by crews in open space aimed at gathering and extending scientific and technological data.

maintenance, including maintenance on the day after the space walk (suit-drying, etc.), took about 3 hours. Operational tasks relating to spacesuit-testing and maintenance and the time needed to carry them out are given in the on-board manuals.

Because of the multi-person use of the spacesuit, the personal hygiene of the crew members and antimicrobial protection are assured by issuing each crew member with a clean outfit before each EVA. Any soiled areas of the suit's internal surface are scrubbed and disinfected, if necessary.

**Table 8.1.** Number of EVA sessions supported by ORLAN-type spacesuits.

| Operational year | Number of EVAs |
|---|---|
| 1977 | 1 |
| 1978 | 1 |
| 1979 | 1 |
| 1980 | 0 |
| 1981 | 0 |
| 1982 | 1 |
| 1983 | 2 |
| 1984 | 7 |
| 1985 | 1 |
| 1986 | 2 |
| 1987 | 3 |
| 1988 | 4 |
| 1989 | 0 |
| 1990 | 8 |
| 1991 | 10 |
| 1992 | 6 |
| 1993 | 7 |
| 1994 | 2 |
| 1995 | 10* |
| 1996 | 9 |
| 1997 | 6* |
| 1998 | 10* |
| 1999 | 3 |
| 2000 | 1 |
| 2001 | 5 |
| Up to 31 December 2002 | 4 |

* Some of these EVA sessions were carried out in the depressurized station module without opening the exit hatch.

Thorough drying of the spacesuit after each EVA period is important in keeping the suit in the correct operational state during its long storage. The internal surfaces of the suit, sublimator, tubes and elements connecting the sublimator with the feedwater tank and elements of the on-board ventilation system all need to be dried. Experience showed that, under conditions of increased humidity aboard the station, insufficiently dried surfaces would start to smell musty.

In the process of lengthy operation of spacesuits certain remarks/observations were naturally made by the cosmonauts regarding the functioning of some of the spacesuit's elements—some pointed out a decrease in crew comfort and some called for replacement of some suit elements or even suit repair. The fact that the spacesuits permanently stayed in orbit, were maintained by crew members and not brought back to the Earth made it difficult to verify the crew's remarks/observations and analyse the post-EVA state of the spacesuits. So, information about how spacesuits

**Table 8.2.** ORLAN-type spacesuit operations.

| Orbiting station | Spacesuit | Spacesuit No. | Number of EVAs | Date of delivery to the orbit | Date of last use | Operational period (months) |
|---|---|---|---|---|---|---|
| *Salyut-6* | ORLAN-D | 33 | 3 | 09.77 | 15 August 1979 | 21 |
| *Salyut-6* | | 34 | 3 | 09.77 | 15 August 1979 | 21 |
| *Salyut-7* | | 45 | 10 | 04.82 | 08 August 1984 | 28 |
| *Salyut-7* | | 46 | 3 | 04.82 | 03 November 1983 | 19 |
| *Salyut-7* | | 47 | 7 | 03.84 | 08 August 1984 | 5 |
| *Salyut-7* | ORLAN-DM | 8 | 3 | 06.85 | 31 May 1986 | ~12 |
| *Salyut-7* | | 10 | 3 | 06.85 | 31 May 1986 | ~12 |
| *Mir* | | 7 | 5 | 03.86 | 30 June 1988 | 27 |
| *Mir* | | 9 | 5 | 03.86 | 30 June 1988 | 27 |
| *Mir* | ORLAN-DMA | 6 | 14 | 06.88 | 27 July 1991 | 37 |
| *Mir* | | 8 | 10 | 10.89 | 20 February 1991 | 16 |
| *Mir* | | 10 | 9 | 08.88 | 21 April 1991 | 32 |
| *Mir* | | 12 | 7 | 10.89 | 20 February 1991 | 16 |
| *Mir* | | 14 | 13 | 04.91 | 22 October 1993 | 30 |
| *Mir* | | 15 | 7 | 04.91 | 18 June 1993 | 26 |
| *Mir* | | 18 | 13 | 10.92 | 8 December 1995 | 38 |
| *Mir* | | 25 | 15 | 03.93 | 13 June 1996 | 39 |
| *Mir* | | 26 | 12 | 10.95 | 20 October 1997 | 24 |
| *Mir* | | 27 | 12 | 02.95 | 20 October 1997 | 32 |
| *Mir* | ORLAN-M | 4 | 14 | 04.97 | 12 May 2000 | 37 |
| *Mir* | | 5 | 12 | 04.97 | 16 April 1999 | 24 |
| *Mir* | | 6 | 10 | 10.97 | 12 May 2000 | 31 |
| *ISS* | ORLAN-M | 12 | 7 | 07.00 | 25 January 2002* | |
| *ISS* | | 14 | 4 | 09.01 | 26 August 2002* | |
| *ISS* | | 23 | 7 | 07.00 | 26 August 2002* | |

* Status at 31 December 2002.

operated was mainly evaluated on the basis of telemetry data, crew reports, post-flight discussions and studies of some of the failed suit elements brought back to Earth. Each remark/observation was thoroughly analysed and then, in an effort to solve the problem, necessary measures were worked out, which included (depending on the cause) additional instructions to the crew, update of the on-board crew manual, replacement of separate elements or introduction of changes in the suits.

If, when getting ready for the EVA session, faults or improper operation of spacesuit systems came to light, additional tests or preventive/repair work were carried out subject to the agreement of the ground mission support team. To support work of this kind, the suit's system set comprised special tools and a limited quantity of spare parts for light repair of the suit (replacements of gaskets, connectors, etc.).

**Figure 8.5.2** Debriefing by J.-L. Chrétien (CNES) at Zvezda 1988 after the first EVA of a non-Soviet cosmonaut; *from left to right (first row)* N.I. Afanasenko, J.-L. Chrétien, G.I. Severin, V.I. Svertshek; *(second row)* I.I. Chistyakov, V.P. Efimov, A.V. Kirdan, M.M. Balashov, F.A. Vostokov, A.Yu. Stoklitsky, I.P. Abramov, P.Kh. Sharipov, A.S. Barer, B.V. Michailov, I.A. Sokolovsky.

From Archive Zvezda.

Possible replacement of the arms and lower "soft" part of the spacesuit aboard the station, in case they are damaged or worn out, is available with the ORLAN-DMA and ORLAN-M spacesuits.

A more substantial or intricate repair, if needed, was carried out under the guidance of the ground support team using available on-board or appropriate equipment specially delivered to the station for such an event.

Analysis of failures that cropped up and critical remarks that were made during operation of the spacesuits aboard the stations fall into several groups:

- failures of spacesuit elements that resulted in the need to repair the suit or shorten the EVA;
- failures of suit elements, which resulted in the reduced comfort of crew members or worsened signalling or telemetry measurements, but did not affect completion of the EVA programme;
- mistakes made by the crew;
- all other events and observations (such as malfunctions of on-board or ground support measurement equipment).

Failures of the first group were very rare and usually presented themselves during pre-EVA suit-testing.

Partial misting of the suit visor, malfunctions of gauges for measurement of the ventilation gas flow rate and medical readings, etc. fell into the second group of failures and remarks/observations. It is important in this regard to point out that significant assistance to cosmonauts while preparing for and carrying out their EVAs is rendered by the Zvezda EVA support team at the Mission Control Centre, TzUP [ЦУП], which monitors the readings from all spacesuit systems, informs crew members about observed deviations and recommends the best way of rectifying them.

Mistakes made by the crew were mostly countered by actions recommended by this very EVA support team. However, there were cases when EVA sorties had to be delayed due to mistakes made by the crew.

Active roles in the work of the Zvezda EVA support team were played by M.M. Balashov and, from 1995, by G.M. Glazov. They maintain direct radio communication with the crews during their work with spacesuits and, if the necessity arises, during EVAs.

It should be pointed out that some of the failures occurred in spacesuits whose service life had expired. Moreover, some malfunctions occurred because the atmosphere aboard the station at certain times was not normal (in terms of humidity or temperature). With long breaks in-between suit operation, such a situation negatively affects the operability of spacesuits after long-term storage (up to 10 months). In particular, failures in the Beta-08 medical apparatus and the electronic units of fans (see below) could have been caused by increased humidity in the station. As a matter of fact, one of the ORLAN-M suits completely froze during storage in the *Kvant-2* module in 1999 (with no crew aboard the station).

Listed below are some of the maintenance and repair activities carried out by cosmonauts both to bring about changes in spacesuits as a result of new modifications and changes in operating conditions and to eliminate any observed malfunctions and defects. Some work concerned the need to use spacesuits whose service life was drawing near to a close or even had run out.

The most substantial maintenance and repair work undertaken aboard the orbiting station was as follows:[2]

- repair of the suit's leg enclosure (1983);
- replacement of the fan unit (1991 and 1998);
- installation of an additional guard for the suit's pressure mode selector (1988);
- replacement of the suit's arm because of introduction of the pressure cuff (1990) and increased leakage (1998 and 2000);
- substitution of the "heat–cool" valve on the control panel of the pneumatic and hydraulic system with a new one with improved performance characteristics (1990);

[2] In the ORLAN-D suit used by Solovyov on EVA No. 15 the water pump failed. The work was undertaken with the two ventilation fans working simultaneously (while the liquid cooling system was not functioning). This suit was not repaired as the suit had reached the end of its life expectancy.

- replacement of the suit's measuring complex (1990);
- replacement of a leaking T-piece in the suit's hydraulic system (1991);
- replacement of the Beta-08 medical data-processing unit (1998);
- replacement of a leaking water valve on the suit's combined service connector (1999).

On completion of all maintenance, repair and replacement operations, the necessary tests were carried out and normal utilization of spacesuits went on. The experience gained in the multi-year operation of ORLAN-type spacesuits shows that these suits feature both high reliability, provided by redundancy of the units and elements of the suit system important for life support, and high maintainability. These features made it possible, in certain cases, to successfully get out of extremely complicated situations. Several examples can be given of ways that were found to get out of dramatic situations.

In 1979, when the third expedition team worked aboard the *Salyut-6* station, on completion of the experiment with the KRT-10 radio antenna (10 m diameter), the latter got caught on one of the protruding elements of the station. The situation called for an urgent EVA session to uncouple the antenna. However, there were circumstances that complicated the situation. To begin with, one of the ORLAN-DM suits was not ready for operation because an earlier failed back-up fan had been removed (it was going to be returned to the Earth with the crew). Then, there were doubts about whether EVAs could be carried out by a crew that had been in orbit in a weightlessness environment for almost 6 months. No experience of carrying out EVA sorties during long-term space missions existed. However, the situation had to be resolved. Several working groups were formed so that they could make the decision. V.P. Glushko of Energia and G.I. Severin of Zvezda were directly involved in their work. As far as the incomplete EVA spacesuit was concerned, the decision was made to return the faulty fan to its place in the suit (to close the ventilation line) and then, after thorough additional checking, use it for operation. Such a decision implied that crew member safety would be provided by the system operating with the injector switched on, even if the main fan failed. The crew of V. Lyakhov and V. Ryumin successfully accomplished the task on 15 August 1979.

Another, arguably unique, case occurred in 1983 on the *Salyut-7* station when cosmonauts managed to repair damage to the main suit's pressure bladder. In 1982, the first shift crew members of the *Salyut-7* orbiting station successfully performed an EVA session using two ORLAN-D spacesuits. The next EVA session was scheduled to be carried out on 1 November 1983 by V. Lyakhov and A. Alexandrov, members of the next space crew, who were to use the same EVA suits even though the suits had been stored for more than a year. However, crew members carrying out pre-EVA checking of the suits on 26 October reported leakage in one of the spacesuits. The suit was leaking under the knee of the right leg. In line with recommendations of the ground support team, the cosmonauts removed the stitches of the longitudinal seam of the restraint enclosure, opened it to find a transverse rupture of the main pressure bladder that ran for two-thirds of the bladder's transverse perimeter. Subsequent analysis showed that the most probable cause of such a rupture was improper stowage (too compact) of the suit for long-term storage by the previous crew. Testing of the suit for leakage showed that the back-up pressure bladder was intact.

If such a defect had occurred during work outside the station, the life of the crew member would not have been in danger. Furthermore, had there been a need to undertake an urgent EVA session, it could have been done with only the back-up bladder effective. However, to avoid any risk, the radical decision was taken to repair the damaged pressure bladder just days before the scheduled EVA period, with help from Zvezda specialists.

Zvezda formed a team headed personally by General Designer G.I. Severin. The team mainly comprised design and test engineers, specialists engaged in production and representatives from NPO Energia. The team was to develop a repair procedure using tools and auxiliary materials available aboard the station, prove the reliability of the enclosure after repair and train the crew how to effect the repair. Such a repair of the suit enclosure not only had never been planned, it had never even been considered.

With the tools and auxiliary materials available aboard the station, various methods of repair were tried. They included both a conventional method using adhesive and a "dry" (without glue) method using a ring (rimming method). The option shown in Figure 8.5.3 was chosen to effect the repair. A ring of aluminium alloy sheet was cut (70 mm wide and 1.3 mm thick). It was made by crew members who cut a strip out of the fan casing using a hacksaw and riveted its ends together. The repair procedure was as follows. Sticking plaster was stuck on the internal surface of the ring and ring edges. Four rings of width equal to that of the aluminium bandage ring were cut out of rubber bags from the waste management system. One rubber ring was glued to the bandage ring.

At the point of the transverse rupture, the restraint enclosure made of nylon was cut into two parts (upper and lower parts) and the fabric edges were melted with a soldering iron. Then the pressure bladder was cut in the same way. The bandage ring was inserted into the cut between the outer and inner pressure bladders, and the lower part of the damaged pressure bladder was pulled over the bandage ring. Then the upper part of the damaged bladder was pulled over the bandage ring, overlapping the lower part.

Two layers of sticking plaster were applied to the butt. The resulting package was wound around the bandage using thread. Two rubber rings were put over the package and again wound around with thread. The resulting package was covered by the fourth rubber ring. The removed longitudinal seam and transverse cut of the restraint enclosure were sewn up over the edges with thread. During in-house development of the repair procedure, Zvezda specialists, including those who were to advise the crew, repaired four fragments of deliberately cut enclosures and legs from pool-training spacesuits whose service life had expired. The reliability of the repaired enclosure was proven in the tests of the mentioned four samples, including pressure, leakage, length of use, and technical tests in a vacuum chamber. Results of the test runs were positive. Moreover, it was demonstrated that such a repair of the

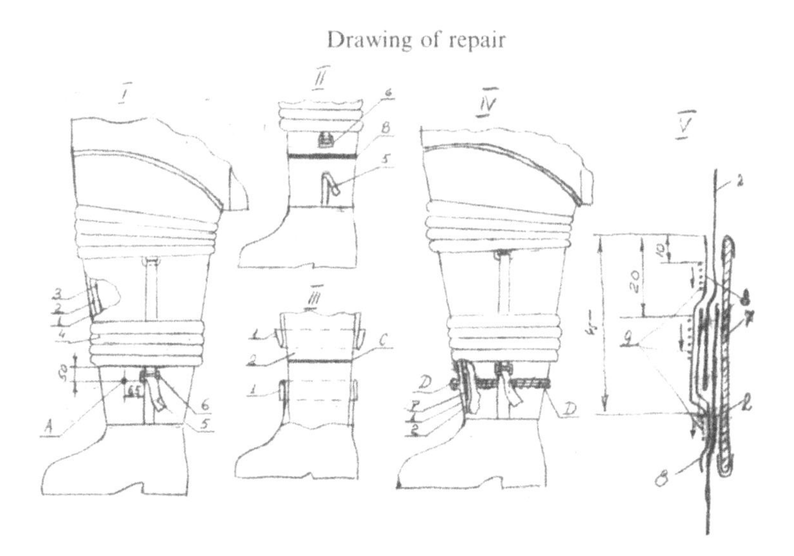


**Figure 8.5.3** The drawings of the ORLAN-D suit's leg repair that were transmitted to the *Salyut-6* OS: I—suit's leg enclosure before repair; II—preparation of the restraint enclosure for repair; III—preparation of the pressure bladder for repair; IV—suit's leg enclosure after repair; V—schematic of band installation; 1—restraint enclosure; 2—primary (external) pressure bladder; 3—redundant pressure bladder; 4—knee joint; 5—adjustment restraint strap; 6—buckle; 7—band (hard ring); 8—plaster; 9—threads; A—position of leakage detected by the crew; B—place where the restraint enclosure was cut; C—place where the pressure bladder was torn and further cut; D—manual stitching of the cut in the restraint enclosure.

From Archive Zvezda.

pressure bladder did not alter in any way the suit-donning/doffing process or operation while in suited mode.

On 29 October the repair procedure and sketches were transmitted to the crew by radio, and the next day (30 October) the crew repaired the spacesuit and carried out all needed tests. On 1 and 3 November 1983, the scheduled EVA sessions were successfully carried out (Figures 8.5.4 and 8.5.5).

There were also occasions when spacesuits were unconventionally used. In 1995 three IVAs were undertaken inside the transfer compartment of *Mir* to shift the position of the docking cone for preparation of docking with *Priroda*. To do this work, electric umbilicals had to be provided for the spacesuits. In 1997 the necessity arose to carry out some work inside the depressurized *Spectr* module, which had been damaged after a collision with a *Progress* spacecraft. However, the suit's sublimation heat exchanger could not be used because the vacuum inside the module was too low. Within a very short time, Zvezda developed, tested and then delivered hose extensions (10 m long) to support the cooling systems of the suited crew members from the on-board system, using the BSS-2M [БСС-2М]. The hose extensions enabled the cosmonauts to get their tasks done.

**Figure 8.5.4** Zvezda experts and cosmonauts V.A. Lyakhov and A.P. Alexandrov discuss the results of the EVA and the suit's leg repair after the mission.
From Archive Zvezda.

**Figure 8.5.5** A.P. Alexandrov and V.A. Lyakhov visiting Zvezda after the in-orbit repair of the spacesuit. *First row (from left to right)*: S.G. Kosichkin, Yu.I. Mishchishin, I.N. Nikitin, V.A. Lyakhov, G.I. Severin, A.P. Aleksandrov, A.A. Leonov, N.I. Afanasenko, V.A. Dubrov. *Second row*: V.P. Yefimov, I.P. Abramov, A.Yu. Stoklitsky, V.B. Razgulin (TsKBEM), A.A. Serebrov, V.I. Svertshek, a representative of the Air Force, F.A. Vostokov, V.I. Kharchenko, F.S. Timokhin, V.E. Tsvetov, Yu.P. Karpov and a representative of the Cosmonaut Training Centre.
From Archive Zvezda.

# 9

# Equipment for the Cosmonaut Transference and Manoeuvring Unit (UPMK)

## 9.1 DEVELOPMENT OF THE FIRST UPMK

Research work on means and methods to support cosmonaut transference and manoeuvring in weightlessness was initiated at Zvezda as far back as the early 1960s. This work ran simultaneously with the design and development of the experimental spacesuit to be used in an EVA sortie—SKV [СКВ]—from the orbiting heavy satellite/station (OTSST) [ОТССт]. Zvezda received a draft of the performance specifications for the suit from OKB-1 (Experimental Design Bureau 1) on 1 December 1961 (see also Section 6.1).

According to the performance specifications, the spacesuit system had to include a UPMK [УПМК]. The purpose of the UPMK was to provide a suited cosmonaut with the capability of undertaking an EVA and of detaching himself from the spacecraft/orbiting station's surface (e.g., to repair a payload accompanying the spacecraft, to travel around the exterior of the spacecraft for its examination and inspection, to transport loads, to perform rescue operations, etc.). To do this, the UPMK had to meet a number of specific requirements.

The UPMK therefore had to be fully compatible with the spacesuit and its design needed to cover all peculiarities and limitations imposed by the spacesuit, especially those pertaining to the field of vision and thus the handling of controls. It seemed natural to develop UPMK controls along the lines of those of the spacecraft, to which the cosmonauts had already got accustomed. The UPMK had to facilitate usage by cosmonauts of different sizes and have such dimensions that it could be transported through the spacecraft hatches, if it needed to be stored inside the spacecraft. Selection of the dynamics and requisite reserve of fuel for the jets were among many other problems associated with the UPMK design. It was essential that the UPMK had the correct dynamics given that its mass and overall dimensions had to be kept to a minimum.

In connection with this, in the initial stage of work in 1962–1963, Zvezda together with NII-2 (Scientific and Research Institute 2) made a theoretical study of the dynamics of man/suit/UPMK system movement at a distance of 50–100 m from the mothercraft. The following parameters were determined:

- rate of movement away from the mothercraft;
- rate of transference under weightlessness conditions;
- probable disturbances at movements;
- translation jet thrust; and
- fuel storage and flow rate.

The movement control system (MCS) for this programme was also developed by NII-2.

In 1964 the study was continued on board the Tu-104 flying laboratory of the LII (Flight Research Institute). For this purpose, an air jet UPMK mock-up was designed and manufactured.

The UPMK engines (12 units) were planned to be liquid jet engines using asymmetric dimethyl hydrazine and, as an oxidizer, nitric acid with an iodic additive. Compressed nitrogen was to be used for pressurizing the expendable tanks with fuel.

The UPMK's main performance characteristics, as applied to the experimental SKV suit, were:

- characteristic velocity: $50\,m\,s^{-1}$;
- total impulse: $15{,}000\,N \cdot s$;
- fuel mass: 6 kg;
- total mass of UPMK: 65 kg.

There was a redundant emergency system with a fuel reserve of 2 kg and total impulse of $5{,}000\,N \cdot s$.

Work on the UPMK for the SKV suit (Figure 9.1.1) was completed by a design review in 1965. When G.I. Severin joined Zvezda in 1964, the programme on UPMK development moved at a faster pace, but this time it applied to the "soft" EVA spacesuits being developed in 1964–1966 (see Chapter 5). Initially this programme was run in accordance with the government directive dated 27 July 1965, concerning the two additionally ordered "3KD" spacecraft (which used the Volga soft airlock). It was planned to use the YASTREB spacesuit on these spacecraft. Upon termination of the "3KD" spacecraft programme, UPMK development was continued by the government directive dated 28 December 1966 for the *Almaz* programme, where it was also planned to use the YASTREB spacesuit and the RVR-1 backpack.

In 1966 two UPMK samples were manufactured for the ground development test programme and testing in the aerostatic support facility at NII-2. In the same year the one-degree-of-freedom rotation test facility was also manufactured at Zvezda.

When the decision was made to use the UPMK in *Almaz*, a TsKBM branch

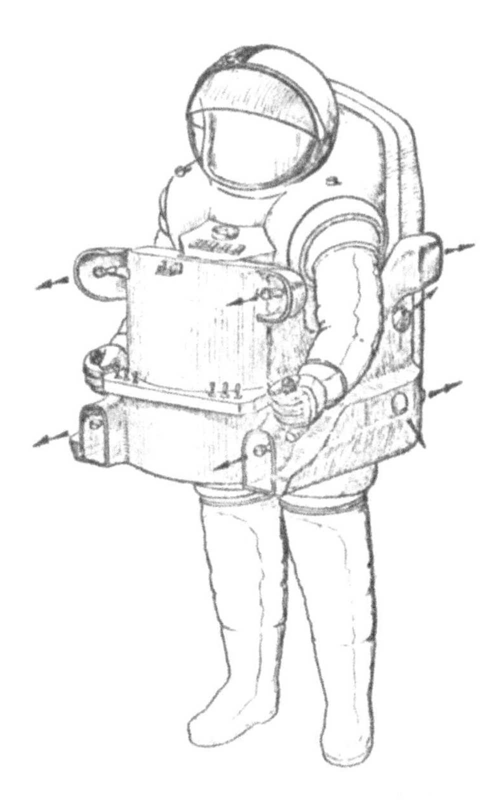

**Figure 9.1.1** General view of the UPMK on the SKV spacesuit (drawing of the preliminary design).
From Archive Zvezda.

developed the UPMK control system and winch. UPMK fit check work was also carried out in the orbital module of *Almaz*. Development of the UPMK prototype, as applied to the YASTREB spacesuit, was completed in 1968. The UPMK passed the full range of development testing and was given the go-ahead to be used for EVA. It was made in the shape of a horseshoe, embracing a suited cosmonaut from the front and included a combination of two systems of actuators (Figure 9.1.2a, b).

The first system consisted of two units with expendable, solid propellant rocket micro-motors (42 in each unit) for linear transference forward and backward, and the second system consisted of 14 air thrusters for linear and angular transference of cosmonauts with six degrees of freedom. The units with single solid propellant motors were arranged in such a way that the thrust line of each motor went through the centre of mass of the suited man/UPMK system. The single solid motors were actuated from the control panel. Operation of each motor changed the linear rate of the system (forward or backward) by $0.2\,\mathrm{m/s^{-1}}$. The total impulse of the UPMK's system of actuators for units of the power motor and for units of the air jet engine was $4{,}000\,\mathrm{N \cdot s}$. The impulse of the single solid motor was $45\,\mathrm{N \cdot s}$. Thrust from the air jet engines was 2.5 N and 5.0 N (eight engines and six engines, respectively). The characteristic speed was $32\,\mathrm{m\,s^{-1}}$ and the mass was 90 kg.

(*a*)

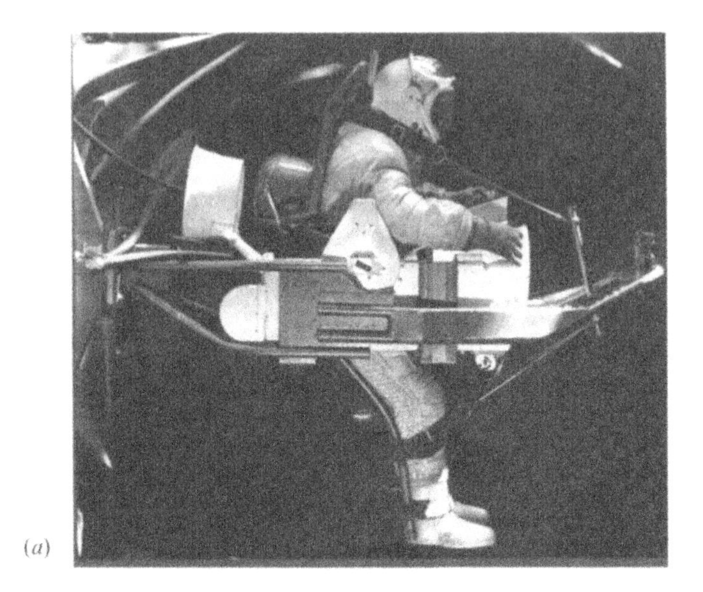

(*b*)

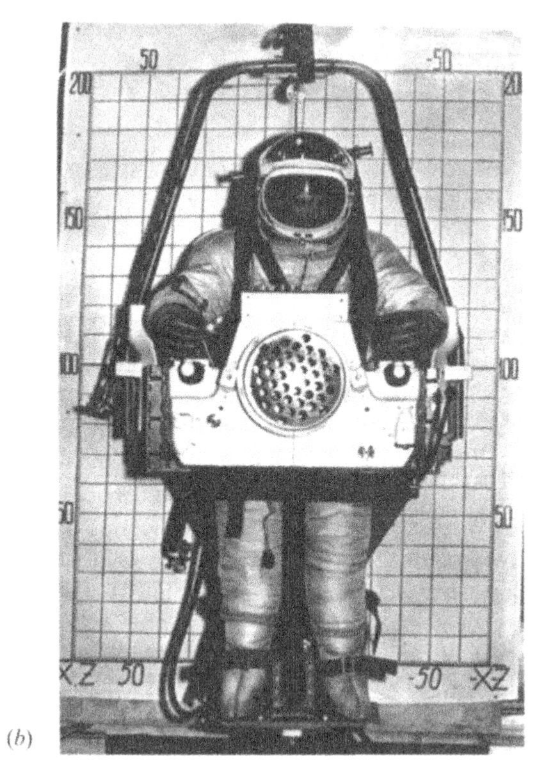


**Figure 9.1.2** General view of the UPMK on the YASTREB spacesuit at the test facility: (*a*) side view; (*b*) front view.

From Archive Zvezda.

When the YASTREB spacesuit was replaced by the ORLAN spacesuit for EVAs from the *Almaz*, Zvezda and TsKBM decided (20 November 1969 and 28 November 1969, respectively) to develop the UPMK for the ORLAN spacesuit. But further work on the UPMK was stopped because at that time there were no specific tasks to be carried out using the UPMK or corresponding equipment on board manned spacecraft and stations. Later this work was resumed for missions to the *Mir* orbiting station, where ORLAN-type spacesuits were also used for EVAs.

## 9.2 *MIR* COSMONAUT TRANSFERENCE AND MANOEUVRING EQUIPMENT

In February 1990, the cosmonauts A.A. Serebrov and A.S. Victorenko carried out in-orbit verification flight-testing of the UPMK (the 21KS [21KC]) during their EVA from the *Kvant-2* module of the *Mir* orbiting station (Figure 9.2.1). The distances travelled were 33 m and 45 m, respectively.

The 21KS unit was a self-contained system (including propulsion). The unit was designed to be used by cosmonauts during space walks. A cosmonaut operating the unit would be able to work and move about the exterior of the spacecraft without using a safety tether, handrails or feet anchor elements. UPMK application in the orbiting station made it possible to increase the efficiency of a cosmonaut's work in space on such things as assembly, repair, preventive maintenance, research, rescue and military uses.

According to a joint decision of the Ministry of Aviation Industry and the Ministry of General Engineering Industry dated 22 March 1984, Zvezda was asked to study this unit along the lines of a proposal made jointly by Zvezda and Energia. It was planned to use the 21KS on *Mir* and in the *Buran* reusable space system.

At the same time work on spacesuit modification began, designed to allow operation of the suit without an electric umbilical connecting it to the on-board power source and radio/telemetry system of the spacecraft (model ORLAN-DMA). Full-scale work was expanded in 1986 as a result of a government resolution dated 25 September 1985. In the same year the Inter-agency Breadboard Approval Review was held, several test samples of the 21KS were manufactured (Figure 9.2.2), working documentation on the unit for in-house and inter-agency testing was issued and adjustment of special test facilities was started.

The above resolution entrusted Zvezda with the function of prime contractor for development and manufacture of the overall system including the self-contained spacesuit with the life support system (LSS), the means of cosmonaut transference and manoeuvring, and on-board means supporting a cosmonaut's work while in self-contained mode.

The Ministry of General Engineering Industry and the Ministry of Defence were appointed the customers. The development was run in line with performance specifications from RSC Energia (Table 9.1).

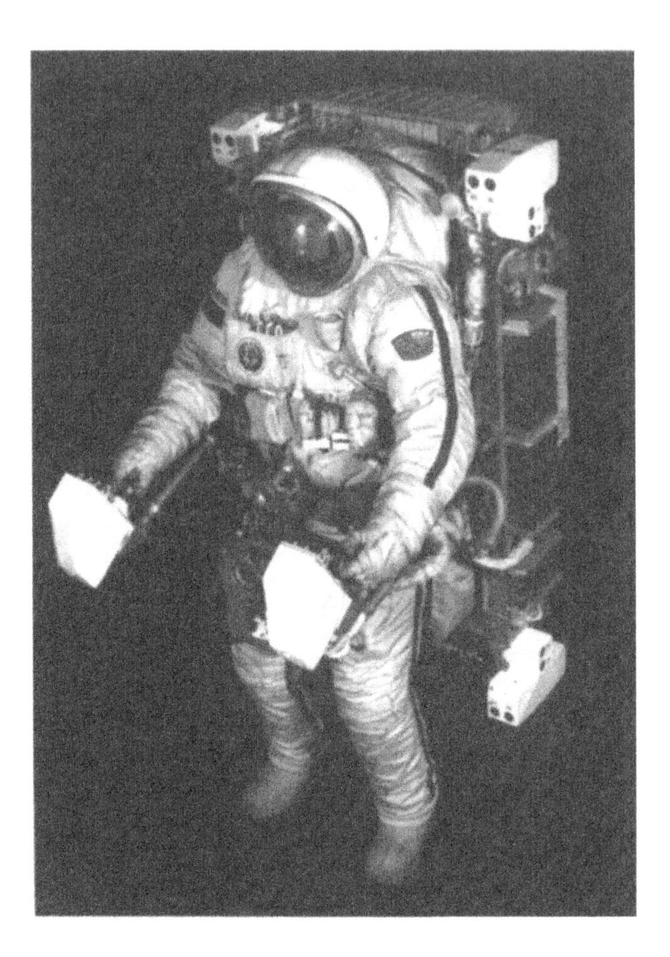

**Figure 9.2.1** General view of the ORLAN-DMA spacesuit with the 21KS.
From Archive Zvezda.

The government directive of 31 October 1985 specified the subcontractors and the main performance characteristics of the system. RSC Energia was charged with development and manufacture of the MCS.

The 21KS unit was made as a backpack that covered the entire back of the suit. At the same time a unique design for attachment to the suit was developed. The design made it possible not only to mount the suit on the 21KS without assistance but also to don and maintain the suit without disconnection from the 21KS (Figures 9.2.3 and 9.2.4).

The suit was attached to the 21KS by restraint elements located on the front of the suit. A rigid belt/frame was fixed on two joints in the front portion of the 21KS. It accommodated the suit/21KS attachment nodes: the central lock, two side rods and nodes for attachment of rotating telescopic booms, on the ends of which were the 21KS control panels for linear transference (left control panel) and rotation

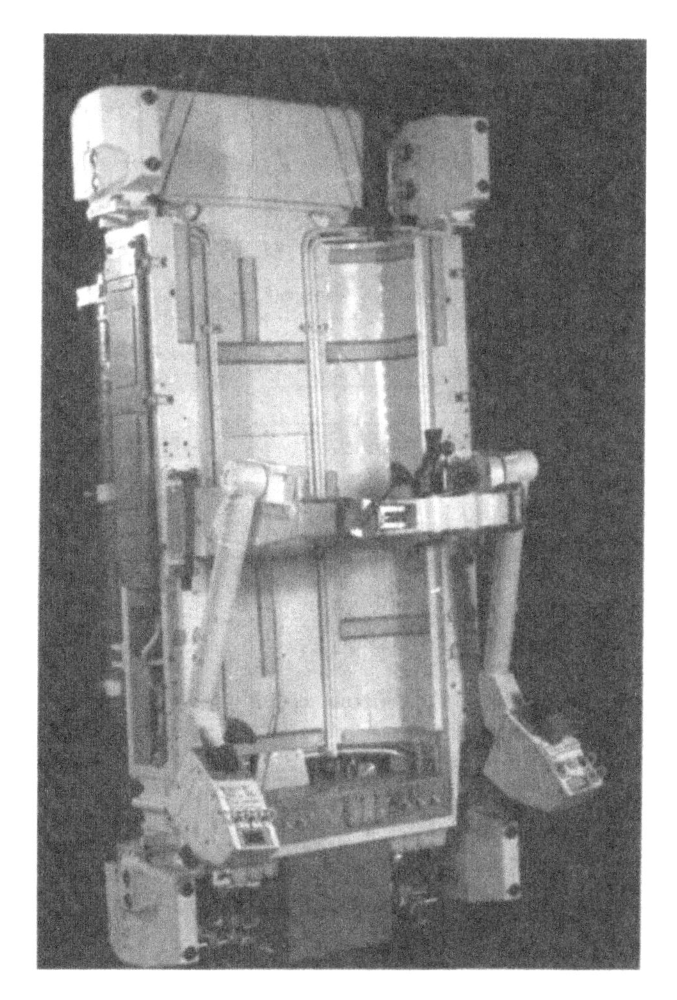

**Figure 9.2.2** The 21KS test sample.
From Archive Zvezda.

**Table 9.1.** UPMK main characteristics.

| | |
|---|---|
| Time of self-contained operation during an EVA without refilling | At least 6 hours |
| Characteristic speed | $30\,m\,s^{-1}$ |
| Total number of EVAs | At least 15 |
| Maximum permissible transfer rate | $1\,m\,s^{-1}$ |
| Maximum angular rate at turns | Up to $10^{\circ}\,s^{-1}$ |
| Accuracy of automatic stabilization | $\pm 0.5 \div 5^{\circ}$ |
| Maximum distance permissible from: | |
| *Buran* space plane | 100 m |
| *Mir* with a safety tether | 60 m |
| Mass of 21KS | No more than 180 kg |

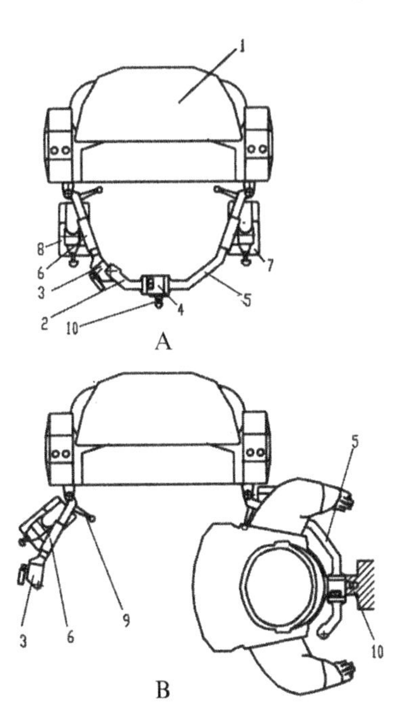


**Figure 9.2.3** Diagram of 21KS/spacesuit attachment: A—top view; B—view with the waist frame open; 1—21KS rear section; 2—waist frame; 3—emergency latch; 4—central latch; 5, 6—left and right segments of the waist frame; 7, 8—control panels with hand controllers; 9—pin for suit attachment; 10—pin for 21KS attachment.

From Archive Zvezda.

(right control panel). The rotation booms of the 21KS provided for two positions of the control panels:

- operation position (while piloting);
- transportation position (while storing the 21KS on board the mothercraft, as well as during work on the mothercraft's surface).

The node for 21KS attachment to the orbiting station's mooring facility was mounted in the central part of the frame. The frame consisted of two unequal portions interconnected by a side lock (Figure 9.2.5). With the lock open, the 21KS backpack together with the smaller portion of the frame rotates relative to the larger portion of the frame so that a suited cosmonaut is able to engage the central lock and a side rod (located on the larger portion of the frame) unassisted by any special facilities in the mothercraft. After returning the backpack to its initial position and closing the side lock, a cosmonaut would be well and truly attached to the 21KS.

As a spacecraft, the 21KS had its own systems: power supply, actuators, control, telemetry, etc. (Figure 9.2.6). The power supply system included two silver/zinc

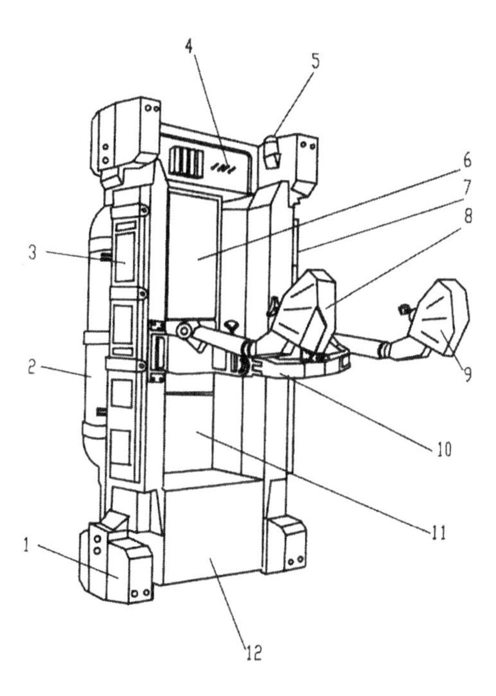


**Figure 9.2.4** General view of the 21KS (design concept): 1—sets of thrusters (4); 2—gas pressure vessel (2); 3—secondary battery; 4—distribution assembly; 5—navigation lights (4); 6—movement control unit; 7—primary battery; 8—rotation hand controller; 9—transfer hand controller; 10—waist frame; 11—angular velocity sensor unit; 12—components of the subsystem of actuators.

From Archive Zvezda.

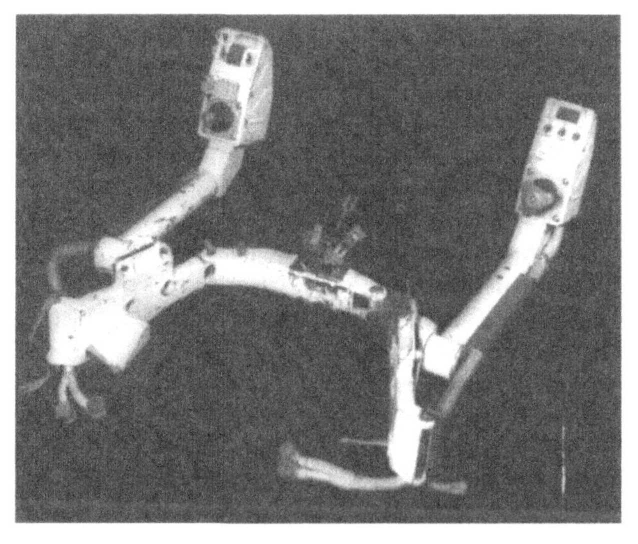

**Figure 9.2.5** The 21KS frame with hand controller modules.

From Archive Zvezda.

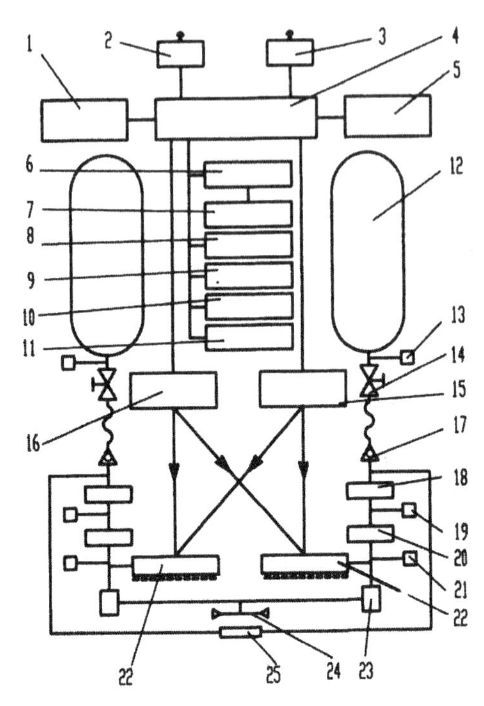


**Figure 9.2.6** Block diagram of the 21KS equipment: 1, 5—primary and redundant battery; 2, 3—control panels; 5—power distribution unit; 6—radio and telemetry assembly; 7—antenna-feeder assembly; 8—TV camera; 9—lamps; 10—navigation lights; 11—other electrical devices; 12—propellant tank with pressurized air (2); 13, 19—high-pressure sensors; 14—shut-off valve; 15, 16—movement control subsystem (primary and redundant); 17—check valve; 18—electric valve; 20—reducer; 21—low-pressure sensor; 22—air jets; 23—safety valve; 24—zero thrust nozzle; 25—cross-feed valve.

From Archive Zvezda.

batteries (primary and redundant). In case of spacesuit power supply failure, the power supply system of the 21KS could also power the suit's systems. The power supply voltage was $27 + 7/-4$ V, and the capacity of the primary and redundant batteries was 18.0 and 8.5 A · h, respectively.

The system of actuators consisted of two similar half-sets. Each half-set included a bottle (a gas tank) with a shut-off valve, check valve, electric valve, reducer, 16 thrusters of 0.5 kg thrust, a relief valve downstream of the reducer and sensors and indicators for the gas pressure in the bottle and in the system. Each cylinder was filled with 28 litres of air under a pressure of 32 MPa. This pressure was reduced by the reducer to the level required for operation of the thrusters: 1.25 MPa.

With the system of actuators functioning, the first half-set would be actuated. It was switched off when air in the cylinder rain out or when a pressure of 11.0 MPa was left; then the second half-set was actuated. The pressure of 11.0 MPa was selected based on the need to provide a gas reserve for safe return of the cosmonaut to the mothercraft from the maximum distance. The half-sets were actuated from the control panels by voltage supply to the electric valve-opening mechanism and were interconnected with the transfer valve. By opening this valve, a cosmonaut connected the bottle of compressed air of the inoperative half-set (in case of failure of the reducer, thruster or some other unit) to the one that was operating.

All 32 jet thrusters were arranged in four units on the backpack in such a way that, first, the resultant thrust of the actuated thrusters went through the centre of mass of the 21KS/cosmonaut/suit system to produce linear transference; second, rotation was effected by pairs of forces acting in the planes perpendicular to the system's axes, about which rotation was done. With such an arrangement of the thrusters, a minimum rotational disturbance at linear transference was generated and translation at rotation was eliminated.

The 21KS's MCS was designed to generate signals for thruster actuation in various operation modes: linear transference, rotation and attitude hold. The system consisted of two similar half-sets. Each half-set included an angular rate gyroscopic sensor, electronic control module and static DC/AC converter to power the gyro motor of the angular rate sensors.

Manoeuvrability was effected by means of linear transference and rotation control handles located on the hand controller modules (HCM) (Figure 9.2.6). The control's subsystem design provided for the following modes: half-automatic control, automatic attitude hold and direct control. There were two operation modes in the half-automatic control: economical and forced. In economical mode with a single engagement of the linear transfer control handle, the thrusters corresponding to the selected direction of transference were actuated and switched off automatically after 1 s; with engagement of the rotation control handle, the 21KS adopted an angular rate of $3^\circ\,s^{-1}$ and rotated at this rate until the cosmonaut placed the control handle in the neutral position. In the forced mode these parameters were 4 s and $8^\circ\,s^{-1}$, respectively. The operation mode was selected depending on the type of work undertaken and distance to the mothercraft. The cosmonaut usually chose economical mode when working near the mothercraft and forced mode when working at a distance.

For ease of control of the 21KS, there was an automatic attitude hold mode in inertial space with an accuracy of $\pm 2^\circ$. To change to this mode, the "null" button on the HCM was pressed. If the cosmonaut needed to change attitude, he engaged the rotation control handle and made the required turn. On disengaging the control handle, the system maintain the attitude that corresponded to the moment the control handle was released.

The direct control mode made it possible for the cosmonaut to preset the time for the thrusters to function and the rotation rate. This would be useful, for example, when approaching a rotating object for the purpose of maintenance. In this mode, control commands from the control handles were sent directly to the thrusters through a timer. With a single engagement of the corresponding control handles, the transference thrusters operated for 2 s, the rotation thrusters for 0.2 s and then they were automatically switched off. To choose the required rates, the cosmonaut pushed the control handles several times in the necessary direction.

The radio and telemetry system transferred about 100 parameters to the Earth, relaying data about the performance of the 21KS's subsystems and enabling the ground support team to monitor operation of the 21KS during EVA. Audible signals about the performance of the 21KS's subsystem were transmitted to the spacesuit's earphones. The navigation lights and the headlight made the complex visible from the mothercraft and illuminated the workstation in the "shade" portions of the orbit.

All the 21KS's systems were arranged in a housing that was covered by shield vacuum thermal insulation, which provided for the thermal mode required for subassembly operation. The 21KS's systems had a high level of reliability. The main systems were backed up, and the cosmonaut could return to the mothercraft without assistance in case of any single failure. The MCS's control handles sent commands separately to both circuits of the half-automatic control module and to the direct control circuit. The 21KS had mechanical and electric quick disconnects that allowed the cosmonaut to jettison the 21KS in an emergency or in case of a side and central lock failure.

To bench-test the reliability of the 21KS and ensure its compliance with the performance specifications, an extensive development test programme was carried out between 1986 and 1989:

- development of a design and engineering mock-up;
- study of the 21KS/ORLAN-DMA arrangement and compatibility;
- determination of mass–inertia characteristics;
- development test in the hydro lab and flying laboratory for compatibility with structural elements of the orbiting station and spacecraft;
- functional tests;
- determination and updating of the main technical characteristics;
- thermal and vacuum tests;
- study of dynamics at aerostatic support test facilities with one and three degrees of freedom;
- study of dynamics at simulation facilities to try and visualize the transference processes;
- preliminary in-house tests for compliance with the performance specifications; and
- development tests of the radio and telemetry systems for compatibility with the orbiting base module.

To carry out the above programme, Zvezda designed and manufactured a series of special test facilities:

- A force–rotation test facility was designed to verify the technical characteristics of the 21KS by applying angular rates to the test facility table while alternating the arrangement of the 21KS's axes relative to the table's axis of rotation, and measuring the corresponding parameters of the 21KS.
- A three degrees-of-freedom test facility with a spherical air bearing was designed

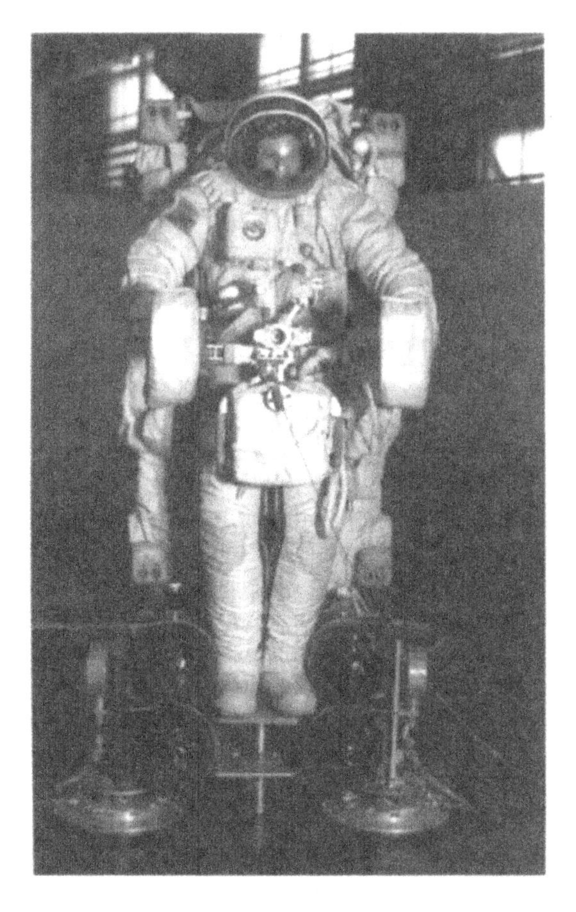

**Figure 9.2.7** A test subject suited in the ORLAN-DMA, wearing the 21KS, on the test facility with the aerostatic supports.
From Archive Zvezda.

to study the 21KS's angular motion relative to three axes. The 21KS was installed on the movable part of the spherical air bearing.

- A test facility with aerostatic supports, ASO [ACO], was designed for comprehensive evaluation of the 21KS as well as for cosmonaut training in 21KS control skills (Figure 9.2.7). The 21KS/ORLAN-DMA/test subject system, secured on the frame, moved with three degrees of freedom (rotation about the vertical and transference in the level plane along two directions) on the precision support floor of dimensions $7 \times 8$ m (sloping no more than 0.1 mm per meter and with a roughness no more than 20 micron). Air for the air supports was supplied from a compressed air line through a flexible hose.
- Zvezda developed a simulator (Polosa) to visualize transfer processes. The simulator was used to develop the methods for 21KS control and to train cosmonauts.

In November 1988, inter-agency testing of the overall system was undertaken with participation of RSC Energia, the Cosmonaut Training Centre, representatives of the Air Force and Ministry of Defence. A.A. Serebrov and A.S. Victorenko, who were training to operate the 21KS on *Mir*, took part in the testing.

The following was bench-tested and verified: the capability of controlling the 21KS by a cosmonaut's own visual evaluation of his attitude and his movement as well as the linear and angular accelerations and rates of the 21KS, accuracy of automatic attitude hold and its performance in various control modes, including abnormal situations. Between 1987 and 1989, a total of 31 hydro lab tests, 32 flying laboratory tests and a large number of cosmonaut training sessions, including those at Zvezda's ASO and Polosa test facilities were undertaken. Cosmonaut training also took place on the Don simulator at the Cosmonaut Training Centre.

On 26 November 1989 the *Kvant-2* module was launched into orbit and the 21KS flight sample with its on-board test equipment was delivered to *Mir*. In order to ensure cosmonaut safety in an unforeseen situation (the *Mir* orbiting complex could not make a cosmonaut pick-up manoeuvre), full-scale testing of the 21KS was carried out by means of a special safety hoist with a rope made of high-strength synthetic material. The rope connected the cosmonaut to the orbiting station. A take-up system for the safety hoist was provided by rope deployment and winding with minimal force, which had little effect on cosmonaut movement dynamics. The safety hoist was attached to the 21KS's frame when the cosmonaut left the airlock. The maximum distance a cosmonaut wearing the 21KS could move away from the orbiting station was 60 m (the length of the rope).

In reality, A.A. Serebrov flew the 21KS for 40 minutes and moved a distance of 33 m from the station (Figure 9.2.8). A.S. Victorenko flew the 21KS for 93 minutes and moved the maximum distance from the station (45 m). All the systems functioned properly.

The cosmonauts were full of praise for the engineering and ergonomic characteristics of the 21KS. They confirmed that it could be used for all sorts of work in open space, including departure from the mothercraft's surface.

## 9.3 SIMPLIFIED AID FOR EVA RESCUE

Unfortunately, further work on *Mir* had no need for the 21KS, but Zvezda used the unique experience gained for development of the Simplified Aid for EVA Rescue (SAFER). SAFER is designed to provide safe return of the cosmonaut, wearing an ORLAN-M spacesuit, to the surface of the *ISS* in case of accidental separation from the station during EVA.

As is known, the safety of crew members wearing the ORLAN-M suit while moving along the station's surface is provided by two safety tethers. When a crew-member disengages one of the safety tethers, he or she grips the handrail additionally. This method has been successfully used for 25 years on board Russian orbiters. In order to improve EVA safety, in case of two failures or mistakes of the crew, the

**Figure 9.2.8** A.A. Serebrov using the 21KS on 1 February 1990 outside *Mir*.

possibility of equipping the *ISS* ORLAN-M spacesuit with SAFER was considered along the same lines as NASA's EMU spacesuit.

In 1998 Zvezda began to develop such a device for the *ISS* ORLAN-M spacesuit under a contract with Rosaviakosmos. Between 1999 and 2000 this work continued with the financial participation of NASA. At the same time the technical performance requirements of the Russian SAFER (Table 9.2) and its operation logic were approximated as much as possible to those of the US EMU SAFER, with the purpose of eventual unification of the system and ease of crew training. Yet the Russian SAFER considerably differs from the EMU SAFER in design, pneumatic flow diagram, means of attachment to the spacesuit, controls, etc. (Figures 9.3.1–9.3.3).

In 2001, NASA's financial participation stopped; nevertheless, between 2001 and 2002 Zvezda completed development of SAFER through its contract with Rosaviakosmos and manufactured the flight samples for delivery to the *ISS*.

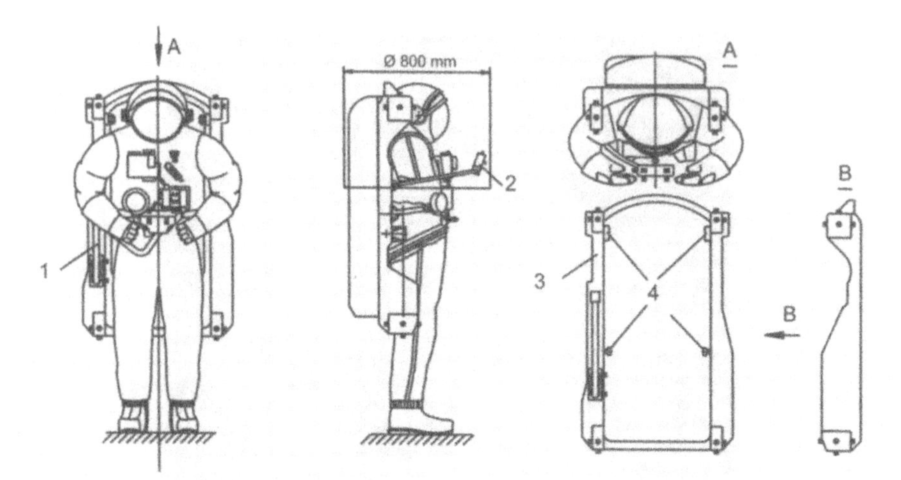


**Figure 9.3.1** SAFER installation on the ORLAN-M suit: 1—HCM in stowed position; 2—HCM in working position; 3—SAFER before installation on the suit; 4—SAFER/suit attachment components.

From Archive Zvezda.

**Table 9.2.** The ORLAN-M SAFER specifications.

| Parameter | Value |
|---|---|
| 1. Working medium | Air |
| 2. Working medium storage (kg) | ≈1.3 |
| 3. Bottle pressure (MPa) at $T = 20°C$ | 35 |
| 4. Number of gas thrusters | 16 |
| 5. Thrust of one gas thruster (N) | 3.5÷4.0 |
| 6. Total increment of velocity ($m\,s^{-1}$) | 3.6 |
| 7. Available linear acceleration ($m\,s^{-2}$) | 0.03÷0.06 |
| 8. Available angular acceleration ($°\,s^{-2}$) | 5.8÷8.7 |
| 9. Doubling up the pneumatic system | Provided by automatic switch-over |
| 10. Number of self-rescue cycles | 5 |
| 11. Number of EVAs throughout its lifetime | 40 |
| 12. Recovery of serviceability after self-rescue | Gas storage assembly changes are foreseen |

SAFER is attached to the spacesuit at four points. At the same time, it is possible to maintain the spacesuit on board the *ISS*, perform suit-donning/doffing, pass through the normal Ø 1-m hatch and effect emergency entry into *ISS* compartments through the Ø 0.8-m hatch. The possibility of detaching SAFER from the suit in case of emergency during EVA (with assistance of the second crew member) is also provided. SAFER is powered from the ORLAN-M battery and actuated by a toggle switch located on the suit control panel. The HCM of SAFER is unlocked and deployed in the working position manually after SAFER actuation.

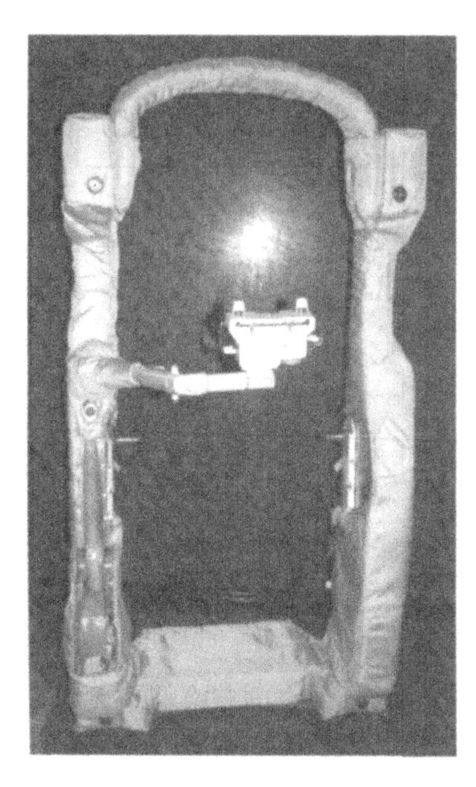

**Figure 9.3.2** General view of the ORLAN-M suit with SAFER.
From Archive Zvezda.

SAFER's MCS was developed by the Ramenskoye Instrument Engineering Design Bureau according to Zvezda's performance specifications. The MCS has three operation modes: half-automatic control mode, direct control mode and emergency control mode.

The half-automatic control mode provides for:

- automatic dissipation of angular rates obtained at the moment of separation;
- attitude hold relative to three axes with an accuracy of $\pm 5°$;
- manual turn relative to one of the axes with attitude hold relative to two other axes;
- manual dissipation or gain of the linear rate along three axes.

The direct control mode provides for:

- manual angular rate dissipation and turn about any axis;
- manual dissipation or gain of the linear rate.

The emergency control mode is similar to the direct control mode, but at the same time the control commands and power supply are sent by independent lines.

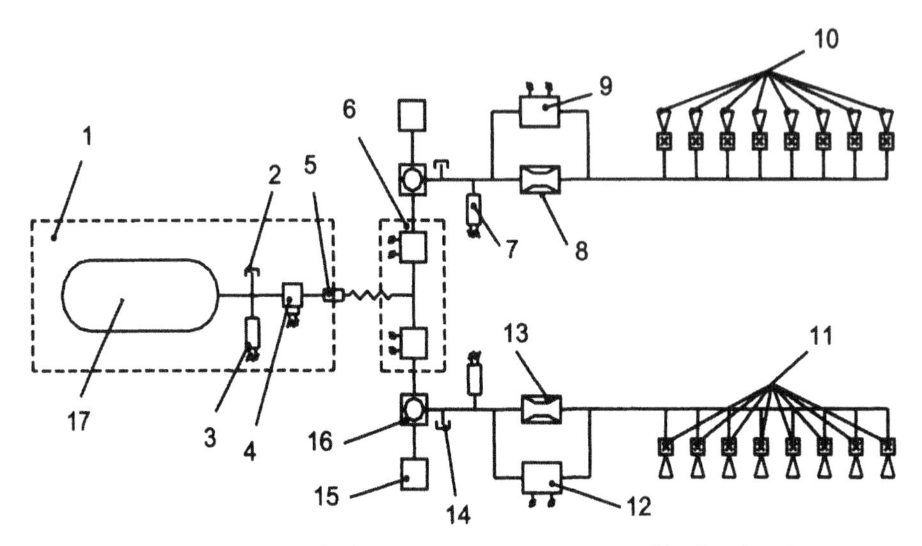


**Figure 9.3.3** SAFER's pneumatic diagram: 1—gas storage assembly; 2—charging connector; 3—pressure transducer; 4—pyro-valve; 5—disconnect; 6—shut-off pyro-device; 7—positive pressure switch; 8, 13—Venturi tube; 9, 12—pressure differential switch; 10, 11—gas thrusters; 14—a disconnect for pneumatic system checkout; 15—relief valve; 16—reducer; 17—bottle with compressed gas.

From Archive Zvezda.

The MCS, besides its control function, monitors the status of SAFER's pneumatic system and generated the commands to change the pneumatic system's configuration and operation in case of failure of a single gas thruster. The gas storage assembly can be replaced by a new one, when necessary after expenditure.

A large team of Zvezda specialists took part in the design and development of means for cosmonaut transference and manoeuvrability. The manager in charge of work on UPMK for the SKV and YASTREB spacesuits as well as on the 21KS was V.A. Frolov and for SAFER it was S.S. Pozdnyakov. Management of the computational and theoretical part of the work was mainly done by A.N. Livshits.

# 10

# *Buran* reusable space system

## 10.1 INTRODUCTION

Development work on the reusable space system (RSS) (called the Energia-Buran system) was initiated by the government resolution of 17 February 1976. The main customer for this system was the USSR Ministry of Defence and the prime contractor was NPO Energia. Prime contractor for the *Buran* orbiting vehicle, a part of the system, was NPO Molnia (G.E. Lozino-Lozinsky was General Director and Chief Designer). Organization of the main subcontractors for the development of the RSS was approved by the government directive of 18 November 1976.

Zvezda received the preliminary specification from NPO Molnia on 22 July 1976 for the preliminary design of a number of subsystems and components for the *Buran* spaceplane including:

- personal life support system (LSS), including a full-pressure rescue suit and on-board LSS;
- EVA spacesuit with a self-contained LSS;
- means for emergency escape from the *Buran* spaceplane for the development test phase;
- personal fire protection means;
- on-board waste removal system (waste management device) in the cabin; and
- equipment for a drinking water supply system in the cabin.

In 1976 Zvezda undertook the preliminary design of these systems, and the technical project was carried out in 1978. The government resolution of 21 November 1977 specified the main phases of the development of the orbiting vehicle (*Buran* spaceplane) and the industrial organizations that were to carry out the development. When the government directive of 27 December 1978 was issued, preparation of the technical documentation for the test and flight samples and later their production and delivery was started. Zvezda developed the hardware for the *Buran* RSS under contracts with a branch of NPO Molnia (for development) and the Tushino Engineering Plant (for hardware delivery). RSC Energia supervised work on the EVA suit.

At first it was planned to build three flight units of the *Buran* spaceplane, but later, in 1983, the number was increased to five. The unmanned maiden flight of the *Buran* spaceplane took place on 15 November 1988. Flight units of all the main items developed by Zvezda were installed in the spaceplane, including two ejection seats, rescue suits worn by manikins and the on-board LSS designed for the rescue suits. It was planned to get the second 7-day unmanned flight to dock with *Mir*, but, because of the political events of 1991, the Energia-Buran RSS was transferred from the military programme to the Government's space programme to carry out tasks for the benefit of the national economy. In 1992 Rosaviakosmos decided to stop work on this programme for financial reasons.

## 10.2 CREATION OF THE PERSONAL LIFE SUPPORT SYSTEM

The personal life support system (PLSS) was designed to provide for the safety and enhanced performance capabilities of the *Buran* RSS's crew in case of emergency depressurization inside the cockpit, on-board LSS failures, smoke in the cockpit, as well as emergency ejection from the spaceplane and subsequent landing on the ground or in the water. The PLSS was also designed to support the crew, wearing EVA suits, in their work in the airlock and docking module of the orbiter by means of the umbilical. During preliminary design, three PLSS versions were considered:

- *Version 1*—use of existing systems or systems under development. In particular, analysis was carried out of a possible application to the SOKOL-KV suit being designed for *Soyuz* spacecraft. This meant that the SOKOL-KV suit would be modified according to the new working conditions: necessity to fly the spaceplane with the suit under positive pressure; change to a closed-type regenerative system; increase of useful life, etc.
- *Version 2*—development of the self-contained on-board system (regenerative type), a universal soft spacesuit and a removable backpack containing the LSS.
- *Version 3*—use of the universal spacesuit and a backpack system similar to Version 2, but using the backpack system as the on-board LSS when the spacesuit was operated as a rescue suit.

Between 1976 and 1977 the above and some other systems and components were analysed and discussed with the customer (NPO Molnia and its branch, managed by V.K. Novikov) and the Nauka enterprise responsible for the whole on-board LSS.

Nauka representatives favoured a version in which the suit would be operated continuously from the cockpit LSS (common to both the cockpit and the spacesuit). This version had the least mass, but had some drawbacks: mainly, its reduced reliability because failure of the LSS could result in crew death. Moreover, a clear split in the responsibilities for the systems between design companies was emphasized during review of the LSSs of the cockpit and suit, as was done earlier for *Vostok* spacecraft.

Some new problems were encountered during preliminary design of the hardware for the LSS. On the one hand, the main requirements of the suit were similar to those of aviation rescue suits. On the other, both the spacesuit and on-board LSS had to operate under spacecraft conditions (where there was no external pressure, inability to land the spaceplane quickly in an emergency, etc.).

Unlike the *Soyuz* programme, the *Buran* programme foresaw a long, multi-hour mission in case of a cockpit depressurization. Another requirement for the suit was to withstand aerodynamic heating, which takes place when a crew member ejects at high altitudes and Mach numbers. Moreover, the suit had to have water supply and waste removal systems because of the long time spent in the pressurized suit. There was also the significant problem, especially for the on-board equipment, of reusability (the initial analysis called for 100 *Buran* RSS missions).

The following prerequisites had to be taken into account in the technical design along with the experience obtained at that time from operating the SOKOL-K rescue suits, the EVA suits for the *Voskhod-2* and *Soyuz* programmes. Results from the first space walk using the ORLAN-D suit on board *Salyut-6* in late 1977 were also considered:

1 The PLSS had to be a self-contained system that only operated at safety critical stages of the mission (with the crew wearing the suits): on the launch pad and entering orbit, dynamic operation in orbit, EVA, re-entry, cockpit depressurization, smoke emission in the cockpit or on-board LSS failure at any stage of the mission. This would simplify the on-board LSS flow diagram and in turn initiate for the first time inclusion of a redundant system, which in turn would improve the general reliability of the spacecraft's LSSs, though it would result in installation of some additional equipment. The PLSS had to be connected to the on-board systems only via the lines supplying the cooling agent, oxygen and power. Taking into account the requirement for suited operation in the depressurized cockpit (at first it was 6 hours, then 8 hours and finally 12 hours) and the metabolic rate of the crew (150–200 W), a closed-type regenerative system to keep crew members cool using ventilating gas was found optimum.
2 The requirement for the suit to be used both as the EVA and the rescue suit would result in a deterioration in performance of both (as already discussed in previous chapters). Therefore, it was proposed to use the ORLAN-D suit as the EVA suit (see Section 8.3) and a "soft"-type, lightweight pressure suit as a rescue suit. The ORLAN-D, a semi-rigid spacesuit with a built-in LSS, had already been operated on board *Salyut* at that time. Use of the operational ORLAN-D suit made it possible to reduce the time and cost of developing the necessary products.
3 Support of the ORLAN-D suits from the on-board PLSS in the course of pre-EVA airlock operations and pre-breathing was to be provided. This would make it possible to increase the time of suit operation outside the mothercraft.

Update of the system flow diagram, item breadboarding, preparation and issuing of performance specifications to the subcontractors and refinement of the drawings for some components of the system were carried out between 1977 and 1979. The first version of the performance specifications for the PLSS as a whole was received in September 1978 and the final performance specifications were coordinated in 1980. The final update of the performance specifications was done in 1981, when the time of closed-loop operation of the suit was increased to 12 hours with the purpose of enabling the spacecraft (in an emergency at any point of the orbit) to land in the given area.

In succeeding years, there was a large number of laboratory, in-house and joint system tests (in the course of which the items underwent considerable modifications). Items were manufactured and delivered for more than 15 various mock-ups of the orbiting spaceplane, which were used for fit checks as well as thermal and vacuum, dynamic, electrical and flight tests. We should especially note the full-scale inter-agency tests of the spaceplane's LSS, which were carried out in the Air Force Scientific Research Institute's (GK NII) vacuum chamber on the 35CT-14 test bench between 1990 and 1991. They included 18 days of man-evaluated tests of the LSS cockpit, when test engineers wore the suits for 18 hours, including 12 hours in the depressurized cockpit's "operation" mode. A series of emergency modes for suited crew operation was also verified. The main portion of in-house testing, including the life tests and mechanical effect tests for five simulated flights in a spaceplane, was undertaken by 1986.

Special on-board equipment to support crew training in the TDK-35 test facilities (in the Gagarin Cosmonaut Training Centre) and the PRSO-2 flight simulator (at NPO Molnia) were developed in parallel with testing of the flight models.

Work on PLSS development was managed by the leading designers: I.P. Abramov (PLSS as a whole), A.Yu. Stoklitsky (spacesuit) and I.I. Chistyakov (electrical equipment). The tests were run under the management of A.S. Oparin.

## 10.3 DESIGN OF THE ON-BOARD ELEMENTS OF THE PLSS

The on-board elements of the PLSS that provided for life support of the suited crew consisted of several modules in which the main components of the system were arranged (Figure 10.3.1). The oxygen supply module BP-1 [БП-1] provided for oxygen receipt from the primary and redundant oxygen stores and reduction of its pressure down to 0.45 MPa. The BRS-1M [БРС-1М] (Figure 10.3.2) contained components that together with the BO-1M (BO-2M) contamination control assembly [БО-1М and БО-2М] (Figure 10.3.3) and the AKhSG [АХСГ] cooling/drying unit, formed a ventilation loop for two suits. The DU-2 module [ДУ-2] provided for remote manual control of the system.

For the four-seater option of the spaceplane, it was planned to install two similar loops including the BRS-1M, the contamination control assembly, DU-2 and AKhSG module. In the initial missions, the use of ejection seats was planned in the two-seater option. The PLSS could operate either in the ventilating mode or in

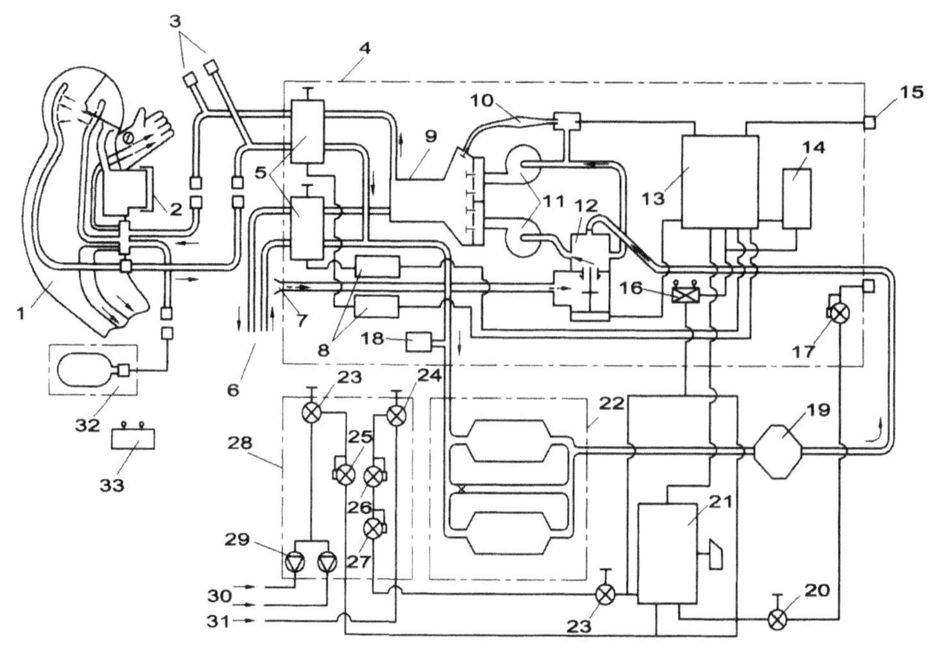

**Figure 10.3.1** Block diagram of the PLSS in the suits for the *Buran* spaceplane (two-seater version): 1—STRIZH suit; 2—suit pressure regulator; 3—connector for the ORLAN-D or -DMA umbilical used in the ShKK airlock; 4—BRS-1M unit; 5—valves to control the rate of suit ventilation and isolation; 6—to the second suit; 7—air leakage from the cabin; 8—compensator for leakage from the system and spacesuits; 9—valve module; 10—injector; 11—primary and redundant fans; 12—a valve switching the system over to closed loop operation mode; 13—oxygen system automatics unit; 14—purge device; 15—oxygen to the ShKK and docking module; 16—electric and pneumatic valve; 17—a reducer to pressurize the drinking water system; 18—pressure regulator (it also serves as a safety valve); 19—cooling/drying unit; 20—a shut-off valve to pressurize the drinking water system; 21—remote manual control; 22—contamination control assembly (removal of $CO_2$ and contamination); 23—a valve for manual actuation of the oxygen supply; 24—shut-off valve; 25, 26, 27—reducers; 28—BP-1 oxygen supply module; 29—check valve; 30—oxygen supply from the electric power generation system; 31—oxygen from the bottles in the equipment compartment; 32—post-ejection oxygen supply unit; 33—barometric relay module. NB. Sensors and annunciators are not shown in the diagram.

From Archive Zvezda.

the regenerative mode depending on whether the ventilating loop was open or closed. With the loop open, crew member life support was provided by ventilation with gas from the cabin. With the helmet open, this gas was released into the cabin from the helmet and, with the helmet closed, it was released through pressure regulators located on the suit and the BRS-1M. With the loop closed, the gas circulating in the system first passed through the ventilation lines of the suit and removed moisture, heat, carbon dioxide and other gases produced by humans. Then it moved to the contamination control assembly to remove the carbon dioxide and

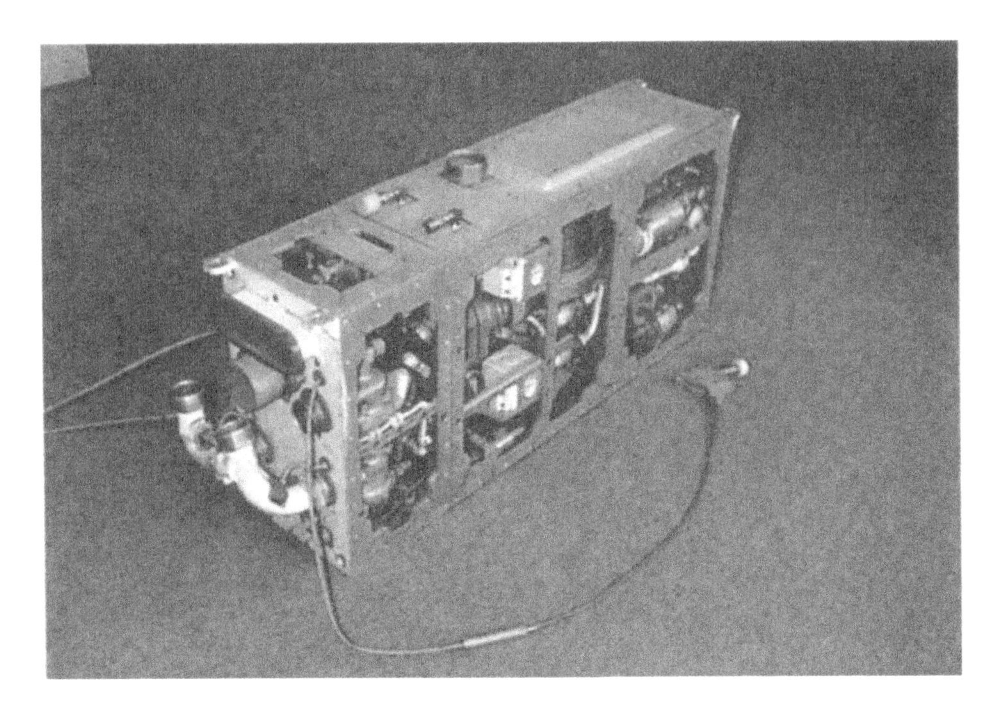

**Figure 10.3.2** General view of the BRS-1M unit (with the panel removed).
From Archive Zvezda.

**Figure 10.3.3** General view of the gas purification unit (removal of $CO_2$ and contamination).
From Archive Zvezda.

other gases. Finally, it reached the cooling/drying unit, which cooled the gas and removed the moisture. In the BRS-1M unit the gas was enriched with oxygen and then flowed back to the suit again. A drop in suit or system pressure, due to leakage from the suit and system and oxygen consumption by the cosmonaut, was compensated in the BR-1M module by oxygen supply into the closed loop. The system switched over to the closed-loop operation mode automatically during cockpit depressurization (when cockpit pressure fell to $600 \pm 40$ hPa) or manually when the crew so wished.

When the PLSS operated in automatic mode, suit-purging was also actuated (to decrease the nitrogen content in the suit's atmosphere). The suit would not be purged if the PLSS was actuated manually. The main operating pressure of the suit, maintained by the pressure regulator, was 440 hPa. Using the same regulator, crew members could maintain a reduced pressure of 270 hPa in the suit.

Thermal control of suited crew members with the loop closed was provided by the AKhSG, which was part of the on-board LSS developed by the Nauka enterprise. The BRS-1M modules were also used for EVA suit ventilation in the *Buran* ShKK airlock (through the closed loop), oxygen supply to the ShKK and the docking module DM [CM], as well as suit-drying after the EVA.

## 10.4 STRIZH FULL PRESSURE RESCUE SUIT

Development of the full pressure rescue suit as part of the *Buran* spaceplane's PLSS was started in 1977 by manufacture of test samples on the basis of the SOKOL-KM, SOKOL-KV and SOKOL-KV-2 full pressure suits. In 1981 the selected experimental sample was named STRIZH (Russian for "swift") (Figure 10.4.1).

Among the other requirements of the RSS suit, some were given priority and called for:

- maximum comfort and performance capability of the crew member with the suit unpressurized;
- maximum performance capability to carry out work in an emergency with the suit pressurized at an operating pressure of 440 hPa;
- easy and correct, unassisted suit-donning in the minimum time and readiness for operation under weightlessness conditions;
- small number of standard sizes and simple personal adjustment of the suit; and
- good compatibility with the ejection seat, in particular with its headrest, restraint and parachute suspension harnesses.

At the first stage of development of the suit enclosure, efforts were concentrated on the built-in restraint harness, which would act as both the seat restraint system and the parachute suspension harness as well as being a function of the suit restraint system. A second task was to provide the required suit/ejection seat compatibility (i.e., crew member capability to carry out the necessary work involved in piloting the

**Figure 10.4.1** General view of the STRIZH suit.
From Archive Zvezda.

spacecraft, including forward bending and assuming the ejection posture under positive pressure in the suit).

When the work on the STRIZH rescue suit started, six fully functional mock-ups of the experimental suits were developed and manufactured, and four more suits followed in 1981.

Besides these tasks, the mock-ups were also used to test several versions of a front don/doff opening (some with zippers and others lacing). However, comparative tests showed that the optimum design, from ease and reliability points of view, was the SOKOL-KV-2 opening, which was then adopted for the STRIZH suits (Figure 10.4.2). In the same year a set of laboratory tests was run: restraint system tests in

**Figure 10.4.2** Front opening of the STRIZH suit.
From Archive Zvezda.

the parachute simulator; thermal protection tests under landing and splashing down conditions in winter; helmet ventilation system and visor misting; wind tunnel tests of the suit with the ejection seat.

In 1982 technical documentation was issued and production of the STRIZH suits for the laboratory tests began. Great attention was paid to development testing of ventilation of the internal suit because the PLSS was of the closed type. The ORLAN suit's ventilation system, which had low hydraulic resistance, was taken as the basis.

The STRIZH suit had to have a provision for drinking with the helmet closed. The on-board water supply system and corresponding device on the suit were developed. The suit assembly also included special hygienic slips that absorbed urine and its smell in case of involuntary urination.

One of the main requirements of the STRIZH suit was safe rescue of the crew member by ejection in the initial part of the mission (ascent) at altitudes up to 30 km and at speeds up to Mach 3. Moreover, because the suit's external surface was rapidly heated due to the high aerodynamic loads, a special heat-resistant fabric for the suit restraint layer was developed and used. This fabric also had low elongation at stretching. To provide additional thermal protection, an external thermal protection layer for the suit enclosure was developed. It was made of elastic

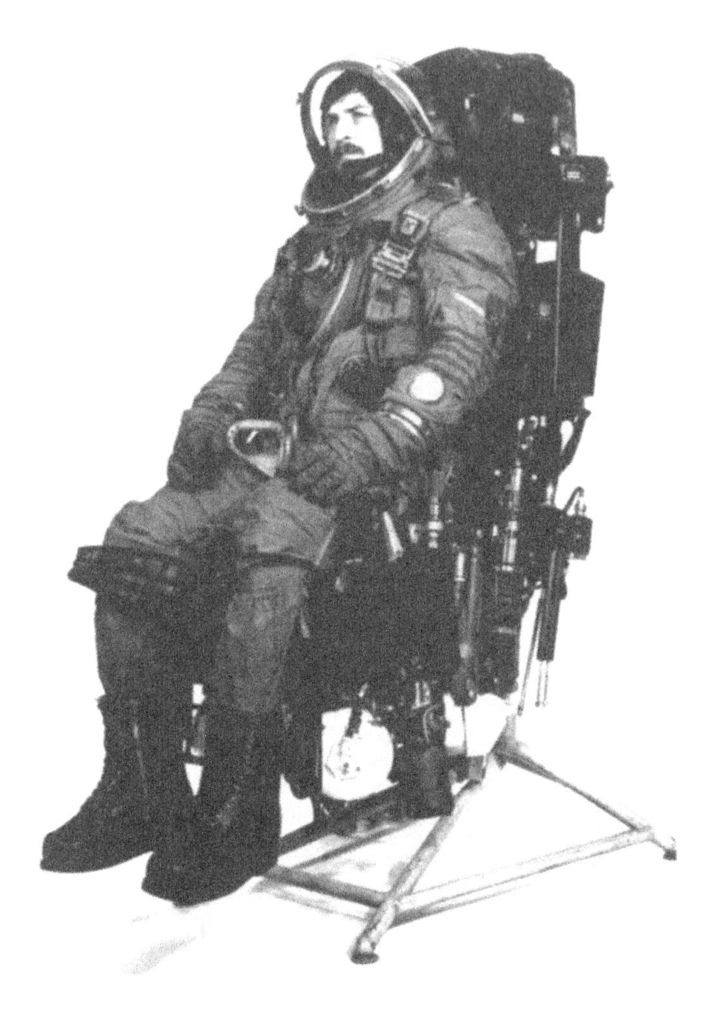

**Figure 10.4.3** A test engineer wearing the STRIZH suit in the ejection seat.
From Archive Zvezda.

leather with an aluminium coating. The helmet visor was converted to a double one to improve reliability, provide additional thermal protection and prevent misting up. To ensure the suit enclosure was warm, thermal protection panels were introduced inside the suit.

The STRIZH suit had a novel restraint system for the upper torso, which provided for both easy donning (by transferring the helmet frame over the head at suit-donning/doffing) and the best possible operator position in the seat. For instance, it was possible to press the nape portion of the helmet against the seat headrest with the suit under operating pressure in case of ejection (Figure 10.4.3).

A new system for selection of the correct suit size, based on an average size (male) for military clothes was developed for the STRIZH suit. In the process of STRIZH suit development, a large amount of laboratory and in-house testing, both

**Figure 10.4.4** Appearance of the STRIZH suit helmet's bubble after the aerodynamic heating test.
From Archive Zvezda.

independent of and together with the on-board PLSS, was run. Besides the standard spacesuit tests (fit checks, strength and life tests, mobility verification, "high-altitude" tests of the pressure regulator, flotation tests, etc.), the following tests were undertaken: water supply system tests, tests of the built-in restraint harness in the simulator and in flight (parachute jumps), confirming the adequacy of the design of the selected restraint harness; wind tunnel tests on the K36M11F35 [K36M11Ф35] ejection seat; physiological low-temperature tests and hygienic slip tests.

Between 1988 and 1990 five K-36M-ESO [K-36M-ECO] mock-up seats and five mock-up STRIZH (ESO) [ECO] suits were produced and five high-altitude experiments were carried out during test flights of the space module's experimental compartments, ESOs, including seat ejection with the suit from that module at altitudes of 35–40 km and speeds of Mach 3.2–4.1 (Figure 10.4.4). In 1989 in-house testing of the STRIZH suit for life expectancy was completed and flight tests in the IL-96 flying laboratory carried out.

# 11

# Evolution of the European EVA spacesuit

## 11.1 INTRODUCTION

European spacesuit work started in 1986 as part of the overall *Hermes* spaceplane project. Between 1986 and 1987 two requirements/feasibility studies were carried out in parallel by Dornier in Germany and BAe in the UK for the European Space Agency (ESA). In 1988 Dornier was awarded the system work under the project title "European Space Suit System" (ESSS). Shortly before that, in 1987, ESA contracted three subsystem studies—EVA Suit Enclosure Module (ESEM), EVA Life Support Module (ELSM) and the EVA Information Communication Module (EICM)—which were subsystems of the planned ESSS. The work was coordinated by the prime contractor for ESSS (Dornier). So, ESEM, ESLM and EICM were not separate suit projects or alternative studies.

After initial studies between 1986 and 1988 and the decision of the 1987 ESA Ministerial Conference in The Hague to support European autonomy in space, the predevelopment phase (C1) took place in 1989–1991, resulting in the system design and considerable technological development. In this phase, all efforts were brought together in one ESA contract, with Dornier as the prime contractor.

During this phase initial contact with the Soviet spacesuit manufacturer RD&PE Zvezda was made (1989), who soon became involved in the ESSS project as a subcontractor to Dornier at the system level. A design phase was envisaged to start in 1992, with planned first delivery of flight hardware in 1998.

At the 1991 ESA Ministerial Conference in Munich, closer cooperation between ESA and the Russian Space Agency Rosaviakosmos (RKA) for the *Hermes* programme was agreed. This raised the possibility of joint ESA/RKA spacesuit development (for *Hermes*, *Columbus*'s Man-Tended Free-Flyer—MTFF, *Mir* and *Buran*), which was analysed in 1992 by Dornier and Zvezda under the study title "EVA-2000" and was found technically and programmatically feasible.

When the *Hermes* project was stopped in 1993, the ESSS was in jeopardy. So, ESA and RKA in 1993 agreed to jointly develop a new generation spacesuit, the ESA/RKA EVA SUIT 2000, to be used on *Mir-2* (later the *ISS*). The feasibility of such cooperation had been verified by the EVA-2000 study! EVA SUIT 2000 was a joint venture between Western Europe and Russia, with Dornier and Zvezda as co-prime contractors and funding from both ESA and RKA. Design and development of the joint suit continued until the end of 1994, when the project was cancelled from the ESA side, due to limited funds for manned programmes in Europe and new priorities set within ESA.

At the time of termination of the EVA SUIT 2000, the first ergonomic model had already been extensively tested. This model, the EVA SUIT 2000 prototype, operated at the full suit pressure of 420 hPa and was built and tested by Zvezda and the Dornier subcontractor SABCA in Belgium. This demonstrator used both West European and Russian manufactured hardware.

Since the end of 1994, no further work on spacesuits has been done in Western Europe. However, some of the new technologies and design features jointly developed for the EVA SUIT 2000 between 1993 and 1994 (increase of torso dimensions, introduction of arm and calf bearings) have been implemented in the Russian spacesuit ORLAN-M, first on *Mir* in early 1997 and now on the *ISS* (see also Section 8.3).

## 11.2 THE EUROPEAN SPACE SUIT SYSTEM (ESSS)

The ambitious European plans for future manned space systems in the mid-1980s covered numerous elements for short- and long-term stay in orbit. The *Columbus* programme included an Attached Pressurized Module (APM) for docking with the *ISS*, an MTFF as an autonomous mini-space station and platforms.

The *Hermes* spaceplane with a crew of three was intended for resupply and repair of platforms and the *Columbus* MTFF. In this context the ESSS was required to carry out MTFF servicing and repair, to support the *Hermes* vehicle in safety critical operations and to undertake external servicing and contingency operations (Figure 11.2.1).

Two initial ESA feasibility studies (starting in May 1986 and continuing until March 1987) were contracted to Dornier and BAe in parallel. Dornier teamed up with Aviation Marcel Dassault and Aérospatiale in France, Nord-Micro in Germany and Microtechnica in Italy, and had for this study consultancy from Hamilton Standard, the US spacesuit manufacturer. BAe selected (as European partners) Matra in France, Sener in Spain and Normalair-Garrett in the UK, with McDonnell-Douglas as the US consultant. The major findings of the Dornier study for a spacesuit for use aboard *Columbus/Hermes* were (Skoog et al., 1991):

- Design of the EVA spacesuit should provide for a 6-hour (plus 1-hour contingency) operation by two astronauts. The spacesuit enclosure was to be of a hybrid type with hard upper torso and backdoor entry.

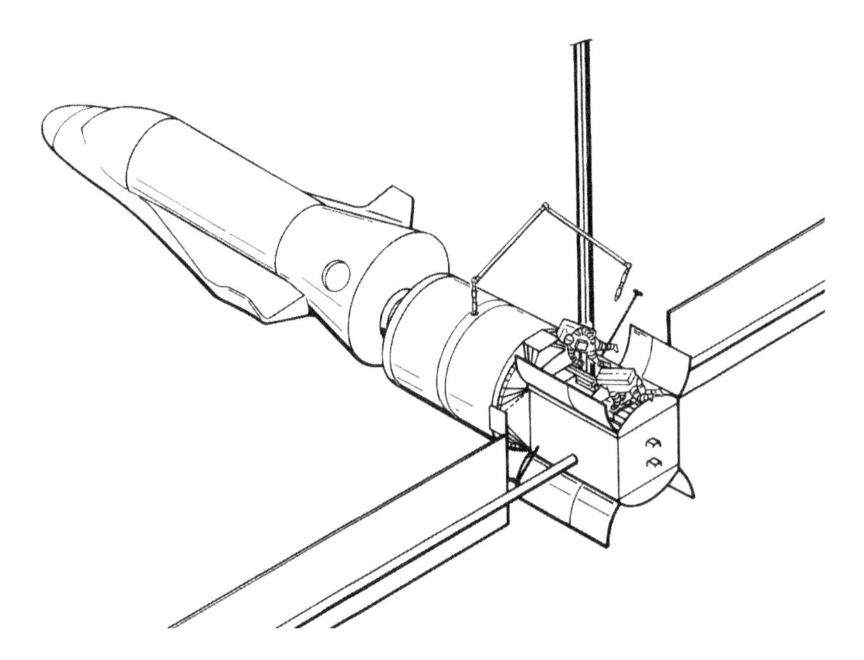

**Figure 11.2.1** *Hermes* docked to *Columbus*'s Man-Tended Free-Flyer.
From Archive Skoog.

- This suit configuration should allow for a design pressure of 500 hPa that, in combination with *Hermes*' reduced cabin pressure (700 hPa), would permit zero pre-breath operation of the spacesuit.
- The life support system (LSS) should be of a non-regenerative closed loop design with $CO_2$ removal by LiOH and thermal control by means of a condensing heat exchanger and sublimator. Metabolic cooling would be achieved by a water-cooled garment and a gas ventilation network with tubing built into the suit enclosure. The oxygen atmosphere would be provided from high-pressure tanks and the LSS integrated within an unpressurized backpack.
- Advanced monitoring and control system, partially providing for hands-free operation at the worksite, due to a display showing operational data.

The following EVA requirements were driven by *Hermes*' operational concept of a crew of three astronauts:

- unassisted spacesuit-donning and doffing;
- shortest possible preparation time for EVA; and
- maximum operational flexibility and crew safety.

ESA's EVA feasibility studies were supported by a number of additional contracts from ESA and national agencies for provision, for example, of man–machine

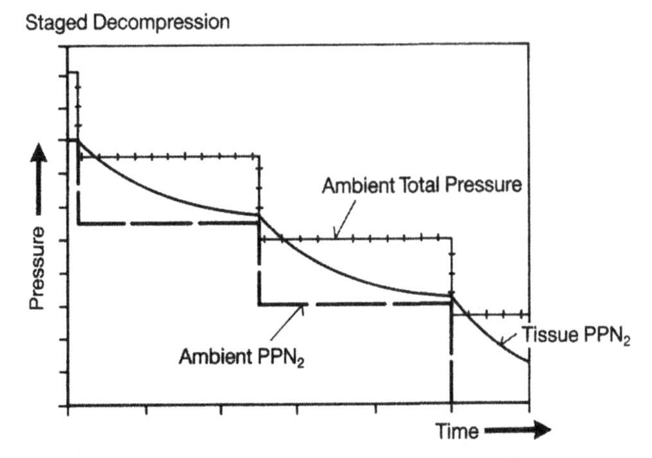


**Figure 11.2.2** Staged decompression procedure.
From Archive Skoog.

interfaces, EVA crew-training (zero gravity aircraft and neutral buoyancy), tools, crew safety and medical compatibility studies.

Zero pre-breathing and the risk of decompression sickness (DCS) was extensively investigated in order to optimize the suit pressure, which was finally selected at 500 hPa, higher than the US EMU (296 hPa) and the Soviet ORLAN (400 hPa) spacesuit pressures (Vogt et al., 1991).

In order to reduce the DCS risk when changing from the cabin pressure to a lower suit pressure (e.g., 500 hPa), one of following general procedures must be applied:

- pre-breathing with pure oxygen for a period that is determined by the pressure differential; or
- reduction of the cabin pressure before EVA for a period to be determined (staged decompression).

Based on *Hermes*' configuration and design of the zero pre-breath suit, the second alternative was selected. By using staged decompression the cabin pressure was reduced to 700 hPa, 12–48 hours before the first EVA sortie (Figure 11.2.2). When the vehicle pressure was set at 700 hPa, assuming nitrogen equilibrium in the body compartments has been established for at least 12 hours, no pre-breathing was necessary for a suit pressure of 500 hPa, corresponding to a ratio of less than $R = 1.22$,[1] which would indicate a 0% risk of DCS symptoms. For the same risk factor ($R = 1.22$) at a suit pressure of 500 hPa and a cabin pressure of 1,013 hPa, approximately 150 minutes of pre-breathing would be required along with consumption of about 100 g of oxygen per crew member per sortie. A ratio of $R = 1.6$ was specified for shorter emergency sorties, corresponding to a 5% risk of DCS

[1] The $R$-factor is defined as the ratio of initial nitrogen partial pressure $p_{N_2}$ to the final total barometric (suit) pressure $p_B$, $R = p_{N_2}/p_B$ (Vogt et al., 1991).

Table 11.1. Summary of major ESSS requirements for phase C1.

| | |
|---|---|
| Sortie duration | 7 hours total, including 6 hours of work outside the airlock |
| Number of sorties (per crew member) | 1 planned, 1 emergency |
| Number of EVA crew members | 2 |
| Re-establishment of fully autonomous capability | Less than 14 hours |
| Operating environment | 500 ± 10 hPa (270 hPa emergency) |
| Suit atmosphere | 95% pure oxygen |
| Emergency life support | 30 minutes minimum |
| Useful life | 15 years with maintenance |
| Anthropomorphic sizing | 5th to 95th percentile of European male population |
| Size | Can pass through a ∅ 900-mm opening |
| Donning/Doffing | Back entry; self-donning and doffing |
| Prebreathing | None from 700 hPa cabin pressure; 150 minutes from 1,013-hPa cabin pressure |
| Leakage rate | 10 g oxygen hour$^{-1}$, maximum |
| Overall cooling capability | 7,600 kJ total metabolic rejection; 962 W maximum total heat rejection rate for 15 minutes |
| $CO_2$ level (oro-nasal area) partial pressure | 10 hPa nominal; 20 hPa maximum for 15 minutes |

symptoms, and required no pre-breathing when the vehicle pressure was set at 1,013 hPa ($O_2$ partial pressure 213 hPa).

The *R*-factors used for normal sorties in the USA are $R = 1.4$–$1.6$ and in the USSR $R = 1.5$–$2.0$; thus, the ESSS concept represented in this respect a conservative approach even for an emergency sortie. In 1990–1991 this approach was extensively discussed with Zvezda's medical experts and became the suit pressure baseline for the subsequent ESSS design work (Vogt et al., 1991).

The overall results of the different EVA and spacesuit studies were then (in 1988) brought together in the ESSS project, under a team led by Dornier. In parallel with and coordinated by the ESSS project, separate studies on the ESEM led by Dassault (France), the ELSM led by Dornier (Germany) and the EICM led by Laben (Italy) were undertaken. By early 1989 the ESSS had come up with the necessary overall ESSS requirements (Table 11.1) and lower level subsystem specifications and reference configurations (Figures 11.2.3 to 11.2.6) to start the pre-development (C1) phase of the ESSS.

When European spacesuit development started in 1986 the only relevant experience in Europe was in aviation pressure suits and LSS development for the Spacelab Laboratory Module flown with the US Space Shuttle. So, as far as Dornier was concerned, the long-standing working relations for LSSs with Hamilton Standard in the USA were of great value; particularly so, as Hamilton Standard was also the prime contractor for the Shuttle spacesuit system Extravehicular Mobility Unit

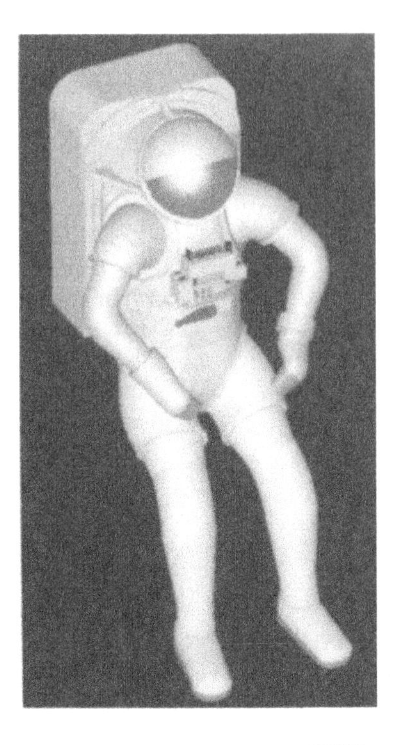

**Figure 11.2.3** Autonomous ESSS.
From Archive Skoog.

(EMU). Hamilton Standard assisted the Dornier spacesuit team in system engineering, and intensive training was undertaken in 1987 and 1988. But, despite this, the ESSS concept defined for the start of phase C1 differed from the US EMU, in particular in its rear-entry concept, higher suit mobility and the higher suit pressure. All this was mainly due to *Hermes*' requirements, which were different from those of the US Shuttle. On the other hand, the LSS and backpack concept showed great similarities to the EMU backpack design.

The ESSS spacesuit design concept progressed through its pre-development phase (C1) between 1989 and 1991 as a part of the overall *Hermes* development programme (Skoog, 1994). The initial phase C1 configuration was based on 3 years of predefinition and conceptual design work undertaken at both system and subsystem level, but without any accompanying technology or other hardware-related activities. However, the need for technology investigations and breadboard-testing had been identified during the predefinition phase, and thus half of the financial budget for phase C1 was allocated to breadboarding the most critical parts of the spacesuit. In some cases (i.e., glove, sublimator and $CO_2$ removal), the purpose was more of an enabling type by building up know-how already existing in the USA and the former Soviet Union. But, in other areas (i.e., rolling convolute shoulders, seals and bearings and voice communication), the selected design concept for the European spacesuit required brand new technologies. The ESSS design was

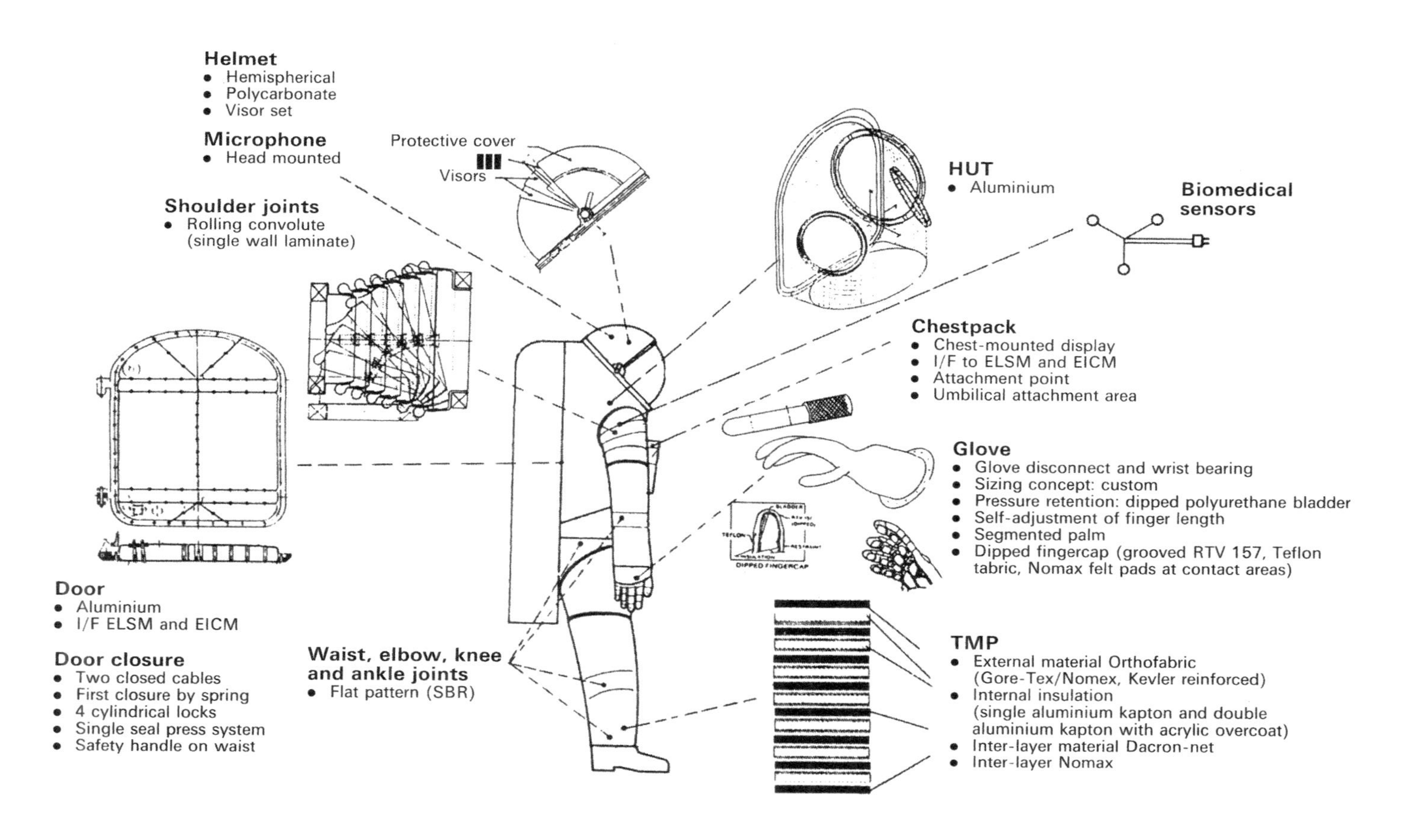


**Figure 11.2.4** Suit enclosure (ESEM) reference configuration.

From Archive Skoog.

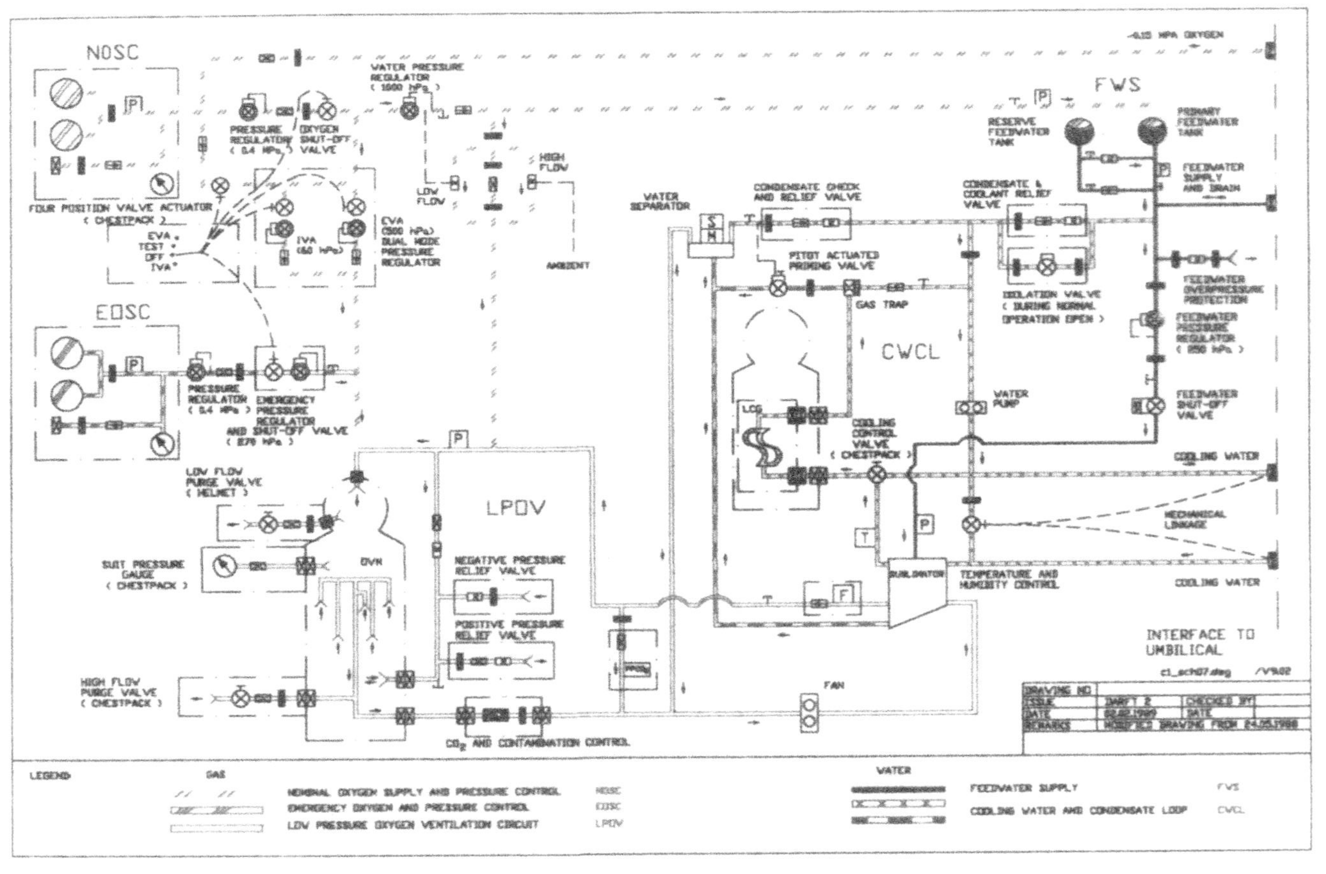


**Figure 11.2.5** Life support system (ELSM) reference configuration.

From Archive Skoog.

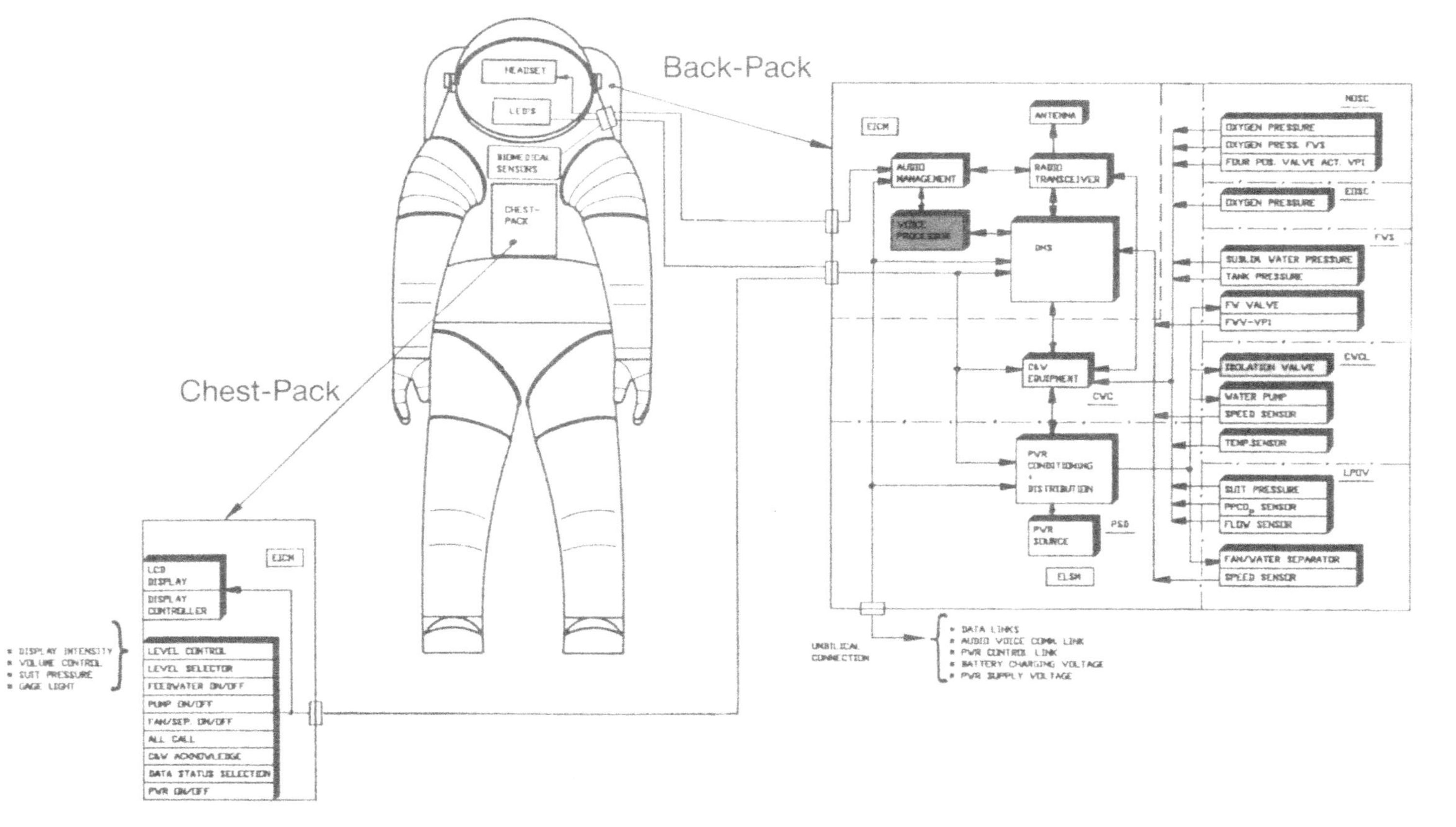


**Figure 11.2.6** Information and communication (EICM) reference configuration.

From Archive Skoog.

based on the above general requirements (Table 11.1) as detailed in the ESA's EVA Requirements Document (ESA-RQ-EVA-001 dated 27 February 1989) later changed to the EVA Design Requirements Document (H-S-2BA-001-ESA, Issue 1 dated 30 November 1989) within the *Hermes*' documentation system.

With such a multitude of tasks in phase C1 a large project team from some 30 European companies was set up at the outset in 1989 (Figure 11.2.7). Around the time of commencement of phase C1, the first official contacts with RD&PE Zvezda took place in 1989, and in 1990 Zvezda was subcontracted by Dornier for system engineering support, in general, and suit enclosure issues, in particular. It should be pointed out that Dassault had its first contact with Zvezda when supporting the second French flight of Jean-Loup Chrétien in 1988. During this flight Chrétien undertook a space walk from *Mir* and thus was the only European cosmonaut/astronaut with EVA flight experience when the ESSS project started. His experience was of great value to the ESSS regarding suit and EVA work (Figure 11.2.8).

The ESSS flight equipment consisted of an autonomous unit and the auxiliary equipment. The autonomous unit allowed the EVA crew member to work outside the manned spacecraft, while the auxiliary equipment provided EVA operations with dedicated tools and the autonomous unit support and interface equipment located on board *Hermes*. The EVA autonomous unit (Figure 11.2.3) consisted of three integrated, differentiable subsystems:

- suit enclosure subsystem (SES), earlier called ESEM;
- backpack subsystem (BPS) with the life support equipment, earlier called ELSM; and
- chestpack subsystem (CPS), earlier called EICM.

During pre- and post-EVA operations the autonomous unit was to be supported by a separate support and interface subsystem (SIS) located in *Hermes*' resource module and airlock, providing the autonomous unit with power, communication, gas and water (via the umbilical) during in-orbit airlock operations. A combined attachment structure and don/doff device and stowage provisions were also part of SIS.

The spacesuit was required to provide the EVA crew member with the best protection against the environment, while allowing him maximum mobility, dexterity and visibility. The suit enclosure is the part responsible for these fundamental functions. The general suit enclosure design for phase C/D (as elaborated in phase C1) is presented in Figure 11.2.9. This design was based on a hybrid concept: soft joints would be mounted on the hard upper torso (HUT), which was closed by a back door on which the life support equipment was mounted. The helmet bubble, visor and sunshade assembly was mounted on the HUT by means of an "Ortman-wire"-type connector. However, no detachment was foreseen in orbit.

Limb mobility was obtained through five pairs of bearings combined with one-degree-of-freedom points. The combination of scye and arm bearings and the shoulder joint provided the three degrees of freedom required for shoulder mobility. The mobility performance foreseen for the waist and brief region posed

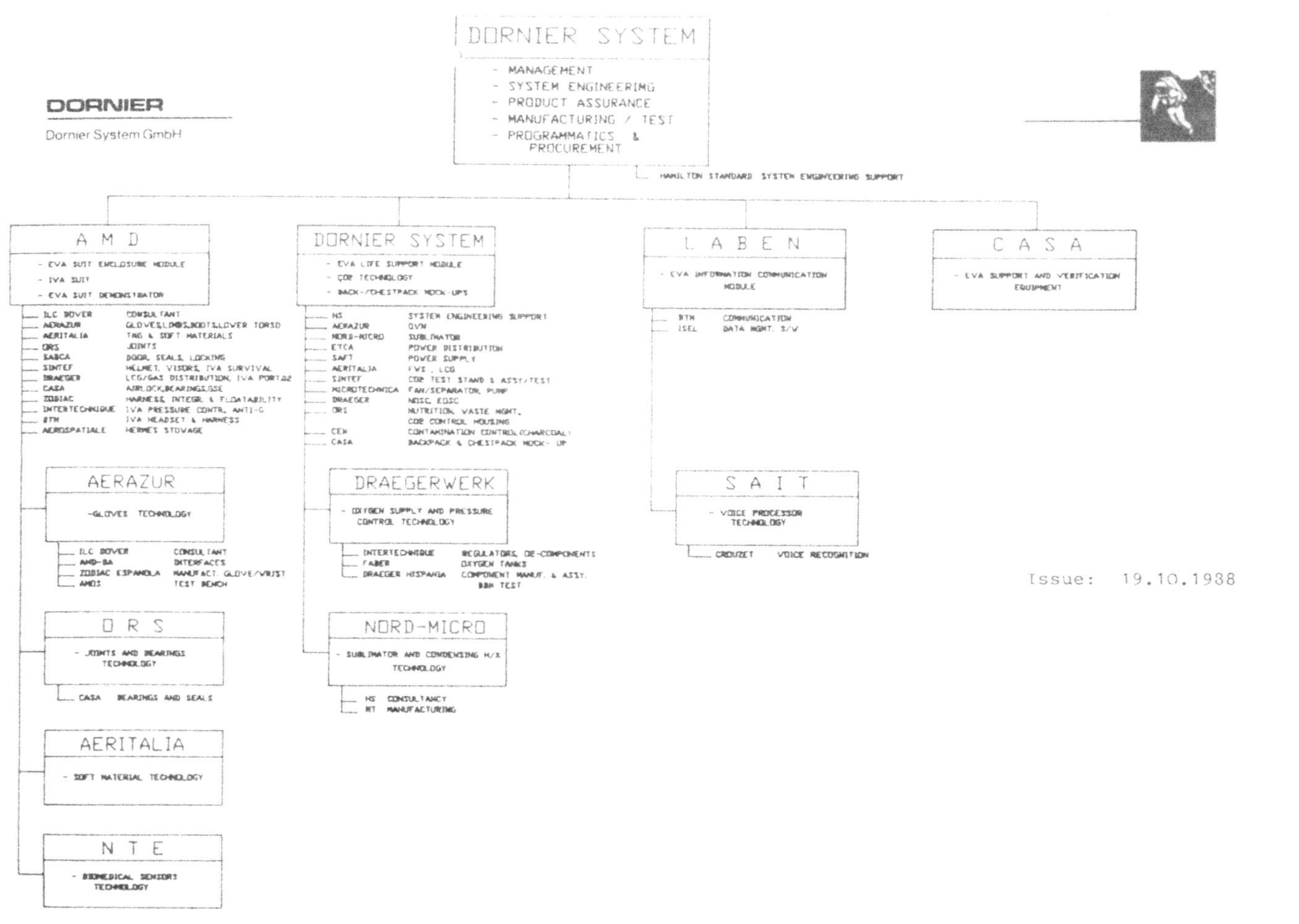


**Figure 11.2.7** The ESSS project team for phase C1.
From Archive Skoog.

**Figure 11.2.8** Jean-Loup Chrétien during a visit to Dornier on 18 January 1990. *From left to right*: R. Loewens (Dornier), S. Berthie (AMD), G. Adami (Laben), G. Cordoni (Laben), A.I. Skoog (Dornier), J.-L. Chrétien, P. Kania (Dornier), J.-R. Chevallier (AMD), S. Mueller (Dornier), Y. Ollivier (AMD).

From Archive Skoog.

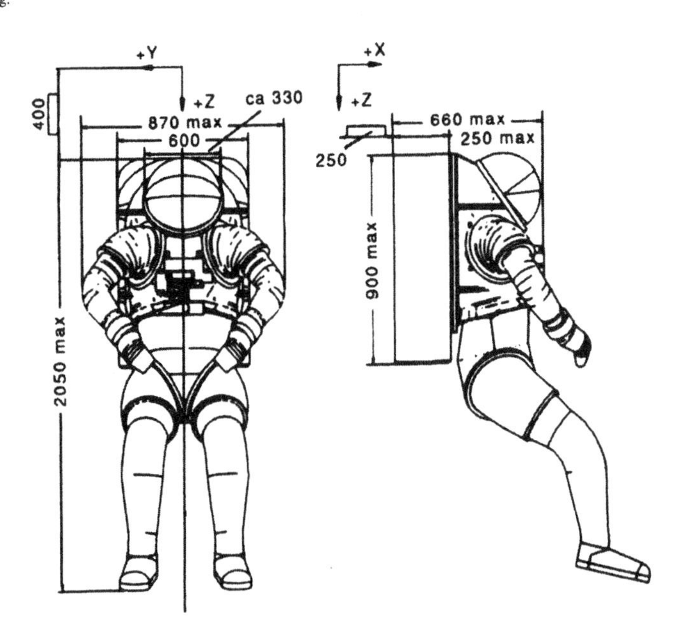


**Figure 11.2.9** ESSS spacesuit overall dimensions.

From Archive Skoog.

many problems and explained the complex design that included a waist and hip joint combined with hip and thigh bearings.

Gloves are among the most important items of a suit, because of the fundamental importance of dexterity and tactility, and were thus custom-sized. Three gloves were worn each on top of the other: the glove bladder for gas retention, the glove restraint for pressure load retention and prevention of ballooning and the thermal protective layer based on multilayer insulation (MLI). The second item considered as critical was the shoulder joint. The chosen design was the "rolling convolute", which included six aluminium rings linked by a mechanical axial restraint and needle bearings. A bladder internally glued to the rings ensured pressure-tightness. The soft joints, distributed along the limbs (elbow, waist, hip, knee and ankle), had their own separate bladders and restraints, completed by axial restraints for transfer of axial pressure loads (the so-called flat pattern concept). The bearing design was based on stainless steel races and balls with dual seals.

Sizing of the suit enclosure to fit each crew member was provided by means of a sizing plan that consisted of a single-size torso, custom-sized gloves and two sizes of each arm, waist and leg. This plan was based on statistical evaluations and analysis that considered the partial length correlation of limbs and known adaptability of the human body to a given suit design in zero gravity.

An open issue at the end of phase C1 was the general design of the lower torso. The complex design with two pairs of bearings (hip and thigh) provided a high degree of comfort and mobility at the cost of high mass. On comparison with American and Russian suit designs, further analyses and tests were required to provide a design adequate to meet the standards required of the EVA suit. Further analyses were also required for bearings because of high torque and design improvements for the glove.

Most of the life support components were located in the backpack, which was to be integrated with the rear entry door of the suit enclosure (Figure 11.2.10). The mode actuation mechanism, switches and manual controls were located in the chestpack.

When the nominal oxygen supply and control (NOSC) pressure regulator was set to "EVA", it maintained the nominal suit overpressure of 500 hPa to make up for $O_2$ consumption and leakage. The NOSC tanks were rechargeable in orbit. The emergency oxygen supply and control (EOSC) pressure regulator was set to maintain a suit pressure of 270 hPa. In case the pressure dropped to 270 hPa (due to leakage), this pressure would be maintained by the EOSC for at least 30 minutes. As soon as the autonomous unit was disconnected from the umbilical, the sublimator would be switched on, leading to a maximum release of 13 g minute$^{-1}$ of sublimated water. The low-pressure oxygen ventilation (LPOV) circuit was driven by the fan in a combined fan/pump/separator unit. $CO_2$ and odours were to be removed by means of LiOH and active charcoal filter. The $ppCO_2$ level would be kept under 10 hPa under all normal conditions. Heat and moisture released by the crew member and the equipment was controlled by the sublimator/condensing heat exchanger, and the ventilated oxygen was cooled down by water condensation wetting the hydrophilic coating in the condensing heat exchanger (CHX) (Figure 11.2.11)

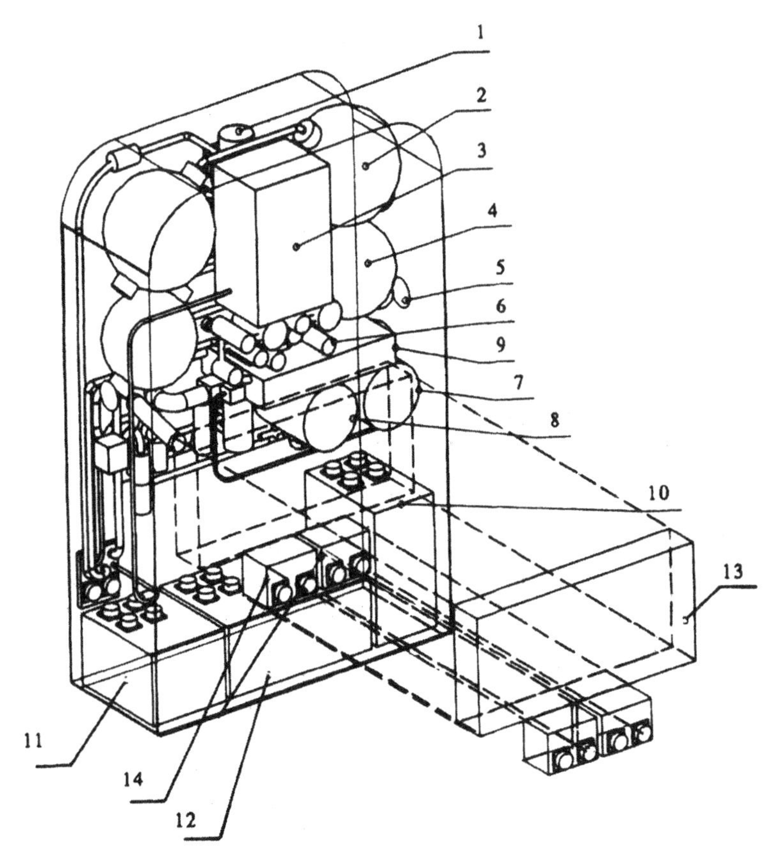


**Figure 11.2.10** Backpack layout: 1—LPOV relief valve; 2—NOSC oxygen tanks; 3—NOSC/EOSC valves and regulations; 4—EOSC oxygen tanks; 5—water vapour vent holes; 6—FWS regulator and valves; 7—fan; 8—redundant fan; 9—sublimator; 10—transceiver; 11—EMS box; 12—DMS box; 13—$CO_2$ and contaminants cartridge; 14—redundant batteries.
From Archive Skoog.

The crew member could adjust the temperature of the water-cooled garment within the comfortable range of 12–23°C by actuation of the cooling control valve located at the chestpack. This valve controlled a bypass of the sublimator/CHX to maintain the requisite temperatures of the water-cooled garment and remove total rejected heat loads.

The CPS consisted of the data management subsystem (DMS), the communication and power supply subsystems located in the backpack, and display and controls on the front of the suit. Information data-handling and data-processing of the ESSS was to be provided by the DMS. It would present the EVA crew member with information in the necessary format and provide the appropriate control functions required to handle the ESSS during all phases of the EVA sortie (Figure 11.2.12). The powerful LCD display (12 × 5 cm) provided graphic and alphanumeric messages to the EVA astronaut, detailing operational procedures and spacesuit status infor-

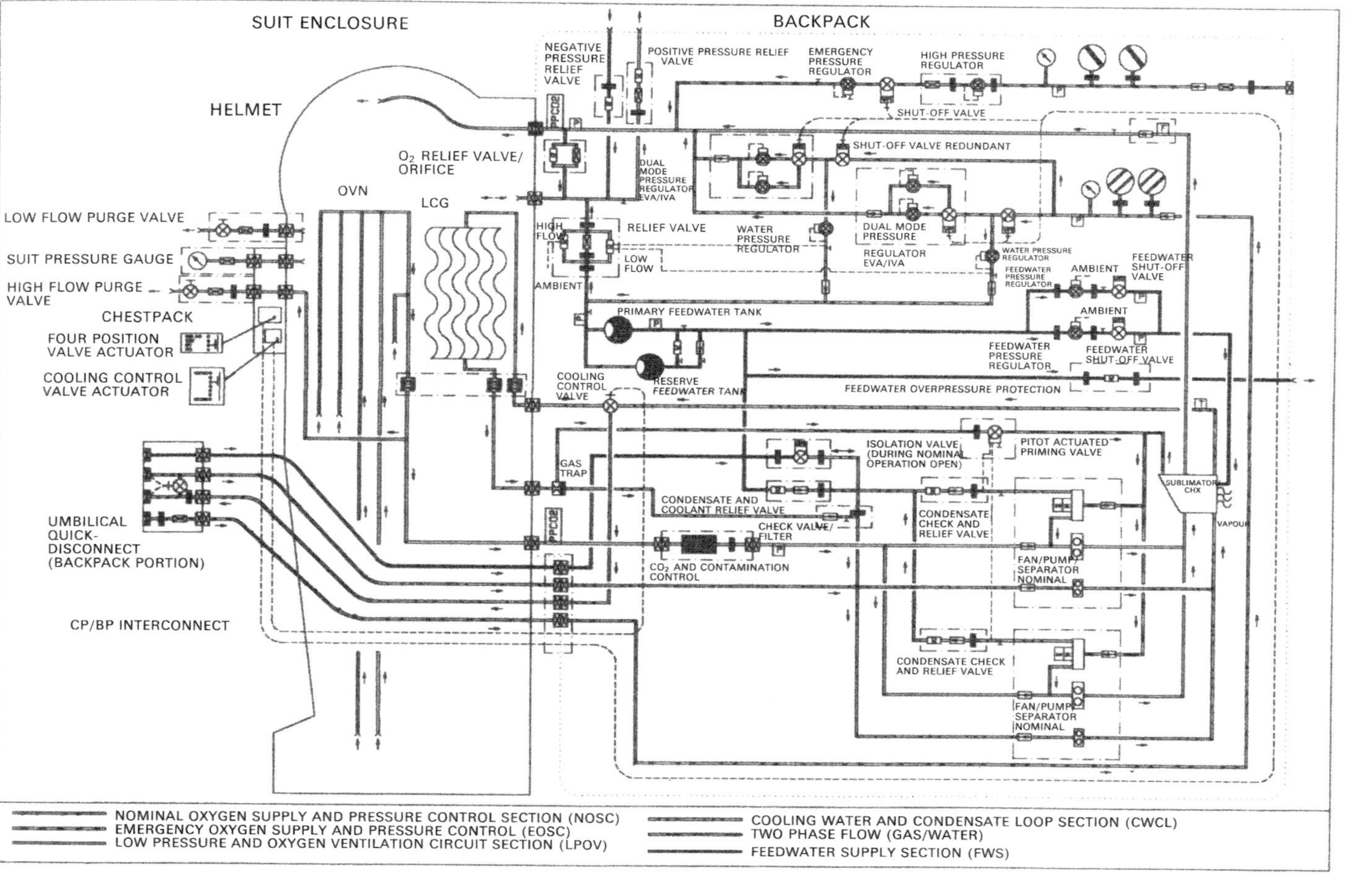


**Figure 11.2.11** Life support system (C1 reference concept).

From Archive Skoog.

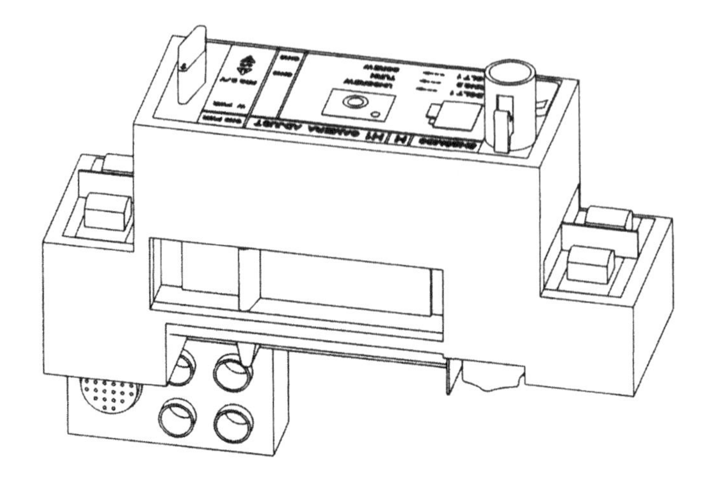

**Figure 11.2.12** Chestpack layout.
From Archive Skoog.

mation. Emergency and warning messages were, in addition to the display on the LCD, accompanied by a buzzer via the headset and by backlight indicators inside the helmet.

The design for power supply and distribution was developed with operational failure and fail-safe requirements in mind. During EVA mode, the power supply consisted of two lithium batteries, each of them providing half (18 A h) the required energy for one sortie at the normal 24 V DC voltage.

Redundancy was built into all communication functions, which were monitored by the DMS. The design for radio frequency transmission was based on the time division multiple access (TDMA) design (in the 400–470 MHz UHF band, with a maximum bandwidth of 500 kHz).

An important part of the phase C1 work was technology development and the breadboarding of critical equipment:

- shoulder joint;
- glove;
- elbow joint;
- thermal and micrometeoroid protection;
- high-pressure components for oxygen supply;
- sublimator/CHX;
- $CO_2$ removal;
- fan/pump/separator; and
- suit demonstrator.

A rolling convoluted shoulder joint breadboard was manufactured (Figure 11.2.13), the hard part of which was very similar to that of the anticipated flight model. The bladder was made of a single layer of laminate (i.e., a fabric coated with

**Figure 11.2.13** Rolling convoluted shoulder joint breadboard unit (manufactured by SABCA and Aerazur).
From Archive Skoog.

polyurethane that served pressure retention and load retention functions). It looked like a series of cones and cylinders beneath the aluminium rings of the joint. The selected flight configuration was to be manufactured using a moulding process, but the breadboard was built by welding together its constituent parts. A glove breadboard was designed and manufactured by Aerazur in France (Figure 11.2.14). Mechanical evaluation of the results indicated the need for extensive development work on the glove:

- the ballooning of the palm was not acceptable and was found to be mainly due to a poor palm restraint design, which had to have its shape and rigidity improved;
- the gripping capability was quite poor and was found to be due to the neutral position of the glove, palm ballooning and the conical shape of the thumb joint.

Thermal test results indicated that the glove performed well, confirming the validity of the multilayer insulation design. However, an improved glove was also manufactured and performed considerably better.

The technology needed to provide the suit's soft joints was a key issue for general, overall mobility. This was pursued by Zodiac Española, who got involved in elbow joint breadboarding (Figure 11.2.15). The breadboard was based on a flat pattern concept, with the bladder and the restraint identically shaped (five gores[2] externally and one internally) and stitched together to avoid any sliding effects. Sizing adjustment of the fabric restraint was obtained by lacing up the two inserts on each side of the elbow. Testing showed good performance (torque below 2 N m)

[2] A "gore" is a triangular or tapering piece of material.

**Figure 11.2.14** Glove restraint breadboard unit (Aerazur).
From Archive Skoog.

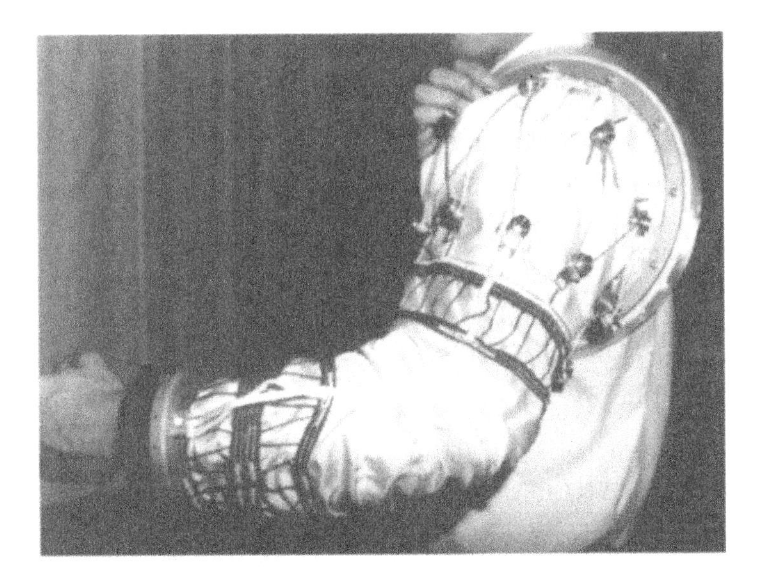

**Figure 11.2.15** Elbow joint breadboard unit (Zodiac Española).
From Archive Skoog.

for low bending angles between 10° and 70°. However, above 70° the torque increased drastically to about 35 N m (at 120°) and required design modification.

The suit's thermal and micrometeoroid protection (TMP) technology was studied by Aeritalia. TMP for the inner layer of the arms was made of Nomex and the external layer out of Orthofabric; an inner reinforcement of Kevlar felt increased mechanical protection. The insulation layers were made of classical, aluminized Kapton, separated by Dacron spacers.

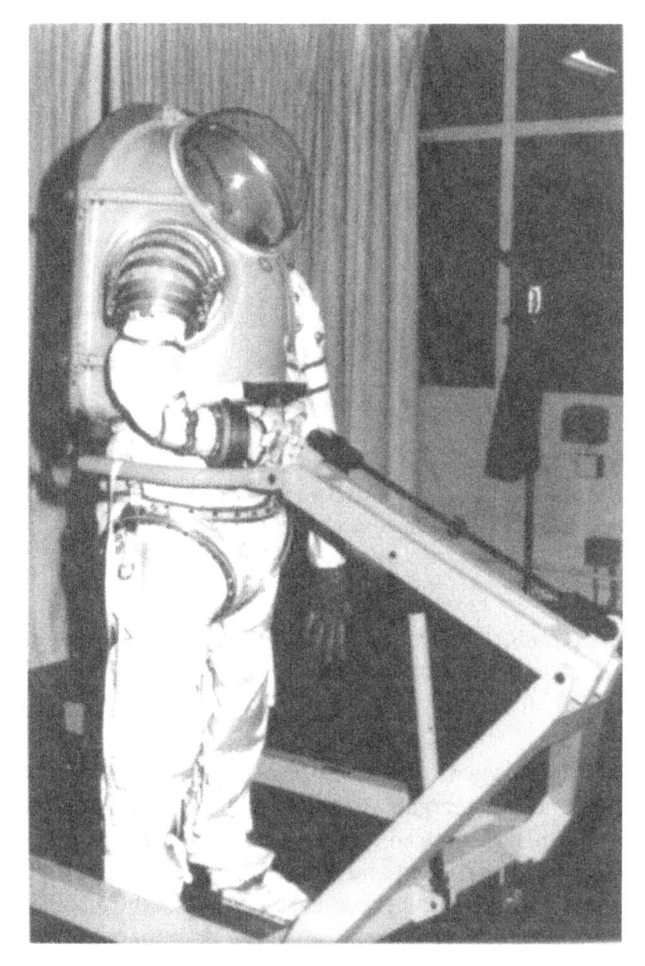

**Figure 11.2.16** ESSS suit demonstrator (Dassault).
From Archive Skoog.

Toward the end of phase C1, a suit demonstrator was assembled and tested (Figure 11.2.16). It included a representative HUT (closed by a door simulator) and a simple hemispherical bubble for the helmet. The right arm illustrated the assembly of the shoulder, the elbow and the glove breadboards. The left arm was simplified, but included a sliding cord shoulder joint, a representative elbow and a simplified glove. The lower torso had no bearings. The main results of ergonomic testing were:

- statistical analysis of test engineers indicated a clear correlation between leg and arm dimensions, allowing simplification of the sizing plan;
- qualitative assessments of mobility confirmed the design of the torso and arm architecture;

**Figure 11.2.17** Oxygen control breadboard unit (Dräger).
From Archive Skoog.

- donning/doffing comfort was proven and confirmed the feasibility of the rear entry design and the torso geometry;
- visibility was adequate.

The original concept for oxygen supply was based on a storage pressure of 20 MPa for both the nominal and the emergency systems. Later the baseline was changed to a 6–20 MPa system, but this was judged to be a minor modification from a technology viewpoint. The components designed and tested were the high-pressure oxygen filler valve, pressure sensor and gauge, the pressure regulator (20/0.4 MPa) and the dual mode pressure regulator (0.4 MPa/500 hPa). These components were individually tested and then attached to the NOSC/EOSC breadboard model (Figure 11.2.17).

The design of the sublimator/CHX for heat and humidity removal was proven in the spacesuit programmes of the USA and the USSR and was also selected for breadboarding in the ESSS project. The design consisted of a stainless steel heat exchanger core with nickel fins and a sublimator porous plate built up of several layers of wire meshes rolled together (Figure 11.2.18). The breadboard model of an all-European-manufactured sublimator/CHX proved the design worked, but heat removal capacity was lower than predicted. Detailed analyses of the test results revealed the likely cause of reduced heat removal, and the unit was modified to give improved performance.

The removal of $CO_2$ from the suit's atmosphere was effected using LiOH, which is used in most spacecraft for this function. Due to its proposed application in a pure oxygen atmosphere at 500 hPa, the behaviour of LiOH was tested for these new environmental conditions.

The original design envisaged separate fan, separator and water pump units, because of the complexity of development and integration. Potential weight savings

**Figure 11.2.18** Sublimator/CHX breadboard (Nord-Micro).
From Archive Skoog.

were however of such importance, due to the overall high mass of the spacesuit, that a combined unit was planned as a technology demonstrator (Figure 11.2.19). This trade-off in design resulted in a peripheral (drag) fan, a centrifugal water pump and a centrifugal water separator with a pitot tube and one brushless DC motor. Initially, the design requirements were met or surpassed for the fan and the water pump, but the requisite water separation efficiency was not reached (only 80% instead of 98%). Design modifications later solved this problem.

Work on the ESSS phase C1 resulted in an improved technical reference concept for the next phase (C/D) that was planned to start early in 1992. The design was backed up by a large database established by work on technology development and extensive breadboard model-testing. The reference concept of phase C1 fulfilled ESA ESSS system requirements, as refined and changed in line with the ongoing design of *Hermes*, in all areas except overall mass. At the end of phase C1 the design was 10% above the given target (125 kg), but options for modifications to reduce the mass had been identified, and the given target was regarded as realistic and achievable.

In 1990 and 1991 (encompassing the ESA Ministerial Conference held in Munich in late 1991), the European autonomy scenario with both *Columbus* and *Hermes* was in question due to the increasing estimated cost for this scenario. Pending decisions for the autonomy scenario to be taken at the next ministerial conference in 1992, phase C1 was extended by a transition phase in 1992, and then extended into 1993, instead of going into full development as planned by the commencement of phase C/D. Thus work on the ESSS design continued at a slower

**Figure 11.2.19** Fan/pump/separator breadboard unit (Technofan).
From Archive Skoog.

pace and breadboarding was extended, pending the decision on the *Hermes* programme.

## 11.3 THE FEASIBILITY OF JOINT SPACESUIT DEVELOPMENT—EVA 2000

With the increasing complexity and prolonged schedule of the autonomous European scenario with both *Columbus* and *Hermes*, new ways were sought to save the programmes, which were becoming unaffordable. In the meantime, initial operation of the ESSS on board *Columbus* and *Hermes* was postponed to the time frame 2002–2004. A similar situation (i.e., a space programme under severe financial constraints) emerged in the new Russian Federation. The ESA Ministerial Conference in Munich decided on closer cooperation with the Russian Federation (in particular for the *Hermes* programme).

In the Russian space programme of 1992, EVAs from *Mir* were undertaken wearing the ORLAN-DMA semi-rigid suit, the origin of which reached back to the 1970s. It had since undergone several modifications to cover the needs of extravehicular work on *Salyut* and *Mir*. At that time, the latest modification dated back to 1988, when the suit was modified to guarantee completely self-contained operation. Thus, Russian spacesuit experts were considering the development of a new suit to meet the future demands of *Buran* and the *Mir-2* station that was under construction.

**Figure 11.3.1** The European–Russian EVA 2000 team on 14 April 1992 at the kick-off of the EVA 2000 study at Dornier in Friedrichshafen. *From left to right*: J. Hernandez (ESA), A. Accensi (ESA, Project Manager), V.I. Svertshek, A. Thirkettle (ESA, Programme Manager), G.I. Severin (Zvezda, General Director), E.B. Ignatova (Zvezda, interpreter), E. Kühnle (Dornier), B.V. Michailov (Zvezda), I.P. Abramov (Zvezda, Project Manager), A.I. Skoog (Dornier, Project Manager), K. Fahlenbock (Dornier), Th. Kleinbub (Interpreter), Rh.K. Sharipov (Zvezda), R. Hause (Dornier), N. Herber (Dornier).
From Archive Skoog.

The past two years of close cooperation with Zvezda led Dornier/DASA to jointly propose to ESA a study to analyse the possibility of joint European/Russian spacesuit development to cover future needs, especially as the two envisaged concepts had a lot in common. In early 1992 the ESA and the newly created Russian Space Agency (RKA) agreed to initiate a requirements analysis and conceptual design study to determine the feasibility of joint spacesuit development—the EVA 2000 (Figure 11.3.1). The feasibility study was conducted in 1992 by Dornier/DASA and RD&PE Zvezda, with Dassault and Laben as subcontractors, as a joint ESA/RKA undertaking (Skoog et al., 1995). If found feasible the new suit would replace *Hermes*' EVA suit's ESSS and be used for the first time on *Mir-2* at the turn of the century.

Merging the future mission scenarios of ESA and RKA resulted in a large scenario of spaceplanes (*Buran* and *Hermes*), space stations (*Mir-2* and *EMSI*) and platforms (Figure 11.3.2), encompassing both short-duration (ground-based) flights and long-duration (space-based) ones, with a minimum of changes in suit designs for both applications. The EVA 2000 spacesuit design had to be compatible

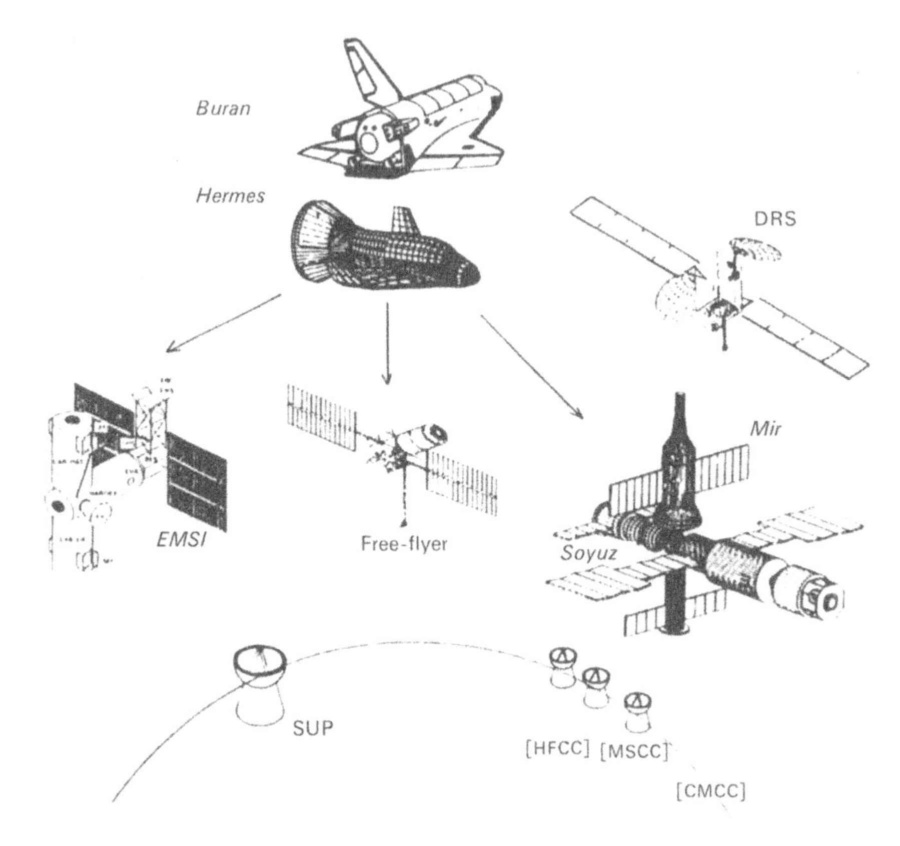


**Figure 11.3.2** EVA 2000 mission scenario.
From Archive Skoog.

with both the ESA EDRD (EVA Design Requirements Document H-S-2BA-001-ESA) and the Russian requirements for *Mir* and *Buran*. The overall EVA 2000 requirements specification (Table 11.2a, b), as defined in the course of the study, shows (besides the varying mission duration):

- a suit pressure (420 hPa) that was lower than original *Hermes* requirements (500 hPa) but slightly higher than for ORLAN-DMA (400 hPa);
- an emergency hatch diameter of 0.8 m; and
- a crew population made up of European and Russian male and female cosmonauts.

Compared with the ESSS design, the design of the EVA 2000's suit enclosure (Figure 11.3.3) was able to dispense with the rolling convolute shoulder joint and the hip and thigh bearings, due to the lower suit pressure. Removal of the rolling convolute joint was also necessary to meet the emergency hatch requirement for the given anthropometric range. Instead, a calf bearing was introduced to maintain leg mobility. Adjustments for individual crew members was achieved by means of axial lacing adjustments of the suit's limbs. The HUT came in one size and spinal growth due to zero gravity was accommodated by adjustment of the waist joint and by the conical shape of the shoulder flanges. These changes also reduced the overall mass to below the given requirement. The LSS (Figure 11.3.4) and the electrical and communication equipment were integrated into the backpack and the chestpack. For the backpack, two alternatives were designed (Figure 11.3.5a, b) for later final selection. In the backpack, the electrical equipment and high-pressure oxygen equipment were held in unpressurized compartments and the ventilation and gas regeneration equipment ($CO_2$ removal, sublimator and sensors) along with the water-cooling equipment in the pressurized part. This was more in line with the ORLAN-DMA design than the ESSS, where all life support and electrical equipment was located in the unpressurized backpack. The overall electrical configuration (Figure 11.3.6) was the same as the ESSS configuration.

The fail operational/fail-safe (FO/FS) requirement and suit reliability of 0.999 were met by a number of redundancies in the suit enclosure (e.g., dual bladder, double visor, double seals for door and pressure bearings, and pressure cuffs in case of glove leakage), life support (e.g., redundant fan, pump, pressure regulator, oxygen supply, feedwater filter, $CO_2$ sensor and moisture collector) and electrical (redundant batteries, power supply lines, emergency caution and warning, and communication) equipment.

The EVA 2000 spacesuit system's mass (minus the spacesuit/mothercraft interface unit) was 121.7 kg and overall power consumption was estimated at

**Table 11.2a.** EVA 2000 mission requirements.

| Title | Short duration (ground-based) | Long duration (space-based) |
|---|---|---|
| Place of maintenance | Ground | Space |
| Life support in spacecraft | Non-regenerative ($CO_2$ removal and $H_2O$ by resupply) | Regenerative (partial) |
| Overboard dumping from spacecraft possible | Yes | No |
| Sorties (two persons) | 2–3 per flight | 50/year |
| Sortie duration | 6 + 1 hours | 6 + 1 hours |
| Cabin pressure | 1,013 hPa | 1,013 hPa |
| Resources pre-/post-flight | On board | On board |
| Operational life of suit (years/no. of sorties) | 15/35 | 5–7/25 |
| Orbit | 8–65°/300–550 km | 8–65°/300–550 km |
| Mass | $<$125 kg | $<$125 kg |
| Failure tolerance | FO/FS | FO/FS |
| Reliability of suit | 0.999 | 0.999 |
| Quantitative safety target | 0.9995 | 0.9995 |
| Hatch diameter | Min ∅ 0.96 m (0.8 m in emergency case) | Min ∅ 0.96 m (0.8 m in emergency case) |

**Table 11.2b.** EVA 2000 requirements specification.

| | |
|---|---|
| Autonomous sortie duration (umbilical disconnected) | 7 hours total; including 6 hours for planned work, plus 1 hour for contingency work |
| Number of EVA crew members (nominal) | 2 |
| Crew sex | Male and female |
| Re-establishment of fully autonomous capability | 2–14 hours |
| Short duration | Replacement of consumables |
| Long duration | Replacement and refill/recharge of consumables |
| Nominal operating pressure | 420 + 10/−20 hPa |
| Redundant bladder actuation | 340 hPa |
| Emergency pressure actuation | 270 hPa |
| Suit atmosphere | 95% pure oxygen |
| Emergency function | 30 minutes minimum |
| Operational life | |
| —short duration | 15 years and 35 sorties |
| —long duration | 5–7 years and 25 sorties |
| Anthropometric height | From 165 to 187.5 cm |
| Anthropometric chest circumference | From 91 to 108 cm |
| Autonomous unit wet mass | <125 kg |
| Donning/Doffing | Back entry; self-donning and doffing |
| $CO_2$ level (oro-nasal area) partial pressure | 10 hPa nominal; 27 hPa maximum for high metabolic rates and short periods |
| Autonomous power (two batteries, half size each) | 27 + 7/−3 VDC (Volt DC); $LiSOCl_2$ |
| Radio frequency | 400–470 MHz TDMA |
| RF bandwidth | TBD |
| Transportation vehicle | *Hermes*, *Buran*, *Progress*, *Mir* |
| Mothercraft for EVA | *Hermes*, *Buran*, *Mir*, *EMSI* |
| Metabolic heat removal from crew member | |
| —average for 60 minutes | 470 W |
| —average for sortie | 300 W |
| —maximum for 30 minutes | 580 W |
| —minimum for TBD minutes | 80 W |
| Mothercraft interfaces | Power 27 + 7/−3 VDC; oxygen for pre-/post-EVA; cooling water for pre-/post-EVA; data and communications provision; storage place |

69 W. The suit enclosure weighed 27.2 kg, life support subsystem 57.6 kg, electrical and communication subsystem 28.3 kg and consumables 8.6 kg.

The short-duration/ground-based design envisaged the use of replaceable LiOH cartridges for $CO_2$ removal, non-rechargeable batteries and replaceable oxygen tanks. This design required further improvements in technology for the glove, combined fan/pump and the battery. The long-duration/space-based design

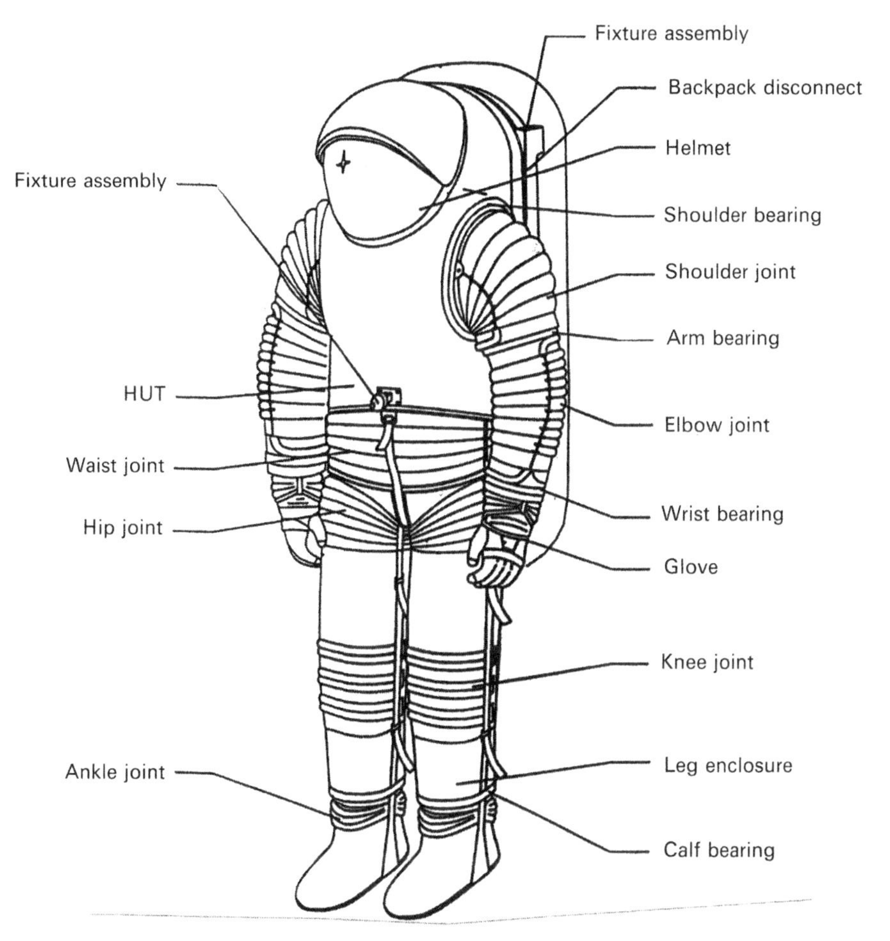


**Figure 11.3.3** EVA 2000 suit enclosure design.
From Archive Skoog.

utilized a metal oxide $CO_2$ removal system, rechargeable lithium batteries and oxygen tanks. This design required advances in technology with respect to metal oxide $CO_2$ removal, sublimator/heat exchanger that had a regenerative heat accumulation capability and rechargeable lithium batteries.

The feasibility study of the EVA 2000 proved the technical viability of a possible joint European/Russian spacesuit design that incorporated both Russian operational experience and the advanced technology of European industry. The need for development and training facilities was also investigated and the best possible use of available Russian facilities proposed.

The constant updating of requirements and generation of a revised reference concept for the spacesuit was accompanied by programmatic evaluation of the joint

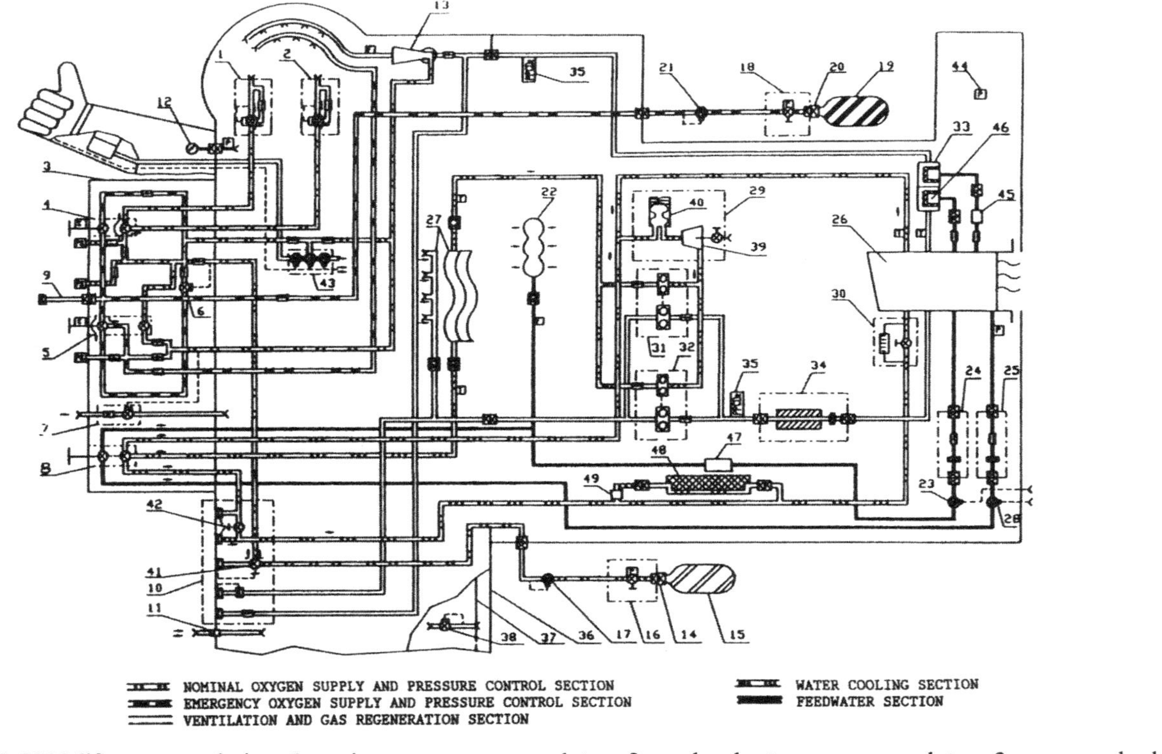


**Figure 11.3.4** EVA 2000 life support design: 1—primary pressure regulator; 2—redundant pressure regulator; 3—pneumohydraulic control panel; 4—pressure switch and reserve oxygen supply actuation assembly; 5— injector and oxygen supply actuation assembly; 6—pneumatic valve for automatic actuation of reserve oxygen supply; 7—back-up pressure valve; 8—handle for thermal control and feedwater supply actuation (combined); 9—emergency hose; 10—combined connector; 11—relief valve; 12—pressure gauge; 13—injector; 14—high-pressure disconnect with a built-in check valve; 15—primary oxygen bottle; 16, 18—shut-off valve with charging connection and transducer; 17, 21—two-stage reducer; 19—secondary oxygen bottle; 20—reserve oxygen supply; 22—feedwater tank (two FW tanks having 1/2 capacity can be used); 23—FW pressure regulator (redundant line); 24—flow rate limiting filter (redundant line); 25—flow rate limiting filter (primary line); 26—sublimator–heat exchanger; 27—Liquid–cooled and ventilation garment; 28—FW pressure regulator (primary line); 29—hydroaccumulator/separator assembly; 30—cooling circulation indicator; 31—fan/pump assembly (primary); 32—fan/pump assembly (redundant); 33—moisture collector (redundant); 34—$CO_2$ and contaminant control cartridge; 35—$CO_2$ sensor; 36—SS primary bladder; 37—SS redundant bladder; 38—aneroid (bypass valve); 39—bubble separator; 40—hydroaccumulator; 41—valve to supply oxygen via an umbilical or to block it during EVA; 42—valve to supply cooling water via an umbilical or to block it during EVA; 43—$O_2$ supply to the cuffs actuation assembly; 44—in-suit pressure sensor; 45—electromagnetic switch; 46—primary moisture collector; 47—electromagnetic switch (redundant FW supply line); 48—ice block (space-based concept only); 49—valve for ice block actuation (space-based concept only).

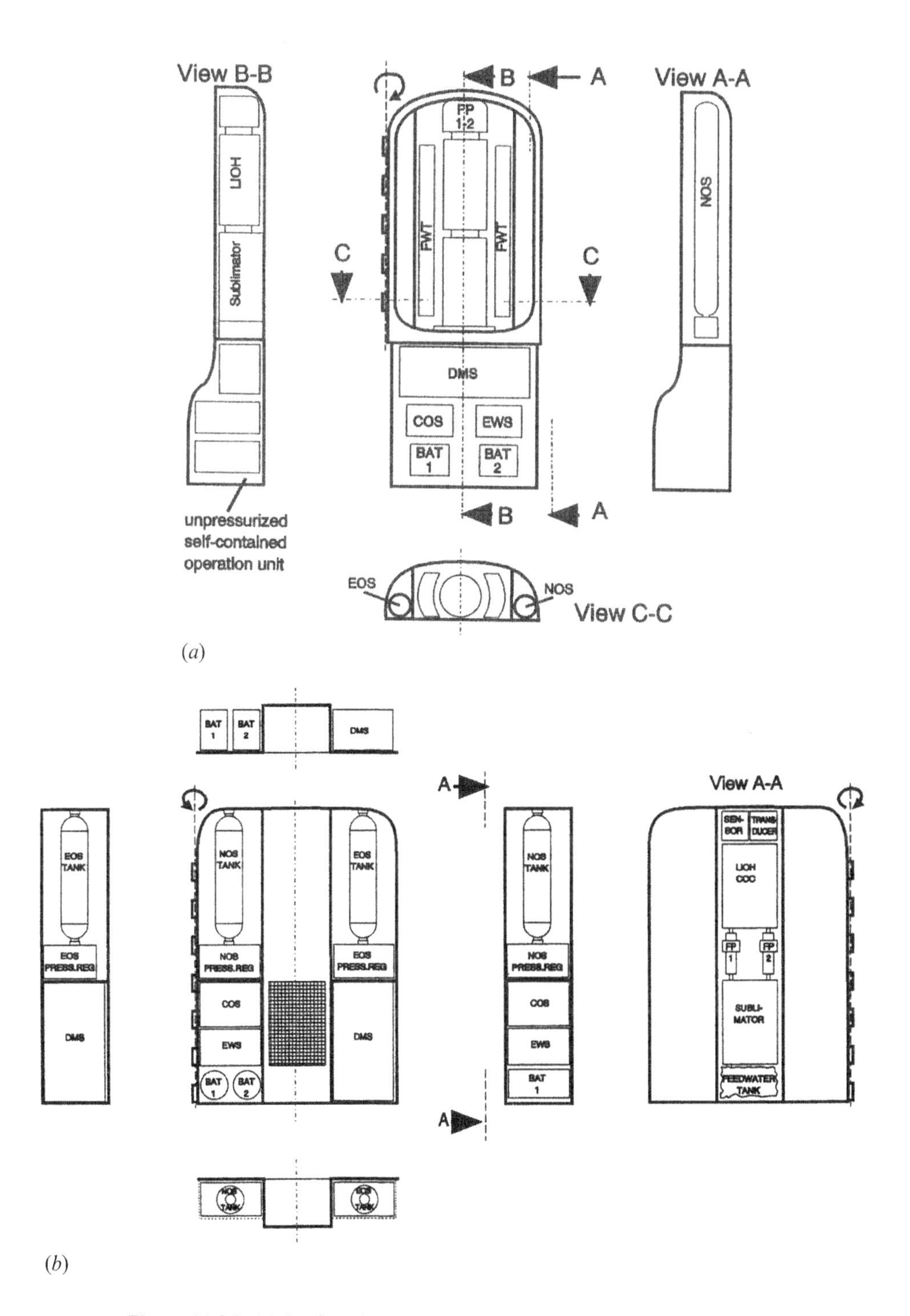


**Figure 11.3.5** (*a*) Backpack configuration alternative I; (*b*) alternative II.
From Archive Skoog.

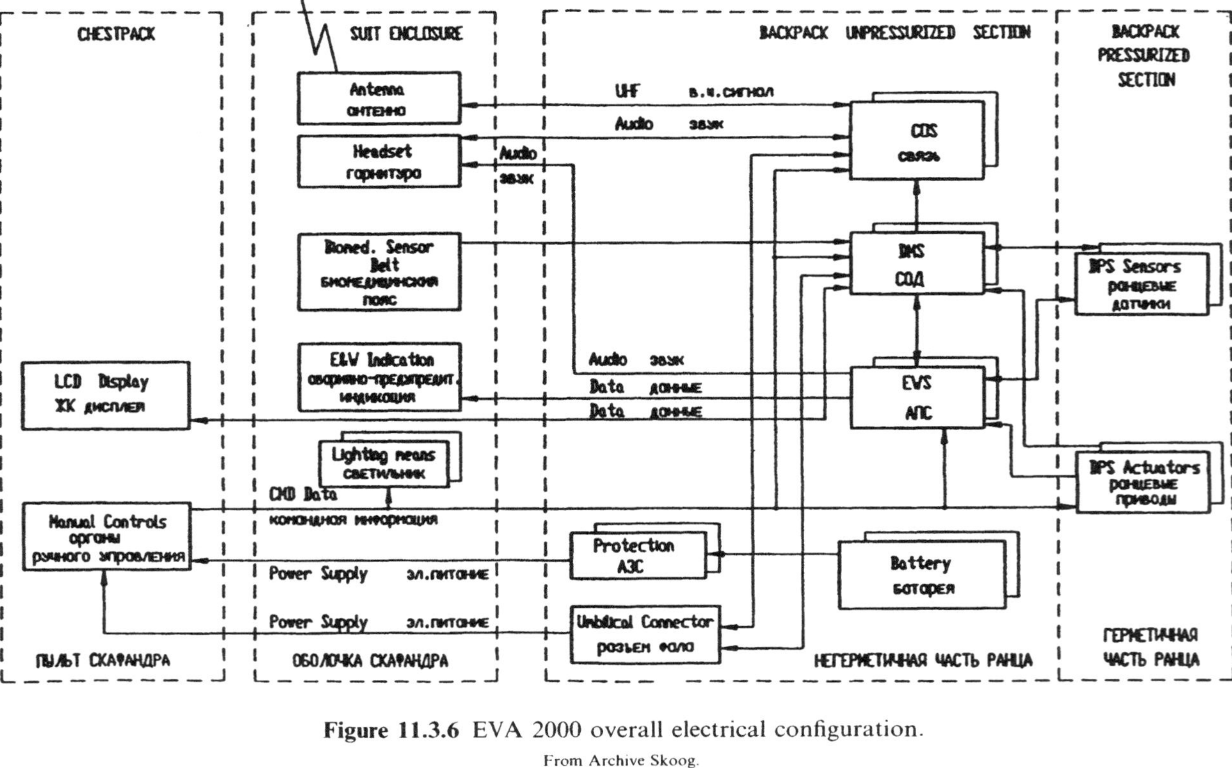


**Figure 11.3.6** EVA 2000 overall electrical configuration.

From Archive Skoog.

undertaking and a formal industrial proposal to make the EVA 2000 a joint European/Russian spacesuit development for the next generation of manned missions (starting around the turn of the century).

With the decisions of the ESA Ministerial Conference in Granada in late 1992 and the results of the ESA/RKA negotiations on future cooperation for spacesuit development completed in early 1993, the EVA 2000 concept had by late 1993 turned into a preliminary development contract with new mission requirements. The joint development was to be undertaken by a European/Russian industrial team under the new project name EVA SUIT 2000. In Europe this replaced the ESSS work, which was transferred along with all its hardware into the EVA SUIT 2000 project.

## 11.4 THE WEST EUROPEAN–RUSSIAN JOINT SYSTEM EVA SUIT 2000

The two years following the 1991 ESA Ministerial Conference in Munich saw several changes in direction for both the *Columbus* and *Hermes* programmes, with consequences for European spacesuit development. After the Munich conference, the *Hermes* programme was examined to see whether there was a chance of possible cooperation with Russia for a future, winged, manned space vehicle. After the Granada ministerial conference in November 1992 the *Hermes* programme was merged into the ESA Manned Space Transportation Programme (MSTP) with the intention of defining a future, manned, space transportation system in cooperation with Russia. However, the envisaged cooperation with Russia on a common winged vehicle became unrealistic, when it emerged that Russia was not financially in a position to participate in the development of a new vehicle. In the meantime the Russian *Buran* programme had also been terminated.

At the Granada meeting, *Columbus* MTFF was abandoned, and in early 1993 (after NASA redesigned the *ISS*) the *Columbus* Polar Platform was transferred to the ESA Earth Observation Programme. For *Columbus*'s remaining Attached Pressurized Module (APM), a bridging phase started in April 1993 to support NASA's space station redesign work. The bridging phase ended in late 1993 with the definition of a new baseline called the *Columbus* Orbital Facility—COF (COF being the only remaining element of the *Columbus* programme).

Therefore, when the joint ESA/RKA spacesuit project EVA SUIT 2000 began in late 1993, the intended use of the suit was on board *Mir-2*, as there were no longer any European *ISS* items of equipment for EVA servicing. The EVA SUIT 2000 was then one of only two cooperation efforts for *Mir-2*, the other being the European Robotic Arm (ERA). In December 1993, the Russian Federation joined the *ISS* programme as a full partner, and as a consequence the items of equipment under development for *Mir-2* were reoriented to the Russian segment of the *ISS* and *Mir-2* was abandoned.

As for the EVA SUIT 2000 system (jointly defined and initiated by ESA and RKA), it was concluded that (ESA Manned Space Programme Board, 1994):

**Figure 11.4.1** The meeting between representatives of ESA, RKA, NASA and industry at RKA in Moscow on 10 February 1994. *From left to right as seated*: B. Kirby (Hamilton Standard), J. Faszcza (Hamilton Standard), F. Morris (Hamilton Standard), S. Kuli (RKA, *turning backwards*), V.I. Svertshek (Zvezda, *standing*), C. McCullough (NASA JSCV), K. Hudkins (NASA HQ), V. Razgulin (Energia, *obscured*), G.I. Severin (Zvezda), I.P. Abramov (Zvezda), G.A. Rykov (Zvezda), S. Chernikov (RKA), A. Thirkettle (ESA), A. Accensi (ESA, *obscured*), D. Isakeit (ESA), A.I. Skoog (Dornier), R. Schaefer (Dornier).
From Archive Skoog.

- the system could efficiently become a joint development with products and work packages being shared between European and Russian industry under joint management of ESA and RKA, *each funding its own activities*;
- the RKA determined that they needed such an advanced suit for the assembly phase of their part of the *ISS*, causing a need date on orbit of April 1998;
- with only a single station to be serviced, the possibility of a common suit was studied. The result of these discussions was an agreement between the nominated representatives of ESA, RKA and NASA (and the respective industrial prime contractors) that the ESA/RKA EVA SUIT 2000 would be the baseline suit for the Russian contribution to the station, and could be the basis for a future interoperable system.

This ESA/RKA/NASA agreement was signed by all parties in a meeting at RKA in Moscow on 10 February 1994 (Figure 11.4.1). This agreement was followed by the industry/agency proposal for EVA spacesuit interoperability capability (Skoog et al., 1995).

The EVA SUIT 2000's development programme was split into two parts. Part 1 covered the first two years of detailed definition studies and pre-development work that included a successful completion of the preliminary design review (PDR). The baseline for the work was the joint ESA/RKA EVA system requirements document (HS-RQ-EV-001-ESA/RKA) and the preliminary concept established in the EVA

2000 study. Zvezda and Dornier/DASA acted as co-prime contractors, with a number of European companies involved as subcontractors (Figure 11.4.2). Responsible for the suit enclosure subsystem was SABCA in Belgium, the avionics subsystem Laben in Italy, the power subsystem Signaal in The Netherlands. Zvezda was, in addition to its system co-prime contractor work, responsible for the life support subsystem and delivery of the suit enclosure's soft parts (Moeller et al., 1995).

The system requirements were the same as those for the EVA 2000, except that the mission scenario was soon narrowed down to the space-based configuration. In a space-based scenario, the suit would typically remain aboard the space station for about 4 years supporting up to 30 sorties, then be retrieved for ground maintenance including replacement of items whose life expectancy had passed and afterwards delivered back into orbit, giving a total useful life of 10 years. Furthermore, anthropometric height was reduced to a maximum of 185 cm, with other requirements remaining unchanged. By summer 1993 Zvezda had already completed manufacturing a suit model with a new flat pattern shoulder and arm bearing for ergonomic testing. This prototype (Figure 11.4.3) was based on the ORLAN-DMA suit's enclosure components as far as possible. One major change was the shape of the HUT. To facilitate better entry in zero gravity and to increase waist mobility, the separation plane of the HUT was slightly inclined upward in the front and the rear entry door was moved up about 5 cm. This prototype helped verify the new shape for improved mobility, which was then implemented in the EVA SUIT 2000 configuration.

The EVA SUIT 2000's overall configuration (Figure 11.4.4 and Table 11.3) also incorporated a backpack with all the life support equipment (Figure 11.4.5) that needed maintenance inside the pressurized upper part (LSS) and batteries and communication equipment in a lower (auxiliary power system—APS) unpressurized compartment (similar to the ORLAN-DMA configuration). The EVA SUIT 2000's system comprised two spacesuits and the station's on-board support equipment, with interfaces to the host vehicle and necessary equipment for servicing, maintenance and checkout of the suits. Displays and control units were chest-mounted. The EVA SUIT 2000 used the rear entry concept, and the HUT was closed by a back door, to which was attached all the essential life support equipment. Following normal space-based procedures, the on-board support equipment provided all specific functions that were only necessary during airlock operations. Oxygen, power, and water for cooling were recharged and the metal oxide $CO_2$ removal cartridges regenerated by on-board support equipment. In case the mothercraft could not provide for regeneration or recharging, the suit design provided the flexibility of replacing these components in orbit with non-rechargeable ones.

The backpack mock-up (Figure 11.4.6) and the ergonomic model of the suit enclosure (Figure 11.4.7) was completed in September 1994 and extensive testing of the ergonomic model with the backpack in position took place in November–December 1994 at SABCA in Belgium. The backpack mock-up served the purpose of the integration and maintenance checkout model at Zvezda and was then shipped to Dornier and SABCA for use in some of the tests with the ergonomic model.

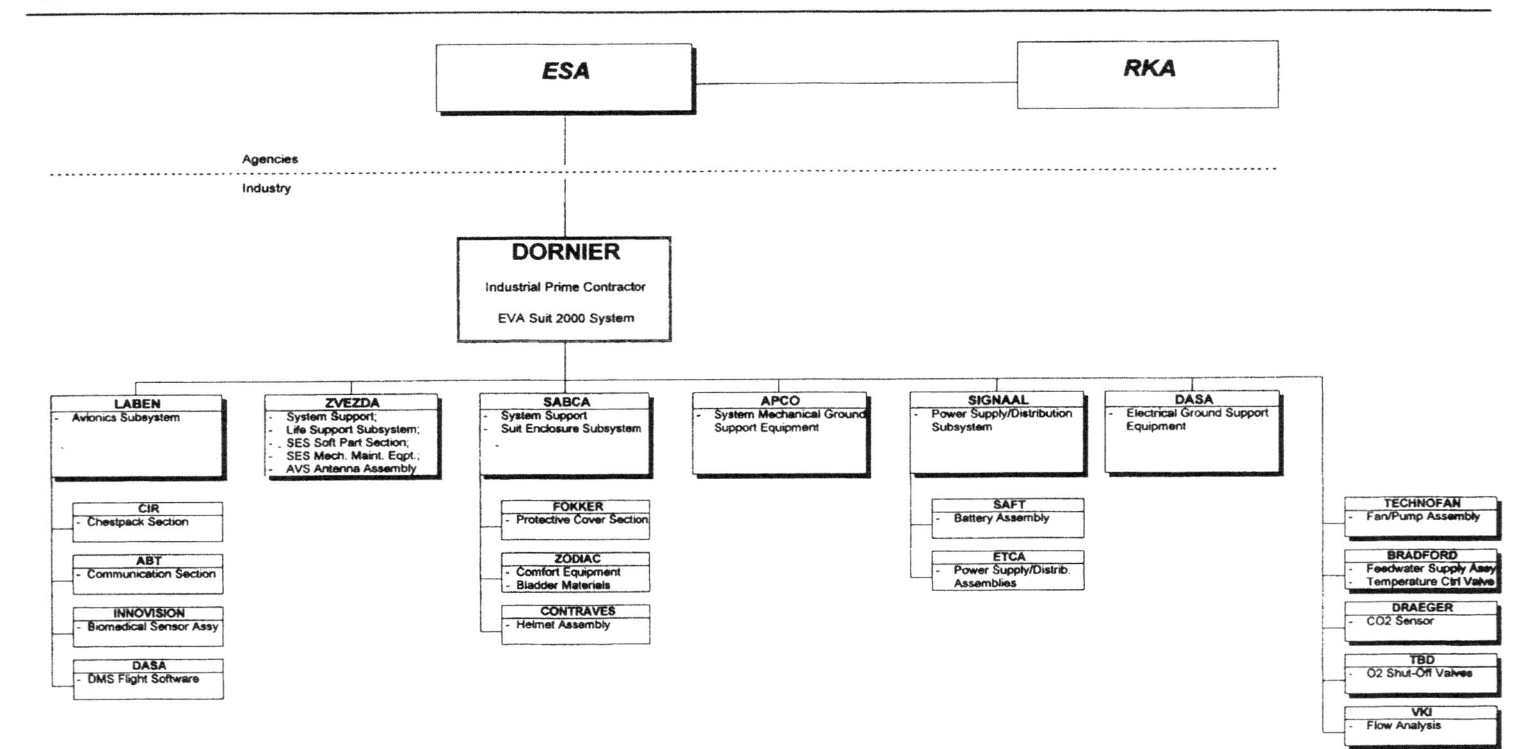


**Figure 11.4.2** EVA SUIT 2000 work distribution.

From Archive Skoog.

**Figure 11.4.3** Initial Zvezda prototype for arms and rear entry testing.
From Archive Skoog.

The ergonomic model comprised: an upper torso made of a composite material and a metallic rear door from SABCA; a polycarbonate helmet from Contraves in Switzerland; arms, gloves and trousers manufactured by Zvezda and a chestpack manufactured by Dornier. The model's soft parts were adapted from the ORLAN-DMA parts with EVA SUIT 2000 bearings. The tests, performed at a suit pressure of 400 hPa, included donning/doffing evaluation, sizing, mobility assessment, arm reach envelope definition, range of vision definition, comfort, control panel and door closure handling (Figure 11.4.8). The adequate design of the test bench and appropriate test procedures allowed the influence of 1-g conditions on the test results not

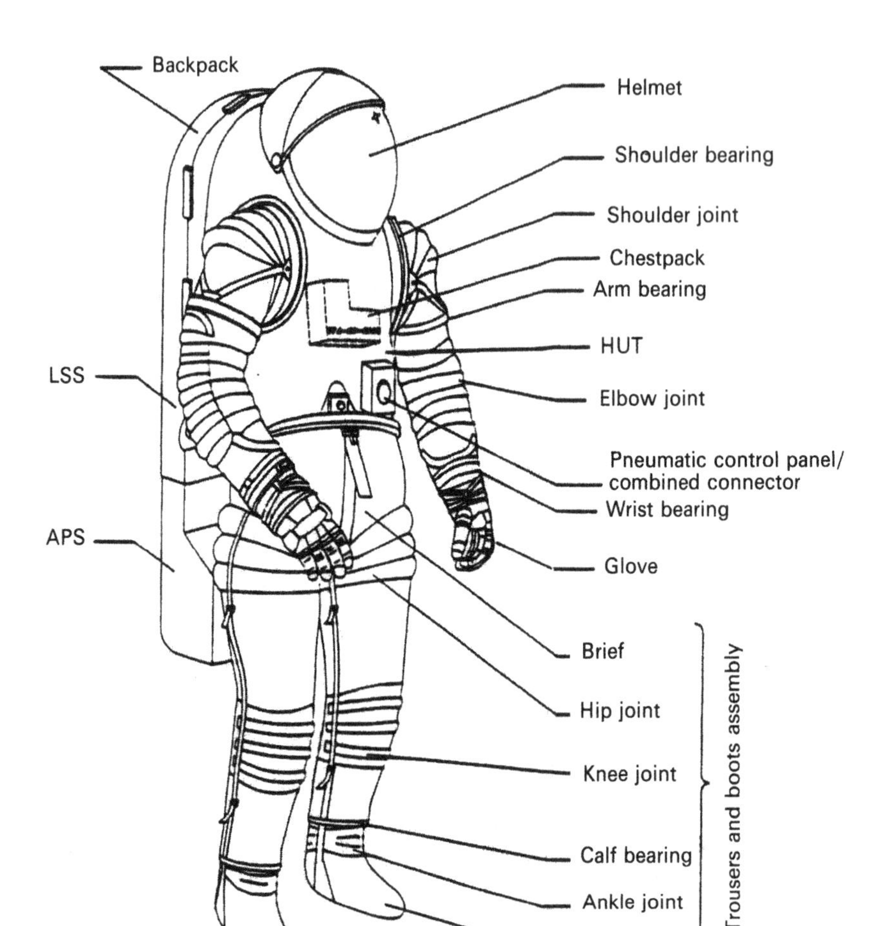


**Figure 11.4.4** EVA SUIT 2000 overall configuration.
From Archive Skoog.

to affect most of the major performances of the suit, with the only significant restriction of manned testing in such conditions being for leg mobility.

Performance testing of the ergonomic model verified the design objectives:

- the EVA SUIT 2000's design allowed for a much wider range of anthropometric sizes than the ORLAN-DMA or ESSS phase C1 suits;
- torso width and backpack length was increased, still allowing passage through the emergency hatch, but the longer backpack would probably reduce leg mobility (especially for bending backward);
- donning/doffing performance was maintained despite the lower inclination of the shoulder bearing;
- visibility performance got very close to meeting the set requirements fully;

**Table 11.3.** The EVA SUIT 2000's reference configuration as of December 1994.

| | |
|---|---|
| Ergonomy | Cosmonaut height from 165 to 185 cm; range of vision ±120° horizontal (90° up, 70° down); visibility of information and suit control; dexterity/tactility >50% of bare hand capability |
| Mobility | Moderate physical effort (soft joints, self-sealing bearings) |
| Protection | Thermal; UV, ionizing radiation, mechanical hazards; $\mu$-meteorites (probability of penetration <0.005) |
| Donning/Doffing | Single crew performance, rear entry |
| Oxygen storage and supply | Pure oxygen, 0.74 kg $O_2$ for one sortie; 1.3 kg oxygen for a 30-minute emergency return |
| Suit pressure control | 420 hPa nominal, in emergency cases $\leq$ 420 hPa |
| Atmosphere revitalization | $CO_2$ and trace contamination removal (regenerative metal oxide cartridge, LiOH cartridge back-up), humidity control 25–70%, suit ventilation, purging of nitrogen |
| Temperature control | Waste heat rejection (up to 950 W), adjustable comfort conditions |
| Crew member support | 0.5 l drinking water, hygienic slips |
| Information management | Failure detection, isolation and recovery (FDIR); caution data; system control data; resources data; telemetry data, biomedical data |
| Crew health monitoring | Active monitoring of critical biomedical functions (i.e. ECG, skin temperature, breathing frequency) |
| Information display | LCD display in chestpack |
| Communication | 3 node TDMA communication network; range for nom/back-up frequency: 400–420 MHz; 2 kbits/s data transmission; 32 kbit voice |
| Power supply | Rechargeable LiC type batteries; 2 × 15 Ah at 29 V average for 1 sortie |
| Power distribution | Unregulated power bus of 28 VDC ± 4 V |
| On-board support | Power, oxygen, cooling water for pre-/post-EVA operation via umbilical; on-board maintenance equipment |
| Failure tolerance | Fail operational/fail safe (by use of redundant equipment) |
| Caution and warning | Independent emergency warning function; caution function part of data management subsystem (DMS) |
| Failure detection, isolation and recovery (FDIR) | Combined approach: DMS-S/W and dedicated hardware |
| System check-out | Semi-automatic C/O under DMS control |
| Maintenance approach | In-orbit maintenance by ORU exchange; ground maintenance after each mission (30 sorties or max 4 years) |

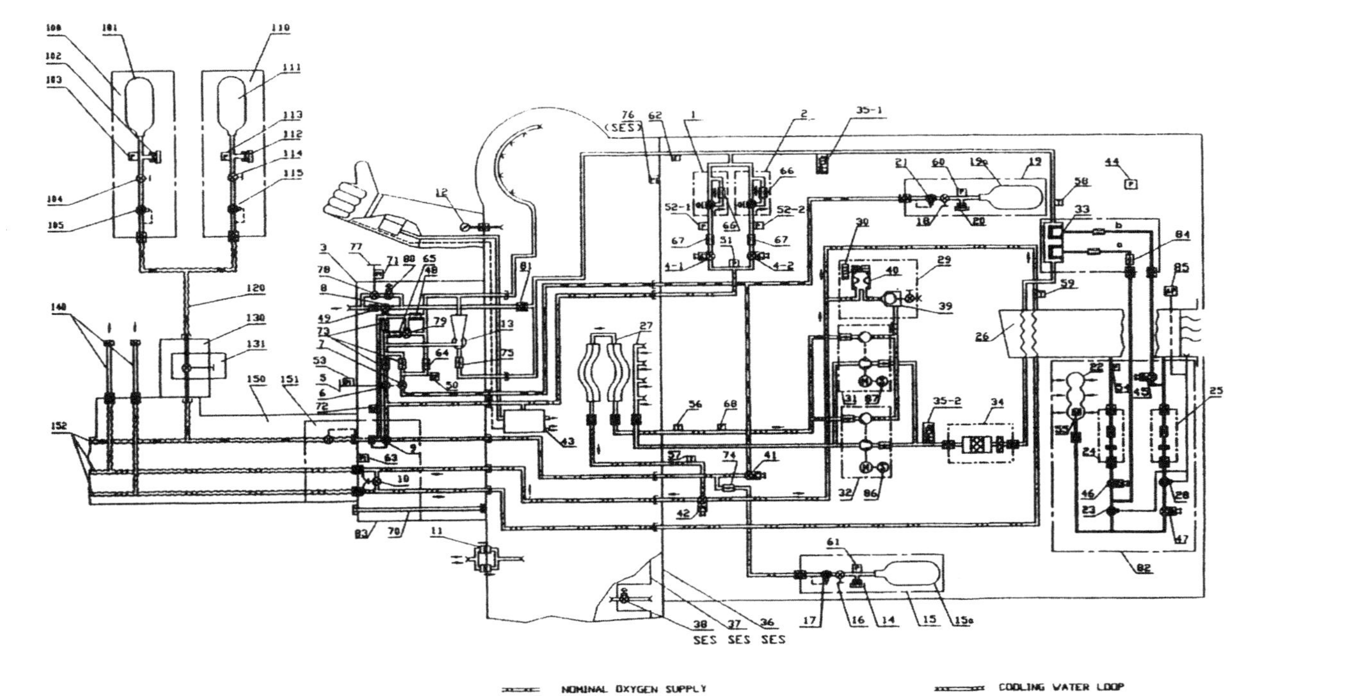


**Figure 11.4.5** EVA SUIT 2000 life support system configuration: 1—suit pressure regulator (primary); 2—suit pressure regulator (redundant); 3—pneumatic control panel; 4-1—electrical oxygen shut-off valve (primary); 4-2—electrical oxygen shut-off valve (redundant); 5—oxygen supply three-position handle; 6—injector actuation valve; 7—emergency oxygen supply valve; 8—pneumatic shut-off valve (dump line); 9—oxygen supply selection valve; 10—cooling water bypass valve; 11—positive/negative overpressure relief valve; 12—suit pressure gauge; 13—injector; 14/20—oxygen fill connector (part of items 16/18); 15—oxygen storage assembly (nominal); 15a/19a—oxygen bottle; 16/18—manual oxygen shut-off valve; 17/21—high-pressure reducer; 19—oxygen storage assembly (emergency); 22—feedwater tank; 23—feedwater pressure regulator (primary); 24—flow limiter/filter (primary line); 25—flow limiter/filter (redundant line); 26—sublimator; 27—cooling garment assembly; 28—feedwater pressure regulator (redundant); 29—accumulator/gas separator; 30—accumulator capacity indicator; 31—fan/pump assembly (redundant); 32—fan/pump assembly (primary); 33—moisture separator/remover; 34—contaminants control cartridge; 35-1—carbon dioxide sensor (at CCC outlet); 35-2—carbon dioxide sensor (at CCC inlet); 36—suit primary bladder (part of SES); 37—suit redundant bladder (part of SES); 38—redundant bladder actuation valve; 39—gas separator; 40—accumulator; 41—emergency oxygen switching valve; 42—temperature control valve; 43—cuff pressurization assembly; 44—suit absolute pressure transducer; 45—electrical water shut-off valve (of the redundant moisture removal line); 46—electrical water shut-off valve (of the primary feedwater and moisture removal lines); 47—electrical water shut-off valve (of the redundant feedwater line); 48—calibrated orifice (for oxygen supply during pre-breathing); 49—relief/dump valve.

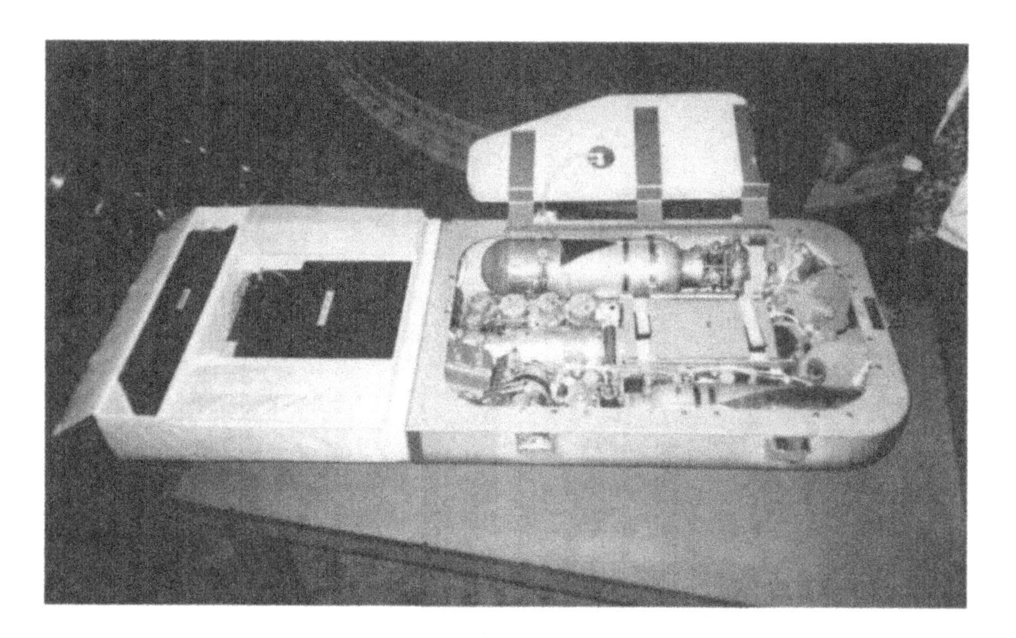

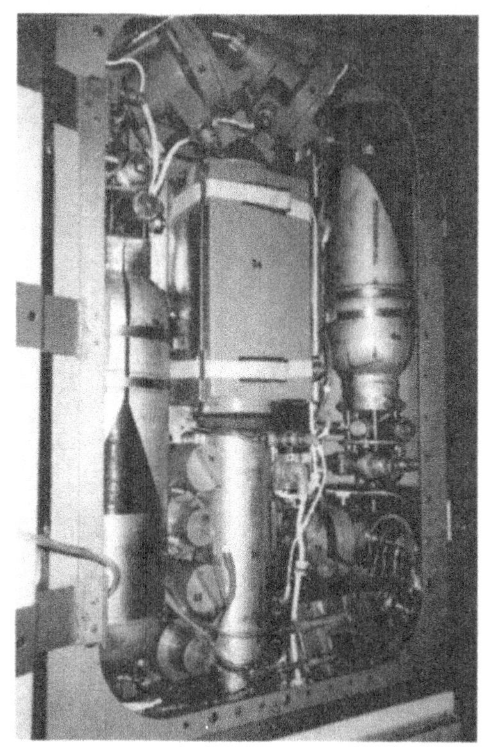

**Figure 11.4.6** Backpack mock-up.
From Archive Skoog.

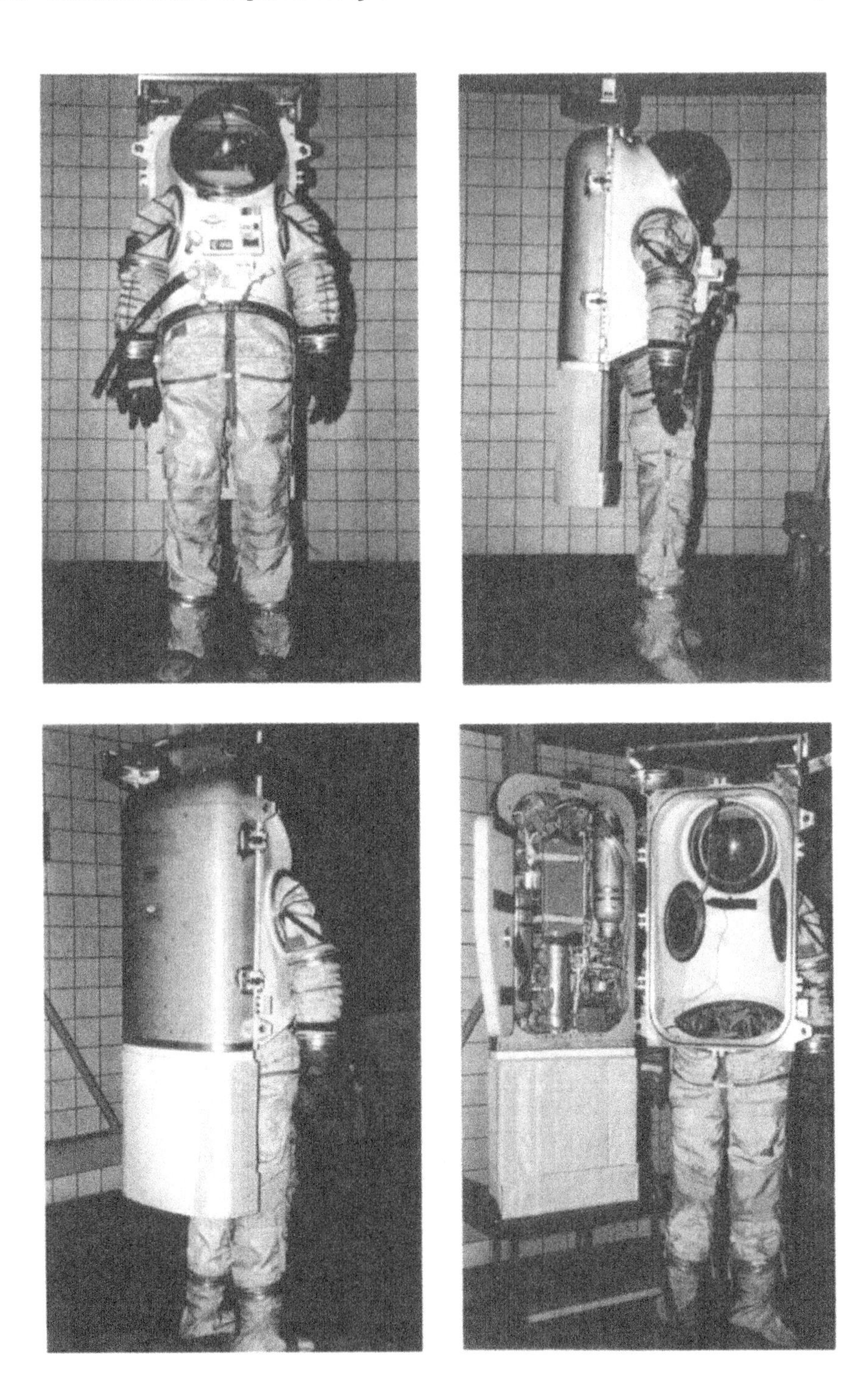

**Figure 11.4.7** Ergonomic model of the suit enclosure.
From Archive Skoog.

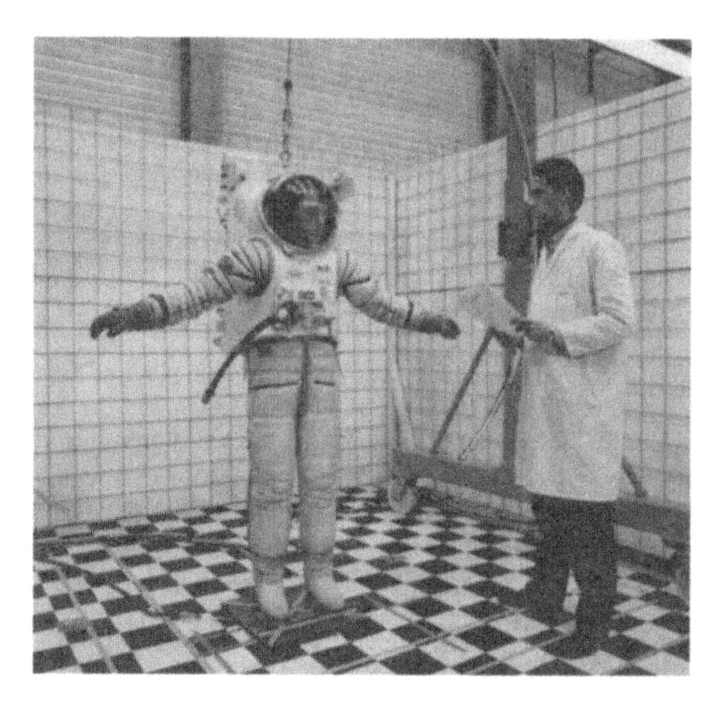

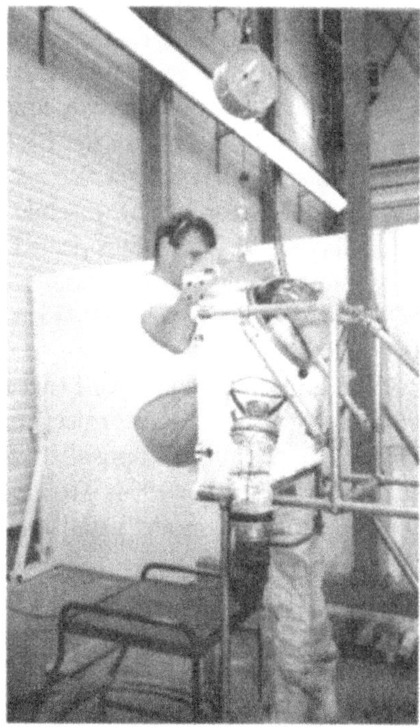

**Figure 11.4.8** Ergonomic model testing at SABCA in November 1994.
From Archive Skoog.

- forward and downward arm mobility was improved compared with the ESSS phase C1 suit, while upward and backward mobility lessened due to the new shoulder design;
- the work reach envelope requirement could be achieved with one hand;
- the handle design for door actuation and its location were assessed, as was the preferred helmet location (less unwieldy than the side handle on the ORLAN-DMA suit);
- performances of the man–machine interfaces were verified;
- the design and location of the chestpack and pneumatic controls allowed good control, visibility and actuation.

Testing of a second ergonomic model was planned to take place at Zvezda with the suit's limbs fully compliant with the EVA SUIT 2000 design, but the decision to end work on the EVA SUIT 2000 and the European spacesuit activates at the end of 1994 prevented these tests.

By the autumn of 1994 not only was the technical design concept complete but also all the necessary programmatic work was prepared to start phase C/D fully. The joint European/Russian industrial team (Figure 11.4.2) had prepared an overall design and development plan, a detailed verification plan, product assurance plan, the phase C/D schedule and the relevant cost estimates. ESA and RKA had signed agreements at the topmost level. The ESA/RKA Joint Project Plan (HS-PL-EV-0001-ESA/RKA, signed 20 May 1994) and the System Requirements Document (HS-RQ-EV-0001-ESA/RKA, signed 20 May 1994) for full development and operation had been generated and agreed. At that time this was the only project in which the RKA provided an equal contribution to that of ESA.

During negotiations on the EVA and ERA projects between ESA and RKA in 1993, it emerged that Russia also needed, as a result of its participation in the *ISS*, an advanced DMS for its *ISS* module. Detailed investigation of the Russian requirements showed that the European DMS, under development within the *Columbus* programme, would suit Russian needs, and it was agreed that Europe would deliver a DMS to Russia as a further item, leaving no financial room for the EVA SUIT 2000 programme.

The ESA Manned Space Programme Board at its meeting on 14–15 September 1994 approved the ERA project and directed a close-out of EVA work. Dornier provided ESA with a proposal entitled "EVA SUIT 2000, Part 1, Including Close-out" on 20 September 1994. Attempts were made to include a bridging phase (to put EVA in "dormant mode") until the next ESA Ministerial Conference in 1995, but these decisions were then confirmed at the meeting of the ESA 5th Manned Space Programme Board (ESA Manned Space Programme Board, 1994). The final contractual close-out meeting took place at Dornier/DASA in Friedrichshafen in December 1994. It is estimated that over €50m had been spent by ESA, national agencies and industry when the EVA work stopped at the end of 1994.

So, instead of entering its full phase (C/D) in 1995, the project was finally stopped at the end of 1994 after the first phase of testing of the ergonomic model. All documentation relating to review of the system's requirements was delivered to

ESA. All the hardware produced for breadboarding, including the ergonomic model, was transferred to ESA/European Space Science & Technology Centre (ESTEC). The ergonomic model was then later displayed at the ESTEC SpaceExpo Noordwijk Museum.

The work done jointly during phase C1 of the EVA SUIT 2000 was later used by Zvezda and RKA to modify the ORLAN-DMA suit into the ORLAN-M for the *ISS* (see Section 8.3). The industrial initiative for EVA spacesuit interoperability (Skoog et al., 1995) also suffered from the withdrawal of the Europeans from EVA spacesuit development. When work on the EVA SUIT 2000 was in question in 1994, development of the US *ISS* EMU continued, and when European activities were finally terminated the work was too advanced to be influenced by any potential joint effort with Russia. Currently, two different EVA suit systems (US EMU and Russian ORLAN-M) are used in the *ISS*. Moreover, the Russian party modified the ORLAN-M/*ISS* interfaces in such a way that the ORLAN-M can be used not only in the Russian airlock, but also in the common airlock of the *ISS*'s US segment (see Section 8.3) along with the EMU. This dual use is neither possible nor foreseen for the US EMU.

# 12

# Human physiological aspects in designing the EVA spacesuit*

The hostile environment of space for human beings necessitates a long list of extra-vehicular activity (EVA) suit design requirements to provide a safe environment for a cosmonaut to carry out work satisfactorily.

## 12.1 SUIT PRESSURE AND ATMOSPHERE COMPOSITION

Principal among these are barometric pressure and gas contents of the suit atmosphere. It is clear that both of these issues are interconnected and their specific parameters depend, on the one hand, on the necessity to maintain the gas exchange between man and the environment and, on the other hand, avoidance of the effects of direct vacuum environments on man. Such effects include:

- dysbarism (gas expansion in partially or completely closed cavities and organs as a result of rapid pressure decrease);
- emphysema in some organs and tissues;
- ebulism ("boiling");
- transformation of dissolved gases into a gas phase, resulting in pain and possible malfunctioning of the vital organs, due to gas embolism;
- decompression sickness (DCS).

The suit's barometric pressure depends on the selected suit operating pressure and is one of the most important parameters to be considered for the suit enclosure and the self-contained life support system (SCLSS) trade-offs. The best version from the physiological point of view is to maintain a suit pressure at the same value as

* This chapter was written by Professor A.S. Barer, Chief of the Aerospace Medicine Department, RD&PE Zvezda JSC.

the spacecraft's cabin. Although current spacecraft and orbiting stations have normal "Earth" atmosphere in the cabins, it is both technically difficult and inexpedient to use such an atmosphere in the suit as an operating pressure. In order to maintain mobility of the suited cosmonaut and improve some characteristics of the LSS, it is advantageous to have the minimum positive pressure, which is determined by man's need for oxygen. But, application of such a pressure, as will be shown below, increases DCS risk. Therefore, spacesuit pressure selection is a trade-off between cosmonaut needs and engineering solutions implemented in the suit design.

As is known, the principal mechanism ensuring transfer of oxygen in the human body is the partial pressure differential over the whole oxygen travel distance. Thus, maintenance of the gas exchange in the lungs and mainly the lung alveoli $O_2$ partial pressure ($PAO_2$) (necessary for saturation with oxygen) at a value close to that at ground level ($PAO_2 = 137$–$147$ hPa) is possible only if the barometric pressure of the suit's monogaseous oxygen atmosphere is kept at values no lower than 253–267 hPa.

The first preliminary calculation of $PAO_2$ versus barometric pressure was made in 1880 by the outstanding Russian physiologist I.M. Sechenov, who analysed the death of French aeronauts on board the *Zenith* balloon at an altitude of 8,600 m. Currently, $PAO_2$ is determined by the formula:

$$PAO_2 = (B - PH_2O)C - PACO_2\left(1 - C\frac{1-F}{F}\right)$$

where $B$ is barometric pressure; $PH_2O$ is the partial pressure of water vapour in the lungs (47 mm Hg at the normal human body temperature of 37°C); $PACO_2$ is the partial pressure of $CO_2$ in the alveoli; $C$ is percent of oxygen content in the environment gas and $F$ is the respiratory factor.

The suit's barometric pressure was set to 360–400 hPa at Zvezda. This was done with the purpose of minimizing unfavourable effects, such as DCS. The decision to use such a value for the suit's operating pressure was made in 1964 when Zvezda got down to developing the flight suit for the first ever space walk. This decision was preceded by a great deal of research and testing, with the participation of many volunteers (Figure 8.2.3).

With this selected suit pressure, the oxygen partial pressure in the alveoli undoubtedly exceeds that in the venous blood coming to the alveoli (53 hPa). This provides the conditions necessary for saturation of the blood with oxygen up to 120–127 hPa ($PAO_2$), owing to its diffusion into the bloodstream through the walls of blood vessels. Oxygen partial pressure levels over the whole path of oxygen transportation from the ambient atmosphere to human body tissues are shown in Figure 12.1.1. Another component of gas exchange in the lungs, also due to the partial pressure differential, is related to removal of $CO_2$, which is generated as a result of metabolism. Under natural conditions, $CO_2$ is removed because of the difference between its partial pressure in the blood coming to the lungs (53 hPa) and that in inhaled air (where its content is very small). Therefore, the requirement for minimum $CO_2$ content in the artificial atmosphere of the suit must also be made. Taking into account the known unfavourable effect of increased $CO_2$ concentration in inhaled gas, the value of its content in the suit atmosphere should not exceed 13.3 hPa, and

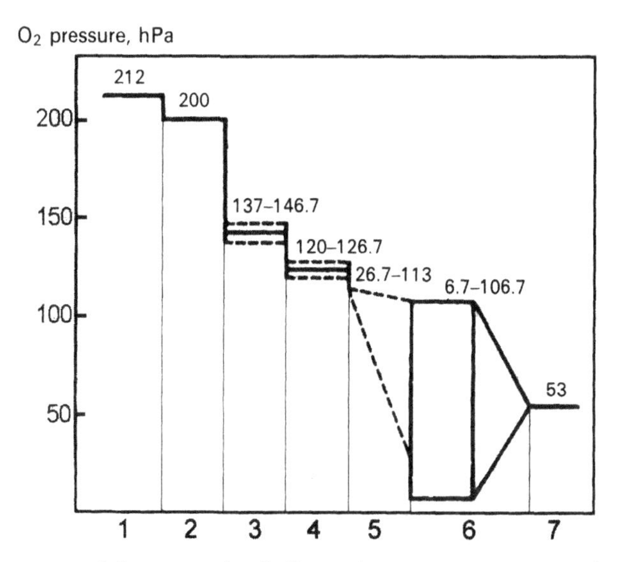


**Figure 12.1.1** Oxygen partial pressure levels for various oxygen transporation phases in the human body: 1—in atmospheric air; 2—in tracheal air; 3—in alveolar air; 4—in arterial blood; 5—in capillaries; 6—in intercellular liquid; 7—in venous blood.

From V.B. Malkin (1975).

only at the end of an EVA (when the $CO_2$ control cartridge—CCC—is almost fully used) should a short-term increase to 27 hPa be permissible.

Because of the damaging effect of low barometric pressure for the pressure range adopted for the Russian EVA suits (360–400 hPa) and its rate of change during airlocking, evaluation of the vacuum effect was tantamount to evaluation of the probability and prophylaxis of DCS. Experience gained from ORLAN suit ground tests and real operations, however, confirmed a high level of safety for the person working in the suit (exceeding 0.97 at the confidence level of 95%). There were no DCS symptoms during real EVAs (more than 200 sorties) and vacuum chamber experiments (about 800) using the adopted suit pressure and short-term pre-breathing.

The concept proposed by Zvezda's physiologists (used as the basis for substantiation of crew member DCS safety) was recognized by practically all scientific laboratories working in this area in various countries. It can be broken down as follows. As is known, one of the main prerequisites for the generation of gas bubble is the process of cavitation in the elastic tissues of humans, which are stretched to a high degree, causing cavities to form in them. Later, the gas bubbles are filled with inert gas (nitrogen in our case), which is transferred from a dissolved to a gaseous state. Because the suit enclosure limits the rates and accelerations of human movements, it sharply prevents the development of these negative processes. This on its own, in our opinion, explains the fact that, at the same barometric pressure, inhaled gas composition, pre-breathing duration and physical activity level, there is such a marked difference between DCS probability at work in the vacuum chamber

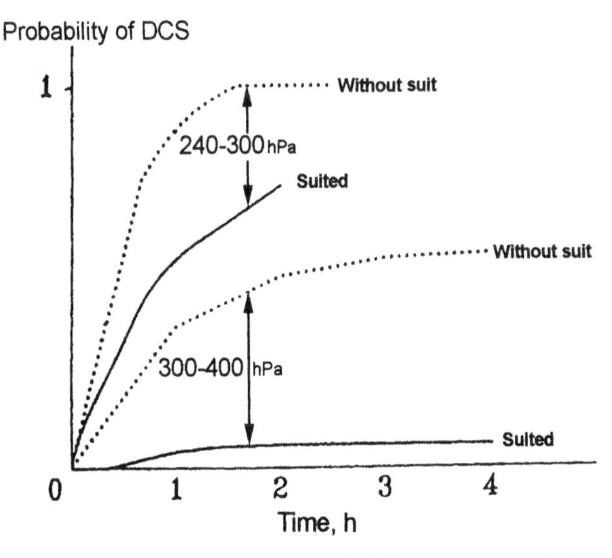


**Figure 12.1.2** Effect of suit application on DCS probability in test runs (without pre-breathing for removal of nitrogen from the body) at various suit operating pressure values.

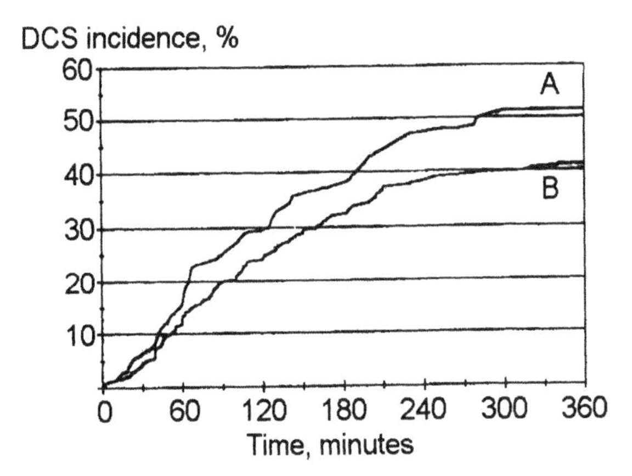


**Figure 12.1.3** Accumulated frequency of DCS occurrence during physical (leg) exercises at various coefficients of tissue over-saturation with nitrogen and various suit operating pressure values.

From Archive V.P. Katuntsev.

using an oxygen mask and DCS probability while wearing a spacesuit. It is 20–30% in the first case and 0% in the second case. This regularity is universal in character and is valid both for the pressures used in modern EVA spacesuits and for lower pressures. This fact is convincingly illustrated by comparative analysis of the Zvezda-run experiments presented in Figure 12.1.2 and supported by Figure 12.1.3.

Another constituent of our concept was the assumption that—with the same factor of over-saturation of the human organism by the inert gas (the ratio of the initial partial pressure of nitrogen in the human organism and the final suit pressure)—the higher the DCS probability the lower the absolute value of the ultimate barometric pressure. This relationship is proven in Figure 12.1.3 (data from V.P. Katuntsev). We can see that when we have a 226 hPa ultimate pressure, 1.73 over-saturation factor and a lengthy (6-hour) pre-breathing for nitrogen removal (A), DCS probability proved to be higher than when we have a relatively higher over-saturation factor (1.91) and no pre-breathing, but with a higher level of the ultimate pressure, 400 hPa (B). These data, in addition to the results shown in Figure 12.1.2, confirm the high probability of DCS in the given pressure range for cases when the pressure suit is not used by the test subject. This assumption is based on the following circumstance. In some body regions or tissues (e.g., in adipose tissue), due to the relatively low vascularity and their higher solution of nitrogen, the nitrogen over-saturation factor at the ultimate pressure in the suit can for a prolonged period of time be relatively higher than its average value for the whole organism, despite pre-breathing. It was for this reason that the suit's barometric pressure, which considerably exceeded the minimum pressure required for human provision with oxygen (360–400 hPa versus 253 hPa), was chosen. Of course, as a result of this the operational conditions for suit joints, which provided the cosmonaut with the required mobility, became harder. The load on the suit also increased due to such a high positive pressure inside it.

It is worth mentioning here that our approach to this task also ensures minimization of the loss of time for pre-breathing during airlocking. This important measure only takes 30 minutes, according to Russian protocol. Note that NASA astronauts have to spend several hours on this procedure because the US suit pressure is 25% lower than that of the Russian suit. Currently, they are trying to reduce this period by intensifying nitrogen removal by getting the astronaut to carry out a set of physical exercises during airlocking.

The capability to switch over to another suit pressure value, close to that in the US suit (about 267 hPa), was envisaged for all Russian EVA suits just in case of an emergency. The cosmonauts could use this pressure value to improve their mobility for a short period of time (about 15 minutes) in critical situations when their life could be at risk (e.g., in case of inability to perform any rescue operations). In the latest modification of the ORLAN-M suit, the manual switch-over to this operation mode was excluded since it had never been used (except once in 1965 by A.A. Leonov, who wore a significantly less mobile suit than the currently available space pressure suits of the ORLAN type). In the ORLAN-M suit, the pressure of 267 hPa can only be activated automatically in case of major leakage from the suit, and the emergency oxygen supply is activated simultaneously. In this case the crew member stops all work and immediately returns to the station.

## 12.2 THERMAL CONDITIONS

The next key physiological and hygienic problem of life support for a suited human was sustenance of the human thermal balance. This task was even more complicated because, on the one hand, a suited human who is working hard is an intensive source of thermal energy and moisture. On the other hand, the environment is characterized by sharp changes of the direction of heat flows to/from the suit due to the frequent change of "day" and "night" in the orbit. This hampered the creation of an efficient system for passive heat exchange between a human and the environment. It should be noted that the heat generated by the suit's LSS components also has to be added to metabolic heat.

Taking into account the complexity and multi-pronged character of the situation, it became clear that the solution could only be found by developing an efficient artificial system to remove heat from the suit combined with a reliable means of thermal isolation of the cosmonaut from the external environment.

In the first EVA suits (BERKUT and YASTREB) the task of counteracting moderate levels of heat generated by the cosmonaut during short-term activities was solved by using ventilation systems. But it was practically impossible to maintain human thermal balance by convection and sweat evaporation alone when the physical load value and duration increased during long-term EVAs. To do this it would be necessary to increase the ventilation flow by several times with corresponding increases in the gas flow rate and thus the power of the gas circulation source. Therefore, more efficient ways of heat removal had to be found at the same time as considering the tasks associated with more intensive and durable operations (assembling and repair of the orbiting station, transfer on the Moon surface, etc.). Such an engineering solution was found that took into account all the specific requirements needed for a human.

In the second half of the 1960s when the spacesuit for the lunar programme was developed, the decision was made to introduce a water-cooling system in the suit, besides the ventilation system. The water-cooling system provided for heat removal by using a conductive constituent. Because this was the primary way of heat exchange and was not natural for humans, evaluation of possible side-effects associated with it was made. In particular, when designing the water-cooled garment, the topography of perspiration was carefully studied, and zones playing a dominant part in heat exchange with the environment were determined with the aim of arranging the water-cooled garment tubes on the human body rationally. Good compatibility between the cosmonaut and the water-cooled garment was established. Cosmonauts are easily trained to control the intensity of heat removal by subjective signs (heat sensations). It should be borne in mind that weightlessness causes a considerable imbalance in human thermal sensations. This can be explained as follows. As is known, the blood takes the leading part in heat and mass transfer in the human body. Under weightlessness conditions, blood is redistributed in the human body due to the lack of hydrostatic pressure. In particular, the blood partially flows away from lower extremities and causes them to cool, resulting in crew members' feet feeling cold. In this case, as our experience shows, lessening the amount of heat removal from the lower extremities by eliminating water-cooling tubes from this area or using warm socks does not fully solve this task. Unfortunately, a practical engineering solution for external heat supply to this part of body has still not been found.

The suit's thermal control system maintains the in-suit microclimate and thermal balance of the cosmonaut at a level that is close to comfort. The relative humidity of the suit atmosphere at 25°C is 30–60%. The crew member can change the inlet water temperature of the water-cooled garment within the range 5–30°C, which is quite satisfactory even at times of intensive physical work, when humans produce up to 13 kcal min$^{-1}$. In this case, as our experience shows, the general enthalpy of the organism differs from the initial value by no more than 120 kJ. The total metabolic rate of a crew member during a 7-hour period of work sometimes amounts to 8 MJ (2,222 kWh). From the viewpoint of thermal control physiology, the Russian system has one more marked advantage. In essence this was the provision of thermal comfort in the head area of the suit, which was maintained owing to the water-cooled garment having a cap and controlled ventilation flow. This provided for temperature control of the brain, and, as a result, for general comfort and for good performance capability even when there was an increase in body temperature. Body temperature is measured behind the ear ($T_b$). The value of this parameter, in terms of thermal control physiology, shows the human "core" and "skin" temperature as 50/50. As a rule, the actual $T_b$ value was within the range 35.5–37.5°C. It depends to a great extent on how the cosmonaut uses the temperature control mechanism (on the LSS control panel) according to how hot or cold he or she feels.

This brings to the fore a specific reaction of the human body. As a result of such an unusual means of heat removal (conduction), a decrease in relative human perspiration (2–4 times) takes place. No doubt, this has to do with a reflex reaction of skin vessels to cooling. To a certain extent, this reduces the criticality of the drinking water problem for a crew member staying in the pressurized suit for a long period of time and simplifies the problem of diuresis. In the course of our multi-year experience, we have had no case of a cosmonaut expressing dissatisfaction regarding these matters. Nevertheless, the latest ORLAN-M spacesuit modification included a drinking water bag (350 ml) and a urine-absorbing garment (for more than 600 ml).

## 12.3 SUIT ERGONOMY

Ergonomic characteristics of the spacesuit affect to a great extent the productivity of the cosmonauts. In particular, the rigidity of the spacesuit enclosure and the space between it and the human body lessen human tactile ability. Despite being able to choose his or her own pressure gloves, the cosmonaut's capability to distinguish the surface character and form of objects, as well as the distance between them, is considerably deteriorated. For example, the threshold for distinguishing the distance between objects by touch is approximately four times worse than the naked hand (about 2 mm). The maximum force exerted by the wrist (dynamometry) decreases 1.5–2 times and amounts to about 225–300 N.

Taking into account the limitations associated with cosmonaut mobility and decrease of his or her tactile ability, a visual analyser can take on special significance in obtaining external information. At the same time, a number of external factors (including the spacesuit) limit the normal functioning of the eyes. The angle of vision through the main helmet visor of the ORLAN-type spacesuit is about 120° in the vertical plane (55° upward and 65° downward) and at least 200° in the horizontal plane (100° to the left and 100° to the right). On top of this, the field of vision in the latest ORLAN-M modification is increased owing to an additional window above the frontal/cinciput area. The field of vision in this direction is increased up to 90°. In order to protect the eyes from the injurious effect of sunlight, the suit visor is equipped with sun visors with the capacity to exclude about 3–5% of sunlight. At the same time, crew members can continue their work at "night" using the lights located on the spacesuit helmet. The lights provide good illumination of the working area with the intensity of 30 lux at a distance of 0.5 m, ensuring object identification of an angular dimension of 1–2 minutes. The frequent "day"/"night" change (typical of an orbital mission) makes the functioning of the human visual analyser more difficult, which has to do with the frequency of visual adaptation to different illuminations from surrounding objects.

Suit ergonomics are of considerable importance for a suited crew member at work (particularly for accessibility to suit controls and adequacy of audio and visual signals generated by the control system). A common problem—inability to recognize readings of the instruments located on the suit control panels—may be caused by long-sightedness. The solution to this problem consists in proper selection of spectacles and contact lenses. Moreover, any evaluation of the ergonomics of suit controls must take into account the overall workload of the crew member during EVA. From time to time we see situations in which the crew member, busy with his or her work, cannot break away from it to readjust the thermal control system, despite the need for this. This highlights the necessity to further automate a number of processes.

It should also be noted that good ground-training, the experience gained by cosmonauts in previous EVAs and good physical and emotional preparation can considerably affect EVA efficiency. The installation of solar panels by one of the *Salyut* crews serves as an example. When this very important and complex operation was undertaken for the second time by the same crew it took half the time and their metabolic rates were 10–20% lower.

Since EVA is still one of the most important and dangerous operations during the mission, a great deal of attention should be paid to evaluating the state of health of the cosmonauts both during EVA and pre-EVA preparation. During the EVA, operational medical monitoring of the health of the crew members (besides talking to them) is undertaken by means of telemetric data about the medical parameters and the performance of the SCLSS.

A medical monitoring system, built into the EVA spacesuit, relays information from an ECG (D-S load) and pneumogram as well as about behind-ear temperature. In addition to this, the cosmonaut's metabolic rate is calculated by such gas exchange parameters as $O_2$ consumption (by change in $O_2$ bottle pressure) and $CO_2$ production (by the difference between CCC inlet and outlet $CO_2$ values), and the amount of heat removal from the body of the crew member (mainly by the water-cooled garment inlet and outlet water temperatures). Through dialogue with the crew, the doctor can analyse the difficulty of the work undertaken and the adequacy of crew member reactions to this work. Note that 40 years of experience in EVA medical ground support shows that the duration of the flight is not a limiting factor in granting a crew member clearance for EVA, as long as the cosmonaut is properly prepared and strictly follows the programme (i.e., working with means that prevent the unfavourable effects of weightlessness). The same applies for a cosmonaut's age. If necessary, it is possible to repeat an EVA after 2–3 days of rest.

As a rule, crew members are highly motivated and look forward to pre-EVA preparation and the EVA sortie itself. A single a case of EVA fear has been known to occur. However, upon completion of the EVA, the crew must be provided with a full-scale rest period for rehabilitation and in some cases for treatment (in case of broken skin, pains in muscles, etc.).

It is important to point out that competent, unobtrusive and friendly medical support of EVA is not only appreciated by crew members, it is instrumental in ensuring fulfilment of this sophisticated and important operation.

# 13

# Potential projects on planetary suits for the Moon and Mars

## 13.1 INTRODUCTION

A study on the feasibility of a planetary suit for a Mars expedition was initiated at Zvezda in the late 1980s when the USSR government included the development of the Mars expedition project in the Governmental Science and Technical Programme of Potential Activities. At that time Zvezda only analysed the operational conditions for a planetary suit because, in accordance with the above programme, the main efforts were concentrated on concept selection for the expedition as a whole and on medical evaluation of human capabilities to undertake such a long-term space mission. Later, due to lack of finance, interest in this study decreased.

In recent years, because of the success of international cooperation on the *ISS* programme, many countries are interested in a manned mission to Mars (principal among them are the USA, Western Europe and Russia). In 1999 a decision was made to establish the International Space Technology Centre, responsible for development of the Mars expedition project (*Novosti Cosmonavtiki*, 2002). The Centre's control committee consists of 21 members from NASA, Russia and ESA. Currently, in Russia, main attention is paid to substantiation of the Mars mission scenario and medical studies (as mentioned earlier). In the USA, besides these issues, NASA and some separate companies are carrying out work on the Mars suit.

Zvezda is also studying some issues associated with development of the Mars suit, despite the lack of finance for this work. This mainly includes analysis of the operational conditions for the Mars suit, development of the requirements for the pressure suit, analysis of design solutions and development of experimental samples of some suit enclosure components. The diagram concepts for the suit's life support system (LSS) depend to a great extent on the system concept for the Mars space vehicle and the Mars expedition scenario. Therefore, Zvezda is currently running only a preliminary study of potential versions of the suit's LSS, which is subject to the amount of finance available.

## 13.2 SOME PROBLEMS OF PLANETARY SUIT DEVELOPMENT

Zvezda has accumulated great experience in development and operations of orbital spacesuits. It has also carried out work on development of the suit for the Moon programme (see Chapter 6). Therefore, this was an opportunity to apply the available experience and components of earlier developed items to the development of the new planetary suit concept. In previous chapters, devoted to spacesuits for extravehicular (EVA) and Moon surface activities (see Chapters 6 and 8), it was shown that pressure suits of the semi-rigid type were optimum for these purposes. These suits (KRECHET and the ORLAN family) had a hard upper torso (HUT) integrated with a helmet, a back entry hatch (incorporating the LSS components) and soft-design arms, lower torso assembly (LTA) and legs.

Almost 30 years of operational experience of these suits on board the *Salyut*, *Mir* and *ISS* space stations was clear proof of their advantages. Therefore, we can assert with full confidence that a suit of the semi-rigid type can be recommended for future planetary suits. But, orbital ORLAN-type spacesuits (without modification) cannot be used as planetary (Moon and Mars) suits because they were not designed to meet requirements necessary for the planetary suit. This mainly refers to suit enclosure mobility.

EVA development called for efforts to concentrate on achieving high performance capability for the cosmonaut's hands, because they are used to move along the station surface by means of snap-hooks, to carry out work with various apparatus, tools, control panels and controls, to transport cargoes, to assemble, to work with hatches, etc. Here, the cosmonaut's legs play a minor role (mostly used to keep the cosmonaut at the workstation or to assist passage through the hatch).

There is the belief that the mass of a suit is of little importance for operation in weightlessness. Quite the contrary, the higher the suit's mass the more stable its attitude and the movements of arms and legs cause fewer disturbances. The arms and gloves of ORLAN-type suits have been improved with each suit modification and now have high-mobility characteristics. In the current configuration, they could be used both for the Moon and the Mars suits. At the same time, the soft LTA of the ORLAN-type orbital spacesuit does not feature any mobility at all and the leg enclosures have minimum mobility (only required to carry out the above-mentioned tasks).

One of the main requirements of the planetary suit is to allow the cosmonaut to walk on the planet surface and to gather ground samples, which calls for bending and kneeling. Other capabilities are required such as walking up and down the vehicle ladder and exit to and from the airlock, Moon rover or Mars rover. This is the reason close attention was paid to formation of the design for the lower portion of the enclosure, its mobility and the suit's footwear at the initial stage of enclosure development. Some enclosure elements, developed for the KRECHET-94 spacesuit during the USSR Moon programme, could be used (see Section 6.2) such as a hip convoluted joint with 2 degrees of freedom and a knee convoluted joint. These could be used as the basis for development of the lower portion of the planetary suit.

Analysis of suit operations on the Moon showed that the Moon suit's LSS design may be close to that of the orbital suit. At the same time, when selecting the LSS flow diagram, it is very important to coordinate it with performance characteristics of the Moon vehicle, airlock system and their interfaces. Many more questions arise when selecting a design for the Mars suit's LSS. In particular, the following main issues should be taken into account:

1 Coordination with the expedition's mission scenario and requirements:

- universality with the orbital (and interplanetary) suit;
- compatibility with airlock systems;
- duration of expedition stay on Mars;
- pressure values for the spacesuit and vehicle;
- duration and character of works;
- number of EVA sorties;
- availability of expendables (water, oxygen, power, etc.) and/or capability to regenerate them;
- methods and time for emergency return to the vehicle.

2 Movement across the planet surface:

- availability of a Mars rover and its design features (pressurized/unpressurized cabin);
- availability of a rover LSS and its coordination with the suit's LSS;
- independent travel across the planet surface in the suit (distance, time, speed);
- dusty surface of Mars.

3 Effect of partial gravity and associated suit mass limitation.

4 Environmental (atmosphere, irradiation, radiation, etc.) effect and protection against it.

5 Provision of thermal control under low temperatures and pressure:

- thermal insulation of the suit;
- type of heat/exchange devices;
- prevention of window fogging;
- materials to work under low temperatures.

The approach used to select the LSS flow diagram, the design of suit elements and materials depends in many respects on solution of the above problems.

We consider the substantial reduction in the suit's mass to be very important. The modern Russian orbital spacesuit weighs 112 kg and the US EMU weighs even more, but this is not acceptable for the Mars suit. Taking the gravity of Mars into account (0.38 of Earth's), the mass of the suit to be used by a crew member walking should not exceed 50–60 kg (to be confirmed in further studies). Development of a suit of such a mass calls for lighter materials and novel approaches to the LSS.

Radiation protection and maintenance of an equable temperature are also going to be difficult to deliver. For example, the heat exchanger/sublimator that is usually used in the orbital suits would not function under the atmospheric pressure of Mars (7–10 hPa) and shield vacuum thermal insulation would not be effective.

In light of these difficulties, the means by which the suited crew member moves across the surface of Mars and the length of time of self-contained operation in the suit become very important. Currently, within the Mars expedition concept, preference is given to robotic concepts with suited crew members moving around the surface of Mars in the rover. However, we believe that it would be a mistake to exclude provision for a suited crew member to remain on the planet surface in case of emergency or repairs. Selection of the LSS flow diagram and mass depends to a great extent on the time of self-contained operation.

We can imagine several ways in which versions of the Mars suit could be used:

1. Suited crew member transfer within a pressurized Mars rover (provision for getting out of the Mars rover and detaching the suit from the rover's systems for 1–2 hours just in case of emergency and return to the mothercraft).
2. Suited crew member transfer within an unpressurized Mars rover (with the use of the rover's LSS). In this case, the time of self-contained operation would be 1–2 hours as in the first case.
3. Suited crew member walking for short distances from the Mars rover, who periodically returns to the rover and changes backpacks containing the LSS (such a version was studied by Hamilton Sundstrand in the USA).
4. Self-contained activities of a suited crew member for 6–7 hours.

For the first two versions, the self-contained LSS can be considerably simplified, and a number of current suit components can be used for it. The third version and, especially, the fourth one call for new engineering solutions. These solutions will depend to a great extent on the selected expedition concept as a whole, especially the concept of the Mars vehicle and the Mars rover (availability of supplies, oxygen in particular, capability to recharge and regenerate expendable elements, the design of interfaces with the mothercraft's systems, etc.). Until the above problems are resolved, Zvezda is only studying and developing enclosure components whose design does not depend on the expedition concept as a whole.

## 13.3 DEVELOPMENT OF CONCEPTS FOR THE MARS SUIT ENCLOSURE AND DESIGN OF ITS LTA COMPONENTS

Zvezda's approach to the concept of some elements for the Mars suit enclosure was laid out within the framework of its recent Mars studies (Abramov et al., 2001a, b, 2002; Hodgson et. al., 2000; Ross et al., 2002).

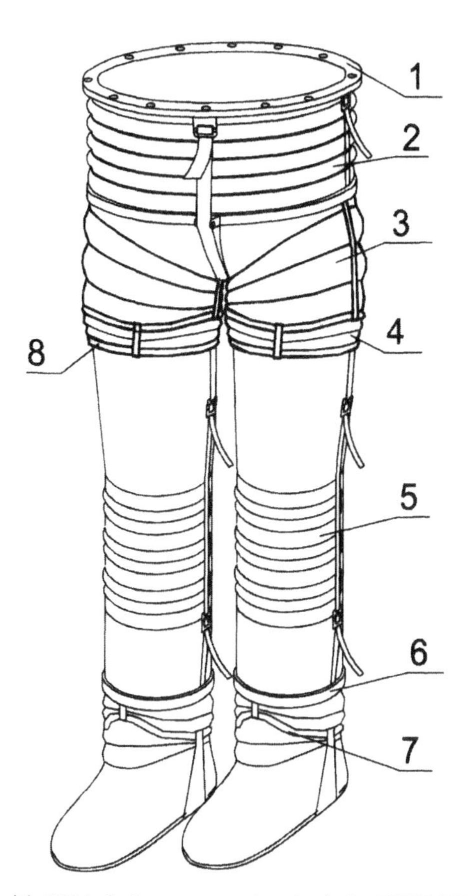


**Figure 13.3.1** Planetary suit's LTA design concept (option): 1—HUT/LTA interface; 2—waist joint; 3—hip flexion/extension joint; 4—thigh abduction/adduction joint; 5—knee joint; 6—ankle joint; 7—two-axis ankle joint; 8—thigh ring or thigh pressure bearing (TBD).
From Archive Zvezda.

The main elements of the Mars suit under study are the leg enclosure, footwear, waist joint and suit/mothercraft (Mars rover) interfaces. Figure 13.3.1 shows one of the LTA concepts being considered at Zvezda. It has the following differences from the concept of the ORLAN-type orbital suit enclosure:

- waist joint that provides for forward bending of the suit body;
- hip joint with two degrees of freedom (flexion–extension and abduction–adduction);
- ankle joint with two degrees of freedom; and
- different knee joint design.

Introduction of the waist joint in the suit enclosure calls for a considerable redesign of the current ORLAN-type HUT. Figure 13.3.2 shows one of the Mars suit's HUT

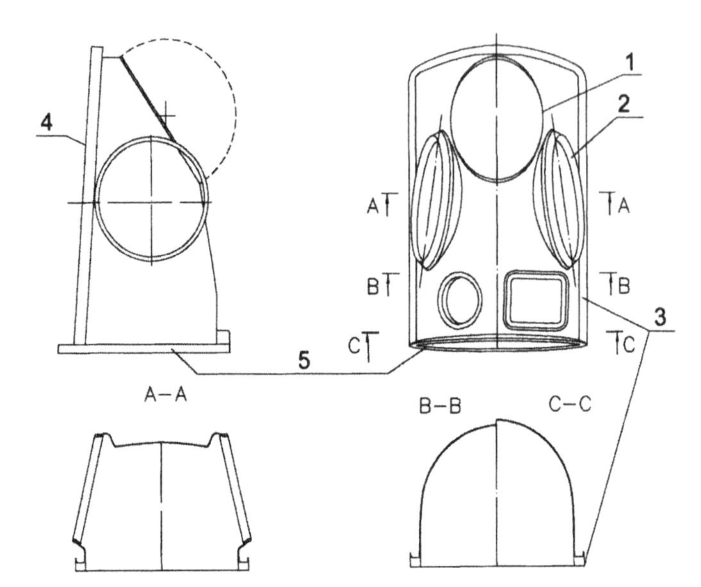


**Figure 13.3.2** HUT concept option: 1—helmet opening; 2—shoulder pressure bearing opening; 3—entry hatch frame; 4—entry hatch; 5—HUT/LTA interface.
From Archive Zvezda.

**Figure 13.3.3** Boot with gimballed ankle joint and calf bearing (concept option): 1—calf bearing; 2—abduction/adduction ankle joint; 3—intermediate gimballed ring; 4—axial restraint tape; 5—flexion/extension joint; 6—transverse restraint tape; 7—restraint layer; 8—protective overboot.
From Archive Zvezda.

concepts, which is remarkable for the shape of the lower edge (flange) and, correspondingly, for the shape and position of the entry hatch. Moved up as far as possible and arranged almost normally to the longitudinal axis of the HUT, the HUT lower flange makes it possible to arrange the soft waist joint on the enclosure.

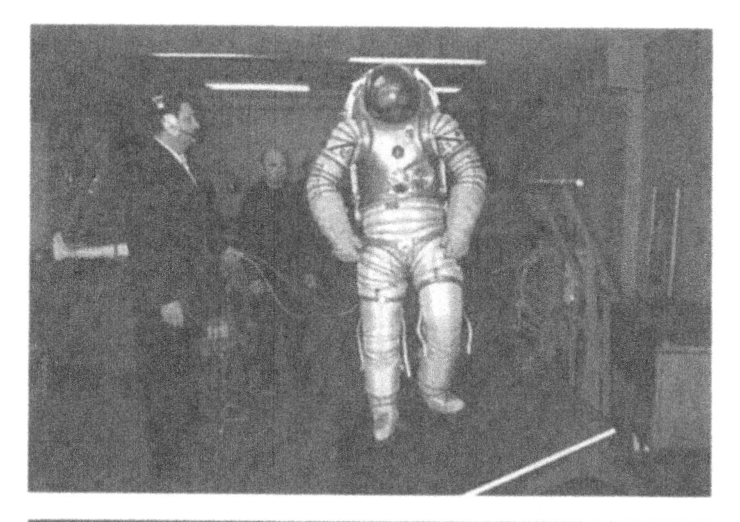

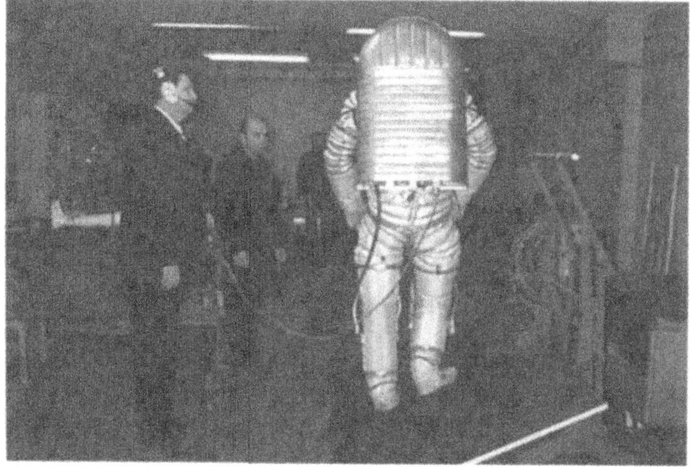

**Figure 13.3.4** A test engineer wearing the ORLAN-type suit with a two-degree-of-freedom ankle joint on an inclined surface.
From Archive Zvezda.

One of the design concepts for the Mars suit's footwear incorporates an ankle joint with two degrees of freedom, a special sole and calf pressure bearing (Figure 13.3.3). The footwear for the Mars suit was developed in cooperation with Hamilton Sundstrand (USA). The need for the ankle joint (with two degrees of freedom) in combination with the pressure bearing was dictated by the requirement to support walking on the rugged surface of Mars (including slopes). The shape and rigidity of the sole is very important for walking. Therefore, several versions of a sole design were developed and tested. Test models of the footwear and sole were tested both at Zvezda in conjunction with ORLAN-type pressure suits (Figure 13.3.4) and at NASA (Figure 13.3.5) in conjunction with the EMU and Mark-III suits.

**Figure 13.3.5** A test engineer wearing the Mark III-type suit with the Zvezda boots on the Martian range in Houston.
Courtesy NASA.

The Mars rover/suit interface is of major importance for the Mars suit. Figure 13.3.6 shows one of the versions (based on the present concept of attachment of the HUT aboard the *Mir* station at three points in the waist region). With such a restraint approach, the crew member on board the Mars rover is rigidly fixed in the standing position. The crew member can drive the Mars rover in the standing position and, if the opportunity to relax arises, he or she can occupy a sitting position inside the suit. Such an approach simplifies the enclosure design and

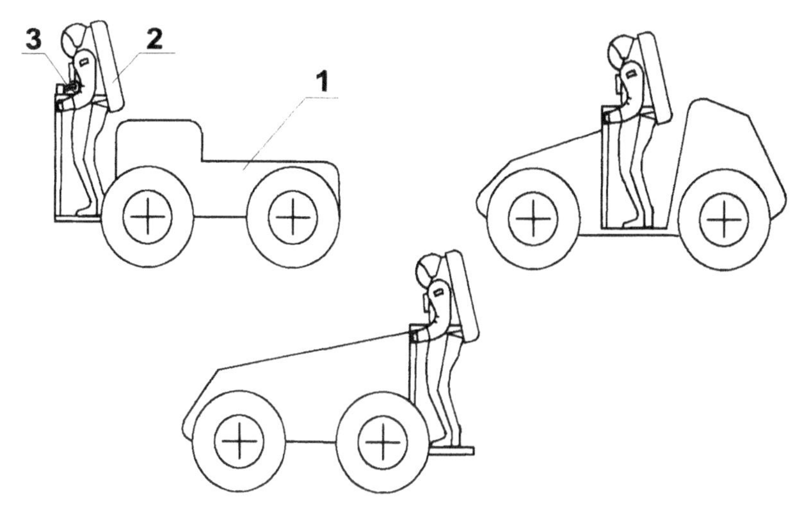


**Figure 13.3.6** Options considered for attachment of the suited crew member to the rover: 1—rover; 2—suited operator; 3—suit/rover interface.
From Archive Zvezda.

**Figure 13.3.7** Option for the lunar suit/rover hatch interface concept.
From Archive Zvezda.

makes the sitting position for the enclosure and, therefore, the seat in the Mars rover unnecessary, improving the field of vision of the surface of Mars.

Entrance to the HUT of the ORLAN-type suit from the rear can be considered as a version of the interface between the suit and the space vehicle's airlock (or that of the Mars rover) provided the spacesuit is fixed to the airlock by its back. Such a version was under discussion at Zvezda as early as the 1960s as part of the Moon programme (Figure 13.3.7).

Currently, A.Yu. Stoklitsky and N.A. Moiseyev are the leading designers of various enclosure elements of the planetary suit.

# 14

# Who's who in Soviet/Russian spacesuit technology

## 14.1 THE COMPANY, RD&PE ZVEZDA

All Soviet/Russian spacesuits and other personal rescue/survival means and protective gear for pilots and cosmonauts are products of the Research, Development & Production Enterprise Zvezda (RD&PE Zvezda), which is today an enterprise comprising specialized scientific and research departments and design and test departments, all equipped with unique experimental know-how. It also includes a department of aerospace medicine, services for scientific and technical operations support of company-developed products, and pilot production. Spacesuit development is just one area of Zvezda's activities.

Zvezda was established on 2 October 1952, as a company belonging to the Ministry of the Aviation Industry. It specialized in safety means for the crews of high-speed and high-altitude aircraft. At that time, jet aviation was growing rapidly and demonstrated enormous potentials to achieve previously unreachable flight speeds and altitudes. At the same time, jet aircraft accidents often resulted in deaths. It became clear to industry experts and management that the only way to minimize the risk for crews was not only to improve the reliability of the aircraft's main systems but also to have special equipment to improve the crew's tolerance to extreme dynamic and climatic conditions, provision for emergency escape from the aircraft at high speeds and for survival on landing or splashing down. The work undertaken by aircraft manufacturers and other companies in this area was not good enough.

In order to provide integrated solutions to this problem and in pursuance of Resolution No. 4325-1715 of the USSR Council of Ministers dated 27 September 1952, Plant No. 918 (Zvezda) (comprising a design bureau, research department and pilot production facilities) was established in Tomilino (Moscow region) by Order No. 1150 of the Minister of the Aviation Industry dated 2 October 1952 (Figure 14.1.1 and 14.1.2). S.M. Alekseyev (Figure 14.1.3), who worked as the head of a

**Figure 14.1.1** First Zvezda building in 1952.
From Archive Zvezda.

**Figure 14.1.2** The construction of the main production building. The laboratory and design building (built 1967) are on the far left.
From Archive Zvezda.

design and production department at the Flight Research Institute (LII), was appointed Manager and Chief Designer of the plant. Early in 1964 G.I. Severin (Figure 14.1.4) became Manager and Chief Designer of the plant. In 1982 he became the General Designer of the company and since 1992 he has been the General Director and General Designer.

During the 50-year period since establishment of the company, it changed its name more than once (Plant No. 918, Post Box A-3927, Machine Engineering Enterprise "Zvezda" and finally "RD&PE Zvezda" Joint Stock Company). The form of company ownership has also changed: from a government-owned company to a private company. Since 1994 it has been the "Research, Development & Production Enterprise Zvezda" Joint Stock Company, with a 38% government

**Figure 14.1.3** S.M. Alekseyev, Chief Designer at Zvezda (Plant No. 918) from 1952 through 1963.

From Archive Zvezda.

**Figure 14.1.4** G.I. Severin, Chief Designer in 1964, General Designer of Zvezda ever since 1982 and General Director in 1992.

From Archive Zvezda.

share in the equity capital. Company employees own 33% of the shares. The remaining 29% belong to institutions and outside shareholders.

"RD&PE Zvezda" JSC is itself a shareholder in nine joint stock companies and a co-founder/founder of five companies. The company's premises cover 16.2 ha, and

(*a*)

(*b*)

**Figure 14.1.5** The main building housing manufacturing facilities (*a*) and the design bureau building (*b*) of Zvezda.
From Archive Zvezda.

a production area of 70,000 $m^2$ (Figure 14.1.5a, b). The company also has a subsidiary in Zhukovsky, the "Zvezdny" boarding house in Kislovodsk, the "Zvezdny" cultural establishment in Tomilino (this is open to the public and houses a theatre) and a mountain skiing base in Chulkovo (Moscow region). On 31 December 2000 the company employed more than 2,000 people.

Currently, the First Deputy General Director and First Deputy General Designer is V.I. Svertshek, deputy general directors are N.I. Afanasenko (Executive Director), I.I. Askerko (Chief Engineer and Deputy General Designer in production) and V.M. Rafeyenkov (Deputy General Designer and Quality Service Manager). Deputy General Designers in subject areas are V.I. Kharchenko (aviation emergency escape and rescue means), A.A. Soldatenko (aviation protective gear and life support equipment) and I.P. Abramov (products for space programmes).

From its inception, "RD&PE Zvezda" JSC (Plant No. 918 at that time) has simultaneously worked on aviation, future space flight safety regarding life support systems and pilot rescue in case of emergency. The first space products of the company were spacesuits and equipment for animals used in early test flights.

From the glory days of the Soviet space programme and the experience obtained during that time, the foresight and astuteness of those who initiated and developed the organizational structure and profile of the future Zvezda was quite simply remarkable. At the dawn of jet aviation, even prior to the advent of cosmonautics, they were successful in providing special safety equipment for pilots (made in the USSR) as integrated systems rather than as separate devices, flying gear and systems, which, although working well as single items, had poor interoperability (as often occurred in Western analogue systems).

Throughout the company's existence, its subject areas have grown and been modernized. However, the scientific and production essence of its work has changed little. Further training and honing of skills of Zvezda's design and production teams, and the bringing in and training of highly skilled personnel, made it possible to effectively solve not only problems in the development and improvement of personal life support systems for pilots and cosmonauts, spacesuits and extravehicular activity (EVA) systems, means of escape for aircraft crew and passengers, but also for other areas such as in-flight refuelling systems, means of protection against aircraft fires and explosion, etc.

Ever since the 1960s, Zvezda has actively participated in many aerospace exhibitions. Pressure suits and other equipment for pilots and cosmonauts always attract the highest attention of both visitors and experts (Figure 14.1.6). Almost all models of Zvezda-developed spacesuits have been exhibited or are currently on display in a number of Russian and foreign museums. Some suits even found their way into foreign private collections.[1] The most complete set of spacesuits on exhibit is in the Zvezda museum (Figures 14.1.7 and 14.1.8).

The Zvezda emblem (Figure 14.1.9) is sewn on all spacesuits and gear currently produced by Zvezda. The name "Zvezda" was not present on earlier emblems, which instead read "SALYUT" and later "MIR".

[1] Currently, it is impossible to give the exact location of a number of exhibits, because some had been written off the inventory (due to falling into disrepair) and some had been transferred or sold to other owners by museums without Zvezda's knowledge (see also Chapter 15).

**Figure 14.1.6** Russian top authorities visit the Zvezda booth at the Zhukovsky air show (MAKS-2001). Seen in the foreground are G.I. Severin, M.A. Pogosian (General Director of Sukhoy), Russian President V.V. Putin, Yu.N. Koptev (General Director of Rosaviakosmos), S.B. Ivanov (Minister of Defence) and I.I. Klebanov (Minister of Science and Technology).

From Archive Zvezda.

Zvezda's 50th anniversary was celebrated in October 2002 (Figures 14.1.10 and 14.1.11). Representatives of many aerospace companies and governmental organizations turned up to congratulate Zvezda's staff.

## 14.2 THE SOVIET/RUSSIAN SPACESUIT CREATORS

The concentration of scientific and production tasks and technologies (including mechanics, aerospace medicine, electronics and many more) in such a comparatively small company as Zvezda called for a specific organizational structure for its work at both formal and informal levels. Zvezda's standard organization diagram for spacesuit development is shown in Figure 14.2.1.

The creative way of working at Zvezda has, as a rule, a collective character, in which representatives of various structural and hierarchical levels generate ideas. This is why it is difficult and often impossible to identify a given individual for this or that technical solution. The hierarchy of formal office subordination in a creative process does not always reflect the actual roles of specific participants in the work. All proposals would go through a process of discussion, improvement and eventually finalization.

The "final" form was not always the best, but sooner or later in the course of the design process the moment would arrive when one of several versions had to be

**Figure 14.1.7** Managers of the development projects for space pressure suits at the Zvezda museum: G.I. Severin (*centre*), I.P. Abramov (*left*) and V.I. Svertshek (*right*).
From Archive Zvezda.

selected. The irony was that others could be seen to be not only competitive but even much better *after* the "approved" version was being embodied in metal or tested. In this regard the history of the development of spacesuits and other Zvezda-designed systems is similar to that of any modern technology, where continuous research results in the birth of new ideas or a return to rejected ideas, the revival of which under new conditions sometimes brings better results.

Zvezda's way of working calls for a "product leader" (i.e., leading designer), who undertakes technical management and coordinates the work on a certain project

**Figure 14.1.8** G.I. Severin demonstrates the recent ORLAN-M spacesuit for the *ISS* programme to Yu.P. Semionov (President and General Designer of RSC Energia).
From Archive Zvezda.

**Figure 14.1.9** Emblem sewn on space gear produced by Zvezda.

or just a part of it. The leading designer is usually a member of the department tasked with development of the item as a whole (Figure 14.2.1), but he or she reports to the corresponding deputy general designer as far as engineering is concerned. At Zvezda the product leader is in most cases the point of all contacts regarding the project, withstanding downward and upward pressure, putting ideas into decisions (the approval of which is sometimes beyond his or her power). Generally, company authorities make the final decision, but sometimes their decisions depend on government body directives.

When we chose the candidates for inclusion in our list of spacesuit creators, we

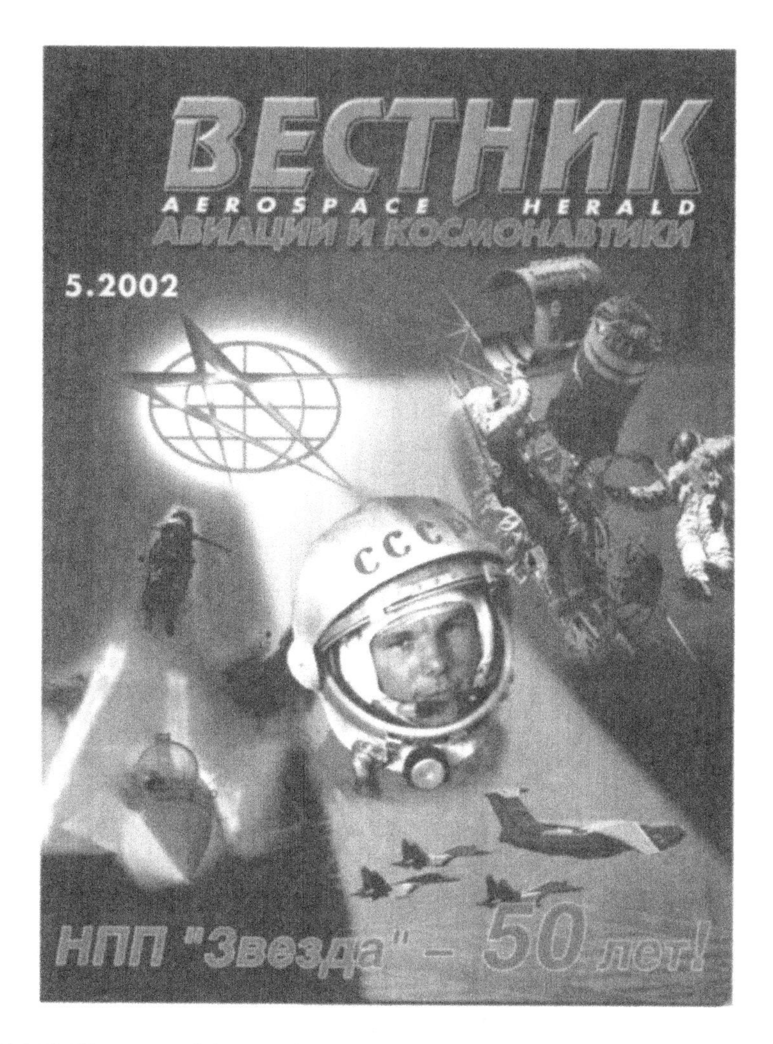


**Figure 14.1.10** The cover of the Russian *Aerospace Herald* dedicated to the 50th anniversary of Zvezda.

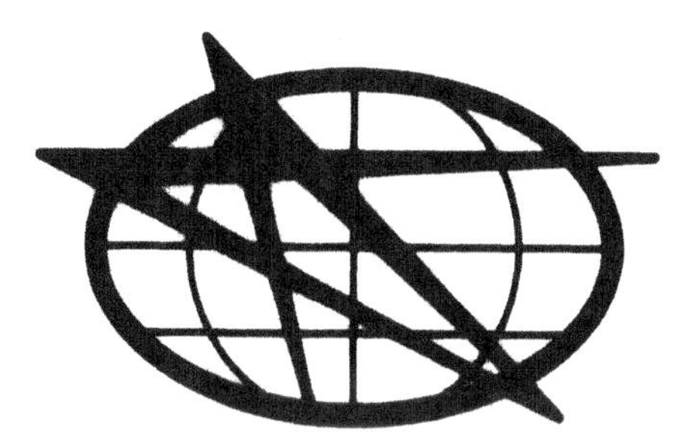

**Figure 14.1.11** Zvezda's golden jubilee pin.
From Archive Zvezda.

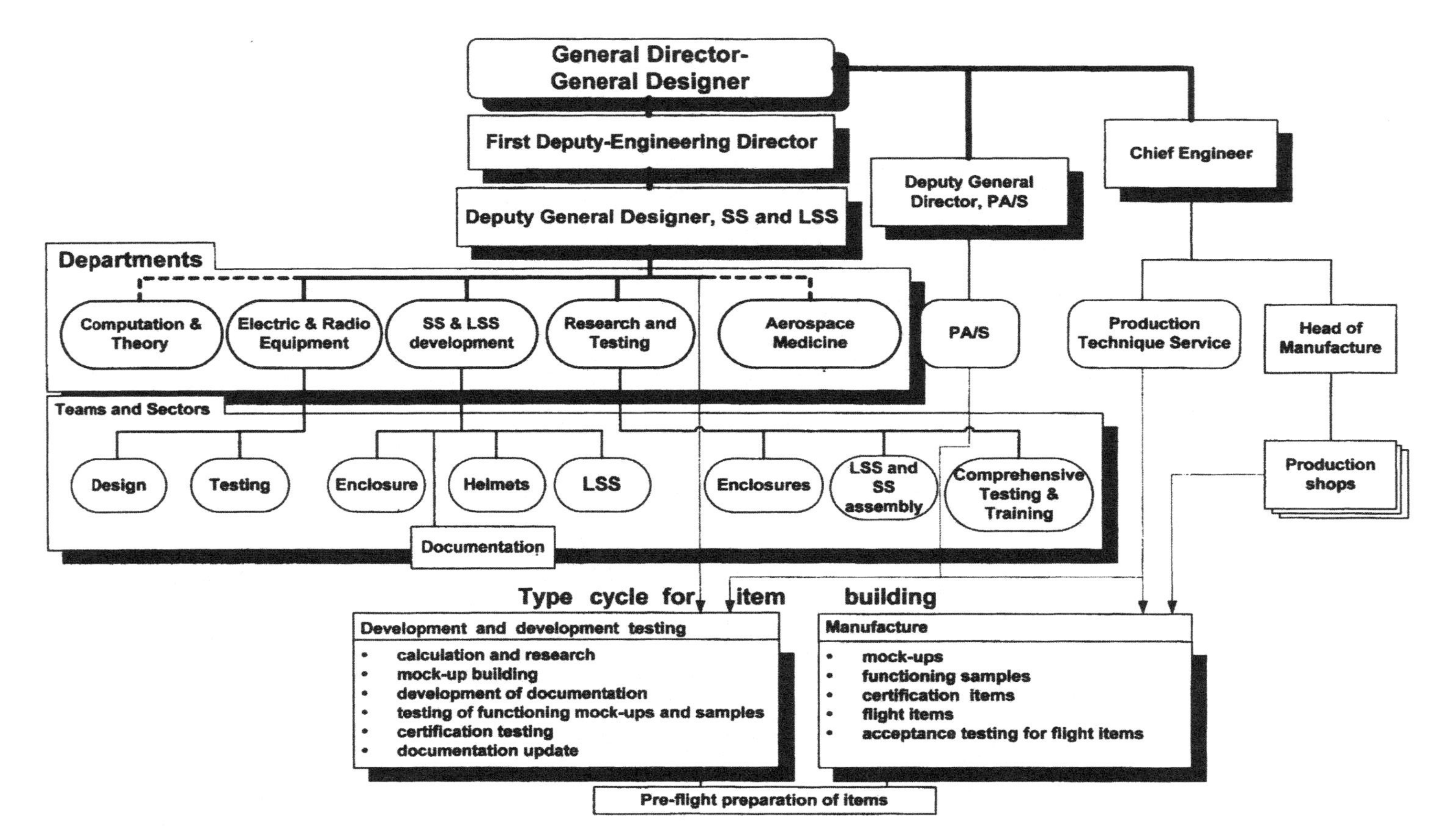


**Figure 14.2.1** Standard structural diagram for spacesuit and LSS development work at Zvezda (financial and economic services are not shown).

From Archive Zvezda.

mainly considered those people who had served as a leading designer (project manager in the West), or were in charge of work run by a separate design or test team. We think it fair to consider these people as the main originators of technical decisions about items that fell within their remit or expertise. In any case, it would be impossible to list all the names of those Zvezda employees who contributed invaluably to the design and development of the many different items.

## 14.3 MOST ACTIVE PARTICIPANTS

*Abramov, Isaak Pavlovich* (born 17.12.1926)
Graduated from the Bauman Moscow Higher Technical School in 1950.

| | |
|---|---|
| 1950–1955 | Respirator plant (manufacturing aircraft oxygen equipment) in Orekhovo-Zuyevo, as design engineer, then the chief of the assembly shop |
| 1955– | Zvezda, a design engineer |
| 1958– | chief of the life support system (LSS) design team |
| 1963– | leading designer and chief expert in spacesuit systems |
| 1995– | Deputy General Designer |

A direct participant and later the manager of activities on the LSS and suit systems as a whole for all spacesuits created at Zvezda as well as the *Voskhod-2* airlocking system. Participated in the preparation of Zvezda space items for launch at the cosmodrome, starting from Yuri Gagarin's flight. He is a Candidate of Technical Sciences and Member of the International Academy of Astronautics.

*Alekseyev, Semion Mikhailovich* (24.12.1909–4.2.1993)
Graduated from the Moscow Aviation Institute in 1935.

| | |
|---|---|
| 1929– | Started work in the aviation industry as a designer at A.N. Tupolev's OKB |
| 1933– | chief of a design team at S.V. Ilyushin's OKB |
| 1939– | leading designer at S.A. Lavochkin's OKB |
| 1942– | chief of a design department at S.A. Lavochkin's OKB |
| 1943– | Deputy Chief Designer at S.A. Lavochkin's OKB |
| 1946–1948 | head of new aircraft development in OKB-21 (in Gorky) |
| 1950–1952 | chief of the design and production complex at the Flight Research Institute (LII) |
| 1952–1964 | Head and Chief Designer at Plant No. 918 |
| 1964–1971 | Deputy Chief Designer at Zvezda |

Under his leadership a series of aviation suits, protective gear, ejection seats and other equipment for aircraft was developed. The ejection seat and spacesuit for the *Vostok* spacecraft, as well as equipment for studying the vital functions of animals during suborbital and orbital flights were developed under his leadership in 1957–1964.

*Alekseyev, Anatoly Vasilievich* (born 3.4.1944)
Graduated from the Moscow Aviation Institute in 1967.

| | |
|---|---|
| 1972– | Zvezda, first as a leading test engineer, later as the chief of a test team and the chief of a team that prepares Zvezda's items for launch at Baikonur and supports cosmonaut training at Zvezda, using the *ISS* spacesuits. |

*Barer, Arnold Semionovich* (born 26.9.1927)
Graduated from the Moscow Medicine Institute in 1951.

| | |
|---|---|
| 1951–1958 | Medical Service Officer, Air Force of the Soviet Army |
| 1958–1960 | chief, Clinicophysiological Laboratory of the Central Science and Research Air Force Hospital |
| 1960–1964 | Zvezda, chief, acceleration research laboratory |
| 1964– | Chief, department of aerospace medicine |

Author of a number of scientific papers in aerospace medicine. He is a Doctor of Medical Sciences and Professor and Member of the International Academy of Astronautics.

*Boiko, Alexander Ivanovich* (20.8.1906–1997)
Graduated from the Moscow Institute of Dirigible Technology in 1936.

| | |
|---|---|
| 1937–1941 | TsAGI, senior engineer |
| 1941–1944 | chief of the team at the aviation plant |
| 1944–1946 | work at the Ministry of the Aviation Industry |
| 1946–1953 | Gromov Flight Research Institute (LII), high-altitude suit development (a pioneer of the development of high-altitude suits in the USSR) |
| 1953–1974 | Zvezda. |

Continued high-altitude suit (VSS type) development, leading designer for the Vorkuta suit. Participant in the *Vostok* spacesuit development, in which components of full-pressure suits developed by him were used.

*Chistyakov, Ivan Ivanovich* (born 18.8.1930)
Graduated from the Moscow Power Engineering Institute in 1959.

| | |
|---|---|
| 1959– | Zvezda, as an electric equipment designer, then the chief of a design team |
| 1966– | leading designer for spacesuit electrical and radio equipment |

He made an important contribution to development of spacesuit electrical and radio equipment. Active participant in the development of the airlock system for the first ever space walk.

*Doodnik, Mikhail Nickolajevich* (born 16.4.1938)
Graduated from the Moscow Chemical Engineering Institute in 1960.

| | |
|---|---|
| 1960–1964 | Zvezda, designer |
| 1964–1965 | leading designer for the *Volga* airlock for the *Voskhod-2* spacecraft |
| 1966–1972 | group manager, chief of the design team for advanced projection |
| 1972– | leading designer in aviation oxygen breathing systems |

During the Soviet Moon suit project, he was one of the key persons in the development of the LSS flow diagram for the KRECHET and ORLAN suits. Between 1973 and 1979, under his leadership, a complex of oxygen equipment for civil aircraft crews and passengers was developed and transferred for mass production. Currently, he manages development of new aviation oxygen systems.

*Elbakyan, Aram Tsolakovich* (born 4.9.1938)
Graduated from the Bauman Moscow Higher Technical School in 1961.

| | |
|---|---|
| 1961–1972 | Zvezda, test engineer for aviation and space equipment, including spacesuits for the *Voskhod*, *Soyuz* and Soviet Moon programmes |
| 1972– | chief of the team for EVA spacesuit tests |

An active participant in research and development of the LSS for all spacesuits. Participated in spacesuit testing as a test engineer as well.

*Gershkovich, Alexander Mironovich* (14.6.1912–1988)
Graduated from the Moscow Aviation Institute.

| | |
|---|---|
| 1938–1941 | TsAGI, engineer |
| 1941–1952 | Gromov Flight Research Institute (LII), where he was involved in work on aircraft cabin air conditioning systems |
| 1953–1983 | Zvezda, chief of a design team, deputy chief of the design department and leading designer |

Under his leadership, a number of LSSs for aviation suits were developed. Leading designer of the spacesuit for *Vostok*. Later, involved in development of the conditioning system for aviation protective outfits.

*Koobar, Felix Vasilievich* (born 23.3.1933)
Graduated from the Bauman Moscow Higher Technical School in 1957.

| | |
|---|---|
| 1957–1965 | Zvezda, test engineer |
| 1965–1972 | chief of a test team |
| 1972– | deputy chief of the test department for aviation and space outfits and LSSs |

Active participant in testing of the systems for all Zvezda-designed spacesuits. He made an important contribution to solving the problem of space suit LSS thermal control.

*Koochevitsky, David Veniaminovich* (8.2.1927–1987)
Graduated from the Moscow Aviation Institute in 1953.

1953–1962 Zvezda, design engineer
1962–1986 chief of a design team for aviation suits and spacesuit LSSs.

Active participant in designing LSSs for all spacesuits developed at Zvezda up to the mid-1980s and the airlock system for *Voskhod-2*.

*Mikhailov, Boris Vasilievich* (born 10.11.1929)
Graduated from the Moscow Aviation Technological Institute in 1952.

1952–1955 Zvezda, designer
1955–1964 chief of the altitude test laboratory
1964– chief of the test department for aviation and space outfit and LSS
1964–1993 manager of pre-launch work with Zvezda-developed systems at Baikonur for almost all manned missions, launches of orbiting stations and their modules.

*Paradizov, Herman Stepanovich* (born 5.1.1938)
Graduated from the Moscow Aviation Institute in 1960.

1960– Zvezda, designer, chief of a design team, leading designer

Active participant in development of aviation and spacesuits, leading designer for development of the Baklan aviation suit.

*Severin, Guy Ilich* (born 24.7.1926)
Graduated from the Moscow Aviation Institute in 1947.

1947–1964 Flight Research Institute (LII), research and flight tests of rescue means for flight vehicle crews, including escape means from *Vostok*
1964– Zvezda, Chief Designer, and then General Director/General Designer

Under his leadership, spacesuits, LSSs, emergency escape means for all spacecraft and orbiting stations (starting with *Voskhod*), the airlock for *Voskhod-2*, equipment for cosmonaut transfer to and manoeuvrability in space and some other systems and items were developed at Zvezda. He is an Academician of the Russian Academy of Sciences, Doctor of Technical Sciences and Professor and Member of the International Academy of Astronautics.

*Sharipov, Rinat Khasanovich* (born 8.9.1938)
Graduated from the Moscow Power Engineering Institute in 1960.

1960– Zvezda, designer, leading designer and chief expert

A participant in development of the LSS for EVA spacesuits for *Soyuz* and all further spacecraft and orbiting stations. He fulfilled a lot of design and research tasks on the development of spacesuit LSS thermal control systems. He is a Candidate of Technical Sciences.

*Skomorovsky, Ilia Israelievich* (1926–1987)
Graduated from the Moscow Aviation Institute.

1953–1987 Zvezda, designer, leading designer, chief of electric equipment laboratory

Made an important contribution in design and testing of the electrical and radio systems for aviation and spacesuits. Active participant in development of the airlock system for the first ever space walk from *Voskhod-2*.

*Smotrikov, Oleg Ivanovich* (22.5.1937–3.8.1988)
Graduated from the Moscow Chemical Engineering Institute in 1959.

1959– Zvezda, designer, participant in development of LSS for *Vostok* and *Soyuz*
1964–1965 coordinator of development work on the complex of systems for the *Voskhod-2* EVA project
1963–1972 deputy chief of the aviation and spacesuit design department
1972–1980 chief of a design department
1980–1988 Deputy General Designer

Tragically died in a car accident.

*Stoklitsky, Anatoly Yudelijevich* (born 10.5.1929)
Graduated from the Moscow Aviation Institute in 1953.

1953–1959 Zvezda, design engineer in aviation and spacesuits
1959–1961 chief of the suit design team
1961– leading designer and chief expert in aviation and spacesuits

Project manager for all spacesuit enclosures for *Soyuz*, *Buran*, *Salyut*, *Mir* and *ISS*. One of the originators of the semi-rigid suit for the Soviet Moon programme, which was used as the basis for all Soviet/Russian ORLAN EVA spacesuits. He is a Candidate of Technical Sciences (1965) and Corresponding Member of the International Academy of Astronautics.

*Svertshek, Vitaly Ivanovich* (born 7.1.1932)
Graduated from the Moscow Power Engineering Institute in 1957.

1957–1972 Zvezda, senior engineer of a test department, then the chief of a test team and deputy chief of the same test department
1972– Deputy Chief Designer, and then the First Deputy General Director/ First Deputy General Designer

Direct participant and manager of research and testing of all spacesuits and their systems developed at Zvezda. Participated in preparation for the flight of Yuri Gagarin as well as other cosmonauts of the *Vostok* spacecraft series. He is a Candidate of Technical Sciences and Member of the International Academy of Astronautics.

*Tsentsiper, Zakhar Borisovich* (born 3.7.1904–1981)
Graduated from the Mendeleyev Moscow Chemistry and Technological Institute in 1930.

1930–1952 chief mechanical engineer, chief engineer and leading engineer at the plants of the Aviation Industry and Scientific Research Institute for Aviation Materials
1953–1978 Zvezda, chief of the team for research and development of new materials for spacesuits and other equipment

He made an important contribution to the development of all the spacesuits and other equipment developed at Zvezda.

*Umansky, Semion Petrovich* (born 9.5.1909)

1933–1952 TsAGI, S.A. Lavochkin's KB, Gromov Flight Research Institute (LII), design engineer, leading designer
1952–1953 Zvezda, Deputy Chief Designer and chief of the design bureau
1954–1977 leading designer of several aviation suits and the ORIOL spacesuit

The author of several publications on space technology.

*Ushinin, Vladimir Vladimirovich* (born 23.6.1935)
Graduated from the Moscow Aviation Institute in 1959.
1959–1965 Zvezda, design engineer
1965– chief of the design team for development of enclosures for space and aviation suits
1974– leading designer for development of the protective gear and oxygen equipment for Air Force pilots.

He was the leading designer of the spacesuit for the *Voskhod-2*. Active participant in the development of spacesuits for the *Vostok*, *Soyuz* and Soviet Moon projects.

*Zelvinsky, Alexander Lvovich* (14.5.1920–1980)
Graduated from the Moscow Aviation Institute in 1942.

1942–1952 Tupolev's OKB in Omsk, later in Moscow
1952–1954 Zvezda, design engineer
1954–1958 chief of the design team on suit enclosure
1958–1972 chief of a design department
1972–1980 leading designer of aviation suits

Under his leadership, the design department developed enclosures for some aviation suits as well as spacesuits for the *Vostok*, *Voskhod-2*, *Soyuz-4*, *Soyuz-5* and the Moon programme.

## 14.4 OTHER PARTICIPANTS

We would like to mention the following people who used to work at Zvezda:

| | |
|---|---|
| *V.G. Galperin* | chief of calculation and theory department |
| *V.I. Streltsova* | chief of the team for research and development of new materials (she took the place of Z.B. Tsentsiper) |
| *V.A. Frolov* | leading designer (UPMK) |

and the people who currently work at the company:

| | |
|---|---|
| *E.A. Albats* | leading designer in space suit LSS |
| *A.N. Livshits* | manager in thermal calculations and Galperin successor |
| *O.N. Bowsin* | chief of the suit enclosure design team |
| *N.I. Dergunov* | chief of the design department for development of aviation and spacesuits |
| *V.A. Kudriavtsev* | chief of the LSS design team |
| *S.S. Pozdnyakov* | leading designer (SAFER). |

We should also mention the contribution made by leading specialists of the test departments, aerospace medicine department and the many test engineers.

An important contribution was also made by the production services managed by:

| | |
|---|---|
| *V.M. Mironov* | chief engineer |
| *A.A. Miskarian* | chief engineer |
| *F.S. Timokhin* | deputy chief engineer |
| *I.A. Miloserdov* | chief of the comprehensive test laboratory, and others. |

Leading experts and active participants in spacesuit development from other companies and organizations working together with Zvezda have been:

- *Zvezda curators at RSC Energia*: I.V. Lavrov, E.N. Zaitsev, E.P. Diomin, B.V. Razgulin, A.P. Aleksandrov, O.S. Tsygankov.
- *VNIIEM*: S.A. Stoma, I.A. Vevyurko, Yu.V. Razumovsky (fan and pump electric motors).
- *Tambov NIKhI*: L.A. Gavrikov, V.N. Shubina ($CO_2$ and contamination control cartridge).
- *SKB AP*: D.M. Sheynin, M.V. Akimov (measuring system).
- *OZKBKO (Orekhovo-Zuyevo Design Bureau for Oxygen Equipment)*: P.I. Zima, E. Nesterov, B.Ya. Tereschenko, E.I. Yakovlev (oxygen equipment units).
- Specialists in suit materials: M.G. Donchenko (fabrics for suit restraint layers), M. Vdovchenkova, V.I. Nosova, S.G. Klavdienko, A.L. Vlasov (rubber parts, bladders and rubberized fabrics), M.M. Gudimov (sun visor), etc.

An important contribution to spacesuit creation was also made by: Air Force personnel (V.A. Smirnov, S.G. Frolov, Yu.D. Kilosanidze, V.D. Dubrov, N.V. Knyazev, B.V. Fiodorov); NIIAM employees (L.G. Golovkin and A.M. Genin); representatives of the Yu.A. Gagarin Cosmonaut Training Centre; many cosmonauts; and other specialists.

The highly professional collective of scientists, designers, production specialists and test personnel were instrumental in setting up the company's unique laboratory facilities and means of specialized production as well as putting together principles of the optimum way of carrying work out. This expertise made it possible to set up the *Zvezda School*, the cornerstone of which is its comprehensive approach to tasks and its strong resolve to get at the very root of problems, even if this calls for the development of untraditional engineering means and methods.

# 15

# Russian spacesuit artefacts

## 15.1 DESTINY OF SPACESUITS

In the 40 years of Soviet/Russian manned space history some 700 spacesuits were manufactured for development, testing, training and flight (Table 15.1). Of these more than half (368) were flight units. Some flight units were returned to Earth, but many were discarded in space or destroyed on re-entry with the spacecraft's habitation compartment.

As for the rescue suits (*Vostok* and *Soyuz*), all launched suits were returned to Earth along with the cosmonauts for whom they were manufactured and size-adjusted.

When it comes to the 41 EVA flight units (involving BERKUT, YASTREB and the ORLAN-family suits), the situation is different. The spacesuits used for *Voskhod-2* were dual-purpose suits (EVA and rescue), and thus all returned to Earth after use. The YASTREB spacesuits for *Soyuz-4* and *Soyuz-5* were only used for EVA and designed for one mission. They were left in the habitation compartment and burnt up, together with the compartment, during re-entry. The suits used during the operation of *Salyut-6* and *Salyut-7* and *Mir* (the ORLAN suits) were designed for a permanent stay in orbit, with size adjusting, maintenance and repair in orbit. At the end of their operational life these suits were placed in a *Progress* capsule for discarding in the Earth's atmosphere or burnt up with the station itself during re-entry into the Earth's atmosphere.

Within the whole *Mir* programme only one ORLAN-DMA suit (No. 18) was returned to Earth (in 1996 aboard the Shuttle). Zvezda and NASA jointly studied its condition, tested components from the enclosure and the LSS and determined the performance characteristics of the spacesuit's components, physical and mechanical characteristics of materials and their rate of wear After these studies, the spacesuit was modified and used as a test model. Otherwise only parts of the spacesuit enclosure were returned for examination, check of wear and deterioration as well

**Table 15.1.** Number of spacesuits produced (as of 31 December 2002).

| Spacesuit model | Suit type | No. of suits |
|---|---|---|
| SK-1 | Test units | 20 |
| | Training units | 11 |
| | Flight models | 9* |
| SK-2 | Test units | 4 |
| | Training units | 2 |
| | Flight models | 2 |
| BERKUT | Test and training units | 9 |
| | Flight models | 4 |
| YASTREB | Test and training units | 18 |
| | Flight models | 6 |
| SKV | Test units | 5 |
| KRECHET | Test units | 3 |
| KRECHET 94 | Test and training units | 22 |
| | Flight models | 0 |
| ORIOL | Test units | 3 |
| ORLAN | Test and training units | 11 |
| | Flight models | 0 |
| ORLAN-D | Test and training units | 27 |
| | Flight models | 7 |
| ORLAN-DM | Test and training units | 5 |
| | Flight models | 5 |
| ORLAN-DMA | Test and training units | 16 |
| | Flight models | 12 |
| ORLAN-M | Test and training units | 17 |
| | Flight models | 7 |
| SOKOL-K | Test and training units | 66 |
| | Flight models | 89 |
| SOKOL-KV | Test and training units | 6 |
| SOKOL-KV-2 | Test and training units | 63 |
| | Flight models | 220 |
| STRIZH | Test and training units | 27 |
| | Flight models | 4 |
| EVA SUIT 2000 | Test units | 2 |

* Two flight units for dummy flights and seven for cosmonauts.

as failed LSS components for examination. Because they were cosmonaut-customized, gloves were generally returned to become souvenirs.

## 15.2 LOCATION OF ARTEFACTS

Spacesuits have always been items of extreme importance for publicity and exhibition, as they were in the Soviet Union era. Today a small number of suits and their locations are known. Many suits, which were available at Zvezda, the Gagarin Cosmonaut Training Centre (GCTC), the Memorial Museum of Cosmonautics, local museums, exhibitions and schools have, after dissolution of the Soviet Union in 1990, been sold at auctions or directly to private collectors. Some spacesuits were given as gifts to foreign cosmonauts or their national museums. At the time of the USSR, artefacts were distributed to some 60 locations all over the USSR.

Furthermore, a number of suits on display in the early 1990s in exhibitions of the former Soviet Union have disappeared and their present whereabouts are not known. This was particularly true of the exhibits in the splendid Pavilion of Cosmonautics at the Exhibition of National Economy Achievements in Moscow. The exact locations of spacesuits exhibited in the various shows, establishments and educational institutions, as well as those in private collections, are today almost impossible to trace.

It goes without saying that those spacesuits and other artefacts that were involved in the breakthrough phases of space exploration, and returned to the Earth are of the highest historical value:

- the two "Ivan Ivanovich" dummies, dressed in SK-1 spacesuits, which were used in the first space mission aboard *Vostok*, prior to Yuri Gagarin's mission;
- the SK-1 spacesuit worn by Yuri Gagarin, the first man in space;
- the SK-2 spacesuit worn by Valentina Tereshkova, the first woman in space;
- the BERKUT spacesuit worn by A.A. Leonov, who was the first man to spacewalk (from the *Voskhod-2* spacecraft);
- the BERKUT spacesuit worn by P. Belyaev, *Voskhod-2*'s commander.

One "Ivan Ivanovich" manikin wearing the SK-1 spacesuit is in the RSC Energia Museum and the other one has been sold at a Sotheby's auction (its location is unknown). The other spacesuits are in Tomilino at the Zvezda Museum, except for Belyaev's BERKUT spacesuit, which is located in the RSC Energia Museum.

# Appendix 1

## Summary of USSR/Russian Federation EVA statistics

| EVA No. | Mission | Date[1] | Crew | Suit type | Suit No. | EVA duration (hr : min) | Accomplishments remarks |
|---|---|---|---|---|---|---|---|
| 1 | *Voskhod-2* | 18.03.65 | A.A. Leonov | BERKUT, LSS KP-55, emergency oxygen and electrical umbilical | | 0:12 | First space walk; assessment of human EVA capabilities; first use of a soft inflatable airlock; distance from the spacecraft up to 5 m |
| 2 | *Soyuz-5/4* | 16.01.69 | A.S. Yeliseyev<br>Ye.V. Khrunov | YASTREB, LSS RVR-IP, electrical umbilical | | 0:37 | First crew transfer from one spacecraft to another via open space (from *Soyuz-5* to *Soyuz-4*) |
| 3 | *Salyut-6* | 20.12.77 | Yu.V. Romanenko<br>G.M. Grechko | ORLAN-D | 33<br>34 | 1:28 | *Salyut-6* docking unit outer surface inspection; first use of semi-rigid spacesuit; Romanenko stayed in depressurized airlock |
| 4 | *Salyut-6* | 29.07.78 | V.V. Kovalyonok<br>A.S. Ivanchenkov | ORLAN-D | 33<br>34 | 2:05 | Retrieved samples exposed to outer space and installed new samples |
| 5 | *Salyut-6* | 15.08.79 | V.A. Lyakhov<br>V.V. Ryumin | ORLAN-D | 33<br>34 | 1:23 | Unplanned EVA with movement to the end of *Salyut-6* to dislodge jammed KRT-10 radio telescope antenna; dismantled research equipment |

(*continued*)

[1] Date of EVA hatch opening (Moscow time) or start of pressure decrease in the airlock or unpressurized module below 400 hPa (for IVA).

| EVA No. | Mission | Date | Crew | Suit type | Suit No. | EVA duration (hr : min) | Accomplishments remarks |
|---|---|---|---|---|---|---|---|
| 6 | *Salyut-7* | 30.07.82 | A.N. Berezovoy<br>V.V. Lebedev | ORLAN-D | 45<br>46 | 2:33 | Dismantled and partially replaced samples and research equipment; performed engineering checks to evaluate different types of mechanical joints |
| 7 | *Salyut-7* | 01.11.83 | V.A. Lyakhov<br>A.P. Alexandrov | ORLAN-D | 45<br>46 | 2:50 | Mounted additional solar cell panel; repaired EVA suit (No. 46) pressure shell (bladder) on board the station |
| 8 | *Salyut-7* | 03.11.83 | V.A. Lyakhov<br>A.P. Alexandrov | ORLAN-D | 45<br>46 | 2:55 | Mounted second, additional solar cell panel |
| 9 | *Salyut-7* | 23.04.84 | L.D. Kizim<br>V.A. Solovyov | ORLAN-D | 45<br>47 | 4:15 | Installed ramp and other equipment and prepared workstation for repair of fuel subsystem of the integrated propulsion system (IPS) |
| 10 | *Salyut-7* | 26.04.84 | L.D. Kizim<br>V.A. Solovyov | ORLAN-D | 45<br>47 | 5:00 | Repaired IPS fuel system |
| 11 | *Salyut-7* | 29.04.84 | L.D. Kizim<br>V.A. Solovyov | ORLAN-D | 45<br>47 | 2:45 | Continued IPS repair |
| 12 | *Salyut-7* | 04.05.84 | L.D. Kizim<br>V.A. Solovyov | ORLAN-D | 45<br>47 | 2:45 | Completed IPS repair |
| 13 | *Salyut-7* | 18.05.84 | L.D. Kizim<br>V.A. Solovyov | ORLAN-D | 45<br>47 | 3:05 | Installed additional solar cell panels |
| 14 | *Salyut-7* | 25.07.84 | V.A. Dzhanibekov<br>S.E. Savitskaya | ORLAN-D | 45<br>47 | 3:35 | First woman to space-walk; worked with a manual welding tool; retrieved samples located on *Salyut-7*'s external surface |
| 15 | *Salyut-7* | 08.08.84 | L.D. Kizim<br>V.A. Solovyov | ORLAN-D | 45<br>47 | 5:00 | Fuel system work at rear of *Salyut-7* assembly module; dismantled solar cell component for return to Earth |
| 16 | *Salyut-7* | 02.08.85 | V.A. Dzhanibekov<br>V.P. Savinykh | ORLAN-DM | 10<br>8 | 5:00 | Installed additional solar panels on third solar cell; modified suit used for EVA |
| 17 | *Salyut-7* | 28.05.86 | L.D. Kizim<br>V.A. Solovyov | ORLAN-DM | 10<br>8 | 3:50 | Tested methods for installing large structures using a hinged lattice truss |
| 18 | *Salyut-7* | 31.05.86 | L.D. Kizim<br>V.A. Solovyov | ORLAN-DM | 10<br>8 | 5:00 | Continued testing methods for installing large structures; removed solar panel sample |
| 19 | *Mir* | 11.04.87 | Yu.V. Romanenko<br>A.I. Laveykin | ORLAN-DM | 7<br>9 | 3:40 | Unscheduled EVA to remove a foreign object interfering with docking of *Kvant* with *Mir* |

| EVA No. | Mission | Date | Crew | Suit type | Suit No. | EVA duration (hr : min) | Accomplishments remarks |
|---|---|---|---|---|---|---|---|
| 20 | *Mir* | 12.06.87 | Yu.V. Romanenko<br>A.I. Laveykin | ORLAN-DM | 7<br>9 | 1 : 53 | Mounted solar cell on truss structure and two sections of photoelectric transducers |
| 21 | *Mir* | 16.06.87 | Yu.V. Romanenko<br>A.I. Laveykin | ORLAN-DM | 7<br>9 | 3 : 15 | Completed installation of solar cell and unfolding the truss; installed test samples |
| 22 | *Mir* | 26.02.88 | V.G. Titov<br>M.K. Manarov | ORLAN-DM | 7<br>9 | 4 : 24 | Mounted an experimental solar cell; installed test samples |
| 23 | *Mir* | 30.06.88 | V.G. Titov<br>M.K. Manarov | ORLAN-DM | 7<br>9 | 5 : 10 | Replaced sensor unit on *Kvant* module's X-ray telescope; work was not completed because of breakdown of device for dislodging the sensor unit |
| 24 | *Mir* | 20.10.88 | V.G. Titov<br>M.K. Manarov | ORLAN-DMA | 6<br>10 | 4 : 12 | Completed replacing the sensor on the *Kvant* module's X-ray telescope; first phase of evaluation of modified spacesuit with power umbilical to station |
| 25 | *Mir* | 09.12.88 | A.A. Volkov<br>J.L. Chrétien<br>(French) | ORLAN-DMA | 6<br>10 | 6 : 00 | Conducted experiments with the Soviet-French program Aragats; mounted and deployed a truss structure; mounted a panel with test samples to study space exposure effect on different materials |
| 26 | *Mir* | 08.01.90 | A.S. Viktorenko<br>A.A. Serebrov | ORLAN-DMA | 6<br>10 | 2 : 56 | Mounted two star-navigation sensors to raise accuracy of the navigation and guidance system of the station |
| 27 | *Mir* | 11.01.90 | A.S. Viktorenko<br>A.A. Serebrov | ORLAN-DMA | 6<br>10 | 2 : 54 | Mounted scientific equipment on *Kvant* module; retrieved samples, prepared transfer for later docking of module *Kristall* |
| 28 | *Mir* | 26.01.90 | A.S. Viktorenko<br>A.A. Serebrov | ORLAN-DMA | 12<br>8 | 3 : 02 | First egress from *Kvant-2* airlock; tested modified self-contained spacesuits; first EVA without power umbilical; installed additional equipment on the module |
| 29 | *Mir* | 0.1.02.90 | A.S. Viktorenko<br>A.A. Serebrov | ORLAN-DMA | 12<br>8 | 4 : 59 | First Soviet test of manoeuvring unit (USPK); Serebrov moved 33 m from the station |

(*continued*)

| EVA No. | Mission | Date | Crew | Suit type | Suit No. | EVA duration (hr : min) | Accomplishments remarks |
|---|---|---|---|---|---|---|---|
| 30 | *Mir* | 05.02.90 | A.S. Viktorenko<br>A.A. Serebrov | ORLAN-DMA | 12<br>8 | 3:45 | Viktorenko performed additional detailed testing of the USPK by moving 45 m away |
| 31 | *Mir* | 17.07.90 | A.Yu. Solovyov<br>A.N. Balandin | ORLAN-DMA | 12<br>8 | 7:00 | Inspected and repaired *Soyuz-TM* damaged thermal insulation; after return to airlock, hatch failed to close completely; *Kvant-2* instrumentation module depressurized to let cosmonauts enter station |
| 32 | *Mir* | 26.07.90 | A.Yu. Solovyov<br>A.N. Balandin | ORLAN-DMA | 12<br>8 | 3:52 | Dismantled equipment on *Kvant-2* module's exterior, inspected damaged hatch, closed it; sealed and pressurized module |
| 33 | *Mir* | 30.10.90 | G.M. Manakov<br>G.M. Strekalov | ORLAN-DMA | 12<br>10 | 2:45 | Attempted hatch repair using special hardware; troubleshooting revealed need to replace one of hatch axis support brackets |
| 34 | *Mir* | 07.01.91 | V.M. Afanasyev<br>M.K. Manarov | ORLAN-DMA | 6<br>10 | 5:18 | Repaired *Kvant-2* module's hatch and replaced bracket and bearing; disassembled instruments; took out solar battery truss to be installed |
| 35 | *Mir* | 23.01.91 | V.M. Afanasyev<br>M.K. Manarov | ORLAN-DMA | 6<br>10 | 5:33 | Installed 14-m (46-ft) boom (Strela) on station core |
| 36 | *Mir* | 26.01.91 | V.M. Afanasyev<br>M.K. Manarov | ORLAN-DMA | 6<br>10 | 6:20 | Lifted solar battery trusses by the boom from egress hatch and installed on *Kvant* module |
| 37 | *Mir* | 25.04.91 | V.M. Afanasyev<br>M.K. Manarov | ORLAN-DMA | 6<br>10 | 3:34 | Conducted experiment to test construction techniques for assembly of large structures in space; inspected damaged antenna located on the end plane of the *Kvant* module, which contributed to the *Progress* M7 supply ship docking problem |
| 38 | *Mir* | 25.06.91 | A.P. Artsebarskiy<br>S.K. Krikalev | ORLAN-DMA | 6<br>14 | 4:58 | Repair damaged antenna at the edge of *Kvant* module |
| 39 | *Mir* | 28.06.91 | A.P. Artsebarskiy<br>S.K. Krikalev | ORLAN-DMA | 6<br>14 | 3:25 | Installed research equipment; prepared Strela boom |
| 40 | *Mir* | 15.07.91 | A.P. Artsebarskiy<br>S.K. Krikalev | ORLAN-DMA | 6<br>8 | 5:56 | Prepared the workstation (based on the *Kvant* module) for the Sofora experiment, designed to development new methods of assembling large structures in space, using thermal mechanical connections |

| EVA No. | Mission | Date | Crew | Suit type | Suit No. | EVA duration (hr : min) | Accomplishments remarks |
|---|---|---|---|---|---|---|---|
| 41 | *Mir* | 19.07.91 | A.P. Artsebarskiy<br>S.K. Krikalev | ORLAN-DMA | 6<br>8 | 5:28 | Installed the work platform and berth; assembled the Sofora truss structure |
| 42 | *Mir* | 23.07.91 | A.P. Artsebarskiy<br>S.K. Krikalev | ORLAN-DMA | 6<br>8 | 5:42 | Continued assembling the Sofora truss structure |
| 43 | *Mir* | 27.07.91 | A.P. Artsebarskiy<br>S.K. Krikalev | ORLAN-DMA | 6<br>8 | 6:49 | Completed Sofora experiment; assembled and mounted 14-m (46-ft) truss |
| 44 | *Mir* | 20.02.92 | A.A. Volkov<br>S.K. Krikalev | ORLAN-DMA | 12<br>8 | 4:13 | Installed scientific equipment on *Kvant* module; collected samples from experiment solar panel; due to malfunctioning of spacesuit cooling garment, commander had to be connected to the airlock cooling system for the better portion of the work period (sublimator was not switched ON) |
| 45 | *Mir* | 08.07.92 | A.S. Viktorenko<br>A.Yu. Kaleri | ORLAN-DMA | 15<br>14 | 2:03 | Installed vacuum lines on station surface for operation of gyro dynes on *Kvant-2* module |
| 46 | *Mir* | 03.09.92 | A.Yu. Solovyov<br>S.V. Avdeyev | ORLAN-DMA | 15<br>14 | 3:56 | Performed preparatory work for the reinstallation of a movable propulsion system (MPS) for the space station complex |
| 47 | *Mir* | 07.09.92 | A.Yu. Solovyov<br>S.V. Avdeyev | ORLAN-DMA | 15<br>14 | 5:08 | Installed a cable on the Sofora truss for the MPS and mated connectors to *Kvant* module |
| 48 | *Mir* | 11.09.92 | A.Yu. Solovyov<br>S.V. Avdeyev | ORLAN-DMA | 15<br>14 | 5:44 | Mated MPS with the Sofora truss and activated it |
| 49 | *Mir* | 15.09.92 | A.Yu. Solovyov<br>S.V. Avdeyev | ORLAN-DMA | 15<br>14 | 3:33 | Dismantled *Kristall* module antenna and scientific equipment |
| 50 | *Mir* | 19.04.93 | G.M. Manakov<br>A.F. Politshuk | ORLAN-DMA | 15<br>14 | 5:25 | Removed and transferred a solar array electric drive from the *Kristall* module and installed it on the *Kvant* module |
| 51 | *Mir* | 18.06.93 | G.M. Manakov<br>A.F. Politshuk | ORLAN-DMA | 15<br>14 | 4:33 | Removed and transferred the second solar array's electric drive from the *Kristall* module and installed it on the *Kvant* module |

(*continued*)

| EVA No. | Mission | Date | Crew | Suit type | Suit No. | EVA duration (hr : min) | Accomplishments remarks |
|---|---|---|---|---|---|---|---|
| 52 | *Mir* | 16.09.93 | V.V. Tsibliyev<br>A.A. Serebrov | ORLAN-DMA | 25<br>14 | 4 : 18 | Installed Rapana experimental truss with scientific equipment on the *Kvant* module; removed metallic and non-metallic material samples; installed specimens of composite materials on truss |
| 53 | *Mir* | 20.09.93 | V.V. Tsibliyev<br>A.A. Serebrov | ORLAN-DMA | 15<br>14 | 3 : 13 | Installed a platform with scientific equipment on the Rapana truss |
| 54 | *Mir* | 28.09.93 | V.V. Tsibliyov<br>A.A. Serebrov | ORLAN-DMA | 25<br>14 | 1 : 51 | Disassembled panel from cosmic ray experiment and installed new scientific cassette with samples of construction materials |
| 55 | *Mir* | 22.10.93 | V.V. Tsibliyev<br>A.A. Serebrov | ORLAN-DMA | 25<br>14 | 0 : 38 | Installed new equipment unit for measuring micrometeorite streams; visually inspected and took video of individual components of station looking for metal corrosion and changes in coloration |
| 56 | *Mir* | 29.10.93 | V.V. Tsibliyev<br>A.A. Serebrov | ORLAN-DMA | 25<br>18 | 4 : 12 | Visually inspected elements status of external station surface and performed set of preventive maintenance operations |
| 57 | *Mir* | 09.09.94 | Yu.I. Malenchenko<br>T.A. Musabayev | ORLAN-DMA | 25<br>18 | 5 : 03 | Repair of thermal insulation on transfer compartment; preparation for installation of the second cargo-carrying boom; removal of samples |
| 58 | *Mir* | 13.09.94 | Yu.I. Malenchenko<br>T.A. Musabayev | ORLAN-DMA | 25<br>18 | 6 : 01 | Installed a platform with solar cell drive; scientific experiments |
| 59 | *Mir* | 12.05.95 | V.M. Dezhurov<br>G.M. Strekalov | ORLAN-DMA | 27<br>18 | 6 : 08 | Preparation to transfer a modernized solar cell from the *Kristall* module to the *Kvant* module |
| 60 | *Mir* | 17.05.95 | V.M. Dezhurov<br>G.M. Strekalov | ORLAN-DMA | 27<br>18 | 6 : 54 | Transfer of modernized solar cell to the *Kvant* module and its partial deployment |
| 61 | *Mir* | 22.05.95 | V.M. Dezhurov<br>G.M. Strekalov | ORLAN-DMA | 27<br>18 | 5 : 15 | Deployed solar cell on *Kvant* module and connected it up; folded second solar cell on *Kristall* module |
| 62<br>(IVA) | *Mir* | 29.05.95 | V.M. Dezhurov<br>G.M. Strekalov | ORLAN-DMA | 27<br>18 | 0 : 20 | Transferred docking cone to "Z" position; worked inside transfer compartment using electrical umbilical |

| EVA No. | Mission | Date | Crew | Suit type | Suit No. | EVA duration (hr : min) | Accomplishments remarks |
|---|---|---|---|---|---|---|---|
| 63 (IVA) | *Mir* | 02.06.95 | V.M. Dezhurov<br>G.M. Strekalov | ORLAN-DMA | 27<br>18 | 0 : 23 | Transferred docking cone to "Y" position; worked inside transfer compartment using electrical umbilical |
| 64 | *Mir* | 14.07.95 | A.Yu. Solovyov<br>A.M. Budarin | ORLAN-DMA | 27<br>18 | 5 : 34 | Inspected and deployed *Spektr* module solar panel No. 4; inspected docking unit surface |
| 65 | *Mir* | 19.07.95 | A.Yu. Solovyov<br>A.M. Budarin | ORLAN-DMA | 27<br>18 | 3 : 08 | Removed Trek equipment and panel with samples |
| 66 | *Mir* | 21.07.95 | A.Yu. Solovyov<br>A.M. Budarin | ORLAN-DMA | 27<br>18 | 5 : 50 | Installed Miras spectrometer on *Spektr* module |
| 67 | *Mir* | 20.10.95 | S.V. Avdeyev<br>T. Reiter (German) | ORLAN-DMA | 18<br>26 | 5 : 16 | Installed European science equipment for material exposure; removal of earlier installed samples |
| 68 (IVA) | *Mir* | 08.12.95 | Yu.P. Gidzenko<br>S.V. Avdeyev | ORLAN-DMA | 25<br>18 | 0 : 29 | Inspected the surface for *Priroda* module docking and installed the docking cone |
| 69 | *Mir* | 08.02.96 | Yu.P. Gidzenko<br>T. Reiter (German) | ORLAN-DMA | 25<br>26 | 3 : 06 | Removed cassettes from European science equipment; installed new samples |
| 70 | *Mir* | 15.03.96 | Yu.I. Onufriyenko<br>Yu.V. Usachev | ORLAN-DMA | 25<br>26 | 5 : 51 | Mounted second cargo-carrying boom on right-hand side of station base module |
| 71 | *Mir* | 21.05.96 | Yu.I. Onufriyenko<br>Yu.V. Usachyov | ORLAN-DMA | 25<br>26 | 5 : 19 | Transferred additional solar cell from the docking compartment and its installation on *Kvant* module |
| 72 | *Mir* | 24.05.96 | Yu.I. Onufriyenko<br>Yu.V. Usachyov | ORLAN-DMA | 25<br>26 | 5 : 43 | Connected and deployed additional solar cell on *Kvant* module |
| 73 | *Mir* | 30.05.96 | Yu.I. Onufriyenko<br>Yu.V. Usachyov | ORLAN-DMA | 25<br>26 | 4 : 20 | Installed MOMS-2 equipment on *Spektr* module |
| 74 | *Mir* | 06.06.96 | Yu.I. Onufriyenko<br>Yu.V. Usachyov | ORLAN-DMA | 25<br>26 | 3 : 36 | Installed Komza science equipment |
| 75 | *Mir* | 13.06.96 | Yu.I. Onufriyenko<br>Yu.V. Usachyov | ORLAN-DMA | 25<br>26 | 5 : 42 | Installed Truss-3 equipment on *Kvant* module |
| 76 | *Mir* | 02.12.96 | V.G. Korzun<br>A.Yu. Kaleri | ORLAN-DMA | 27<br>26 | 5 : 57 | Ran electric cable to additional solar cell on *Kvant* module; installed Rapana equipment on Truss-3 |

(*continued*)

| EVA No. | Mission | Date | Crew | Suit type | Suit No. | EVA duration (hr : min) | Accomplishments remarks |
|---|---|---|---|---|---|---|---|
| 77 | *Mir* | 09.12.96 | V.G. Korzun<br>A.Yu. Kaleri | ORLAN-DMA | 27<br>26 | 6 : 38 | Installed antenna for Kurs system on the docking compartment; mated electric connectors to additional solar cell |
| 78 | *Mir* | 29.04.97 | V.V. Tsibliyev<br>J. Linenger (American) | ORLAN-M | 5<br>4 | 4 : 57 | Installed US optic characteristic monitor and radiation sensor (OPM); removed space radiation recording instrument |
| 79 (IVA) | *Mir* | 22.08.97 | A.Yu. Solovyov<br>P.V. Vinogradov | ORLAN-DMA | 27<br>26 | 3 : 16 | Mated electric connectors of unpressurized *Spektr* module's solar cell and base module; inspected places of possible loss of pressure-tightness (entry from transfer compartment to unpressurized *Spektr* module using extension hoses) |
| 80 | *Mir* | 06.09.97 | A.Yu. Solovyov<br>M. Foale (American) | ORLAN-M | 5<br>4 | 6 : 00 | Inspected damaged *Spektr* module (from outside); installed additional handrails |
| 81 (IVA) | *Mir* | 20.10.97 | A.Yu. Solovyov<br>P.V. Vinogradov | ORLAN-DMA | 27<br>26 | 6 : 38 | Connected electric cables to solar cell control units in *Spektr* module to provide for solar cell orientation (entry from transfer compartment to unpressurized *Spektr* module using extension hoses) |
| 82 | *Mir* | 03.11.97 | A.Yu. Solovyov<br>P.V. Vinogradov | ORLAN-M | 5<br>4 | 6 : 04 | Folded and removed modified solar cell from *Kvant* module, transferred it to the base module; launched a mock-up of the first Earth satellite; plugged *Vozdukh* system. Airlock hatch was not pressure-tight after closing (returned to PNO), inner compartment of *Kvant-2*, used as a redundant airlock |
| 83 | *Mir* | 06.11.97 | A.Yu. Solovyov<br>P.V. Vinogradov | ORLAN-M | 5<br>4 | 6 : 12 | Secured and deployed modified solar cell on *Kvant* module |
| 84 | *Mir* | 09.01.98 | A.Yu. Solovyov<br>P.V. Vinogradov | ORLAN-M | 5<br>4 | 3 : 06 | Removed US OPM equipment from *Kvant* module; inspected airlock's EVA hatch |
| 85 | *Mir* | 14.01.98 | A.Yu. Solovyov<br>D. Wolf (American) | ORLAN-M | 5<br>4 | 3 : 52 (in pressurized suit 6 : 47) | Unloaded and worked with US SPSR equipment; access time (2 : 55) was required to open and close damaged hatch |

| EVA No. | Mission | Date | Crew | Suit type | Suit No. | EVA duration (hr : min) | Accomplishments remarks |
|---|---|---|---|---|---|---|---|
| 86 | *Mir* | 03.03.98 | T.A. Musabayev<br>N.M. Budarin | ORLAN-M | 5<br>6 | | EVA sortie was not performed due to a failure to open airlock hatch lock; were in pressurized suits for 2.5 hr |
| 87 | *Mir* | 01.04.98 | T.A. Musabayev<br>N.M. Budarin | ORLAN-M | 5<br>6 | 6 : 26 | Getting ready for *Spektr* repair; preliminarily secured damaged solar cell |
| 88 | *Mir* | 06.04.98 | T.A. Musabayev<br>N.M. Budarin | ORLAN-M | 5<br>6 | 4 : 23 | Installed fixing beam on damaged solar cell of *Spektr* module; began work on changing the portable power plant (PPP); removed Rapana equipment |
| 89 | *Mir* | 11.04.98 | T.A. Musabayev<br>N.M. Budarin | ORLAN-M | 5<br>6 | 6 : 25 | Removed and threw portable power plant (PPP) away; put a cap on *Electron* valve |
| 90 | *Mir* | 17.04.98 | T.A. Musabayev<br>N.M. Budarin | ORLAN-M | 4<br>6 | 6 : 33 | Prepared PPP delivered by *Progress* for installation on Sofora; folded Truss-3 |
| 91 | *Mir* | 22.04.98 | T.A. Musabayev<br>N.M. Budarin | ORLAN-M | 4<br>6 | 6 : 21 | Attached new PPP to Sofora |
| 92<br>(IVA) | *Mir* | 15.09.98 | G.I. Padalka<br>S.V. Avdeyev | ORLAN-M | 4<br>5 | 0 : 30 | Mated *Spektr* electric connectors with feed-thru board in docking module (DM) |
| 93 | *Mir* | 10.11.98 | G.I. Padalka<br>S.V. Avdeyev | ORLAN-M | 4<br>5 | 5 : 54 | Installed French scientific equipment; launched artificial Earth satellite mock-up |
| 94 | *Mir* | 16.04.99 | V.M. Afanasyev<br>J-P. Haignéré<br>(French) | ORLAN-M | 5<br>4 | 6 : 19 | Ran CNES science experiments; launched second artificial Earth satellite mock-up |
| 95 | *Mir* | 23.07.99 | V.M. Afanasyev<br>S.V. Avdeyev | ORLAN-M | 6<br>4 | 6 : 07 | Removed samples from French equipment; began opening reflector antenna |
| 96 | *Mir* | 28.07.99 | V.M. Afanasyev<br>S.V. Avdeyev | ORLAN-M | 6<br>4 | 5 : 22 | Completed reflector opening and pushed it off into free space; picked up samples of Sprut equipment |
| 97 | *Mir* | 12.05.00 | S.V. Zalentin<br>A.Yu. Kaleri | ORLAN-M | 6<br>4 | 4 : 52 | Ran experiment with the use of a sealing tool; worked with Panorama equipment; inspected solar cell damage; removed experimental solar cell |
| 98(1)[2] | *ISS* | 08.06.01 | Yu.V. Usachiov<br>J. Voss (American) | ORLAN | 23<br>12 | 0 : 20 | Transfer of the docking cone |

(*continued*)

[2] EVA number for *ISS* is given in parentheses.

| EVA No. | Mission | Date | Crew | Suit type | Suit No. | EVA duration (hr : min) | Accomplishments remarks |
|---|---|---|---|---|---|---|---|
| 99(2) | *ISS* | 08.10.01 | V.M. Dezhurov<br>M.V. Tyurin | ORLAN-M | 23<br>12 | 4:58 | Mating of the Transit system electric connectors; installation of additional handrails; Strela assembly and check at DM-1 |
| 100(3) | *ISS* | 15.10.01 | V.M. Dezhurov<br>M.V. Tyurin | ORLAN-M | 23<br>12 | 5:51 | Installation of Kromka samples; removal of Flag samples |
| 101(4) | *ISS* | 13.11.01 | V.M. Dezhurov<br>F. Culbertson (American) | ORLAN-M | 23<br>12 | 5:05 | Mating of the Kurs-P electric connectors of the DM-1 and SM;[3] checking Strela (GStM1) for proper operation |
| 102(4a) | *ISS* | 03.12.01 | V.M. Dezhurov<br>M.V. Tyurin | ORLAN-M | 23<br>12 | 2:45 | Inspection of the SM docking port; removal of an incidental object from the port |
| 103(5) | *ISS* | 15.01.02 | Yu.I. Onufriyenko<br>C. Walz (American) | ORLAN-M | 14<br>12 | 6:02 | Transfer and assembly of the Strela (GStM2) on the DM-1 |
| 104(6) | *ISS* | 25.01.02 | Yu.I. Onufriyenko<br>D. Bursch (American) | ORLAN-M | 14<br>12 | 5:59 | Installation of containers with samples SKK No. 1 on DM-1, SKK No. 2 on SM, protective device (GZU) on the SM engines, Platan-M, RLSWA4 antenna for amateur radio communication |
| 105(7) | *ISS* | 16.08.02 | V.G. Korzun<br>P. Whitson (American) | ORLAN-M | 14<br>23 | 4:23 | Carrying out and installation of the panel to protect from meteoroid penetration on the SM module |
| 106(8) | *ISS* | 26.08.02 | V.G. Korzun<br>S.E. Treschev | ORLAN-M | 14<br>23 | 5:21 | Operation with NAZDA; removal of Kromka-1 samples; installation of Kromka-2 samples; installation of the antenna for amateur radio communication |

[3] SM—service module.

# Appendix 2

## The main characteristics of ORLAN-type spacesuits

| Characteristic | Spacesuit model/modification | | | | | | |
|---|---|---|---|---|---|---|---|
| | ORLAN | ORLAN-D (*Salyut-6*) | ORLAN-D (*Salyut-7*) | ORLAN-DM (*Mir*) | ORLAN-DMA (*Mir*) | ORLAN-M | |
| | | | | | | *Mir* | *ISS* |
| Beginning of suit operation (year) | 1969* | 1977 | 1982 | 1985 | 1988 | 1997 | 2001 |
| Guaranteed number of EVAs | 2[1] | 6 | 10 | 10 | 10 | 12 | 12–15 |
| Total work duration in one cycle (max hr) | 5 | 5 | 7 | 8 | 9 | 9 | 9 |
| Duration of the $CO_2$ removal cartridge operation (hr) | 5 | 5 | 5 | 6 | 7 | 7 | 9 |
| Duration of autonomous operation in one cycle (hr) | 5 | 5 | 5 | 6 | 6+1 | 6+1 | 6+1 |
| Life time (years) | 2.5 | 3.5 | 4 | 4 | 4 | 4 | 4 |
| Wet mass (kg. max) | 59 | 73.5 | 73.5 | 88 | 105 | 112 | 112 |
| Stored $O_2$ (kg) | | | | | | | |
| primary | 0.5 | 1 | 1 | 1 | 1 | 1 | 1 |
| reserve | 0.2 | 1 | 1 | 1 | 1 | 1 | 1 |
| Feedwater capacity (kg) | 2.5 | 2.9 | 2.9 | 2.9 | 3.6 | 3.6 | 3.6 |
| Heat removal (W) | | | | | | | |
| average | 250 | 300 | 300 | 300 | 300 | 300 | 300 |
| maximal | 600 | 600 | 600 | 600 | 600 | 600 | 600 |
| Electric power supply | Umbilical | Umbilical | Umbilical | Umbilical | Battery since 1990 (umbilical) | Battery (umbilical) | Battery (umbilical) |
| Consumed power (W) | 30 | 32 | 32 | 32 | 42 | 54 | 54 |
| Number of measured parameters | 3 | 14 | 14 | 17 | 23 | 26 | 29[2] |

(*continued*)

[1] Without refilling or replacement of consumables (total time of two cycles was 5 hours).
[2] Including two SAFER parameters.
* State of readiness for beginning of flight operation.

| Characteristic | Spacesuit model/modification | | | | | | |
|---|---|---|---|---|---|---|---|
| | ORLAN | ORLAN-D (*Salyut-6*) | ORLAN-D (*Salyut-7*) | ORLAN-DM (*Mir*) | ORLAN-DMA (*Mir*) | ORLAN-M | |
| | | | | | | *Mir* | *ISS* |
| Spacesuit pressure (hPa) | | | | | | | |
| primary mode | 400 | 400 | 400 | 400 | 400 | 400 | 400 |
| reserve mode | 270 | 270 | 270 | 270 | 270 | 392 | 392 |
| Amount of $O_2$ emergency supply, manually activated (kg hr$^{-1}$) | 1 | 2 | 2 | 2 | 2 | 2 | 2 |
| Spacesuit pressure during emergency $O_2$ supply automatic activation (hPa) | 220 | 220 | 220 | 220 | 220 | 270 | 270 |
| On-board system | | BSS-1 | BSS-2 | BSS-2M | BSS-2M | BSS-2M | BSS-4 |

# Appendix 3

## Technical details of spacesuits

# VOSTOK SUIT SK-1 AND SK-2

**Official name** SK-1 spacesuit for male cosmonauts.
SK-2 spacesuit for female cosmonauts.

**Description** The pressure spacesuit was of the soft type, with an open-type ventilation system. The enclosure incorporated two layers: the outer restraint layer was made of Lavsan and the inner pressure bladder of sheet rubber. A protective coverall was put over the enclosure. Under the enclosure, a thermal protection suit with hoses and ventilation system panels was worn. The helmet was rigid and unremovable, with a double visor and a system for emergency visor slide-down. From a design point of view the SK-1 and SK-2 suits were identical. The SK-2 suit differed from the SK-1 in the enclosure tailoring, which took into account specific features of the female body.

**Utilization (operations)** The SK-1 (SK-2) pressure suit was designed for *Vostok* cosmonauts. The suit was first used on 12 April 1961 by Yuri Gagarin. The suit was then used in all the five following *Vostok* missions by cosmonauts of the first team up to 1963.

**Development and operation dates** Development and tests of SK-1: 1960–1961 (SK-2: 1962). Operations aboard *Vostok*: 1961–1963.

**Technical characteristics** The suit in conjunction with on-board ventilation system, on-board and parachute oxygen supply systems provided a cosmonaut with:

- normal hygienic conditions in a pressurized cabin of a spacecraft for a time period up to 12 days;
- safe stay in a depressurized cabin for 4–5 hr;
- protection during ejection from the cabin from altitudes up to 8 km;
- oxygen supply during parachute descent from a 10-km altitude;
- survival during stay in cold water for 12 hr;
- spacesuit positive operating pressure of 270–300 hPa;
- ventilating gas flow rate from the fan of 150 l $min^{-1}$;
- gas supply from the on-board bottles: 50 standard l $min^{-1}$;
- spacesuit mass of 20 kg.

**Quantity of manufactured spacesuits**

| Test models | Training models | Flight models |
|---|---|---|
| SK-1: 20 | SK-1: 11 | SK-1: 2 (for dummies) |
| SK-2: 4 | SK-2: 2 | SK-1: 7 (for five male cosmonauts) |
| | | SK-2: 2 (for female cosmonauts) |

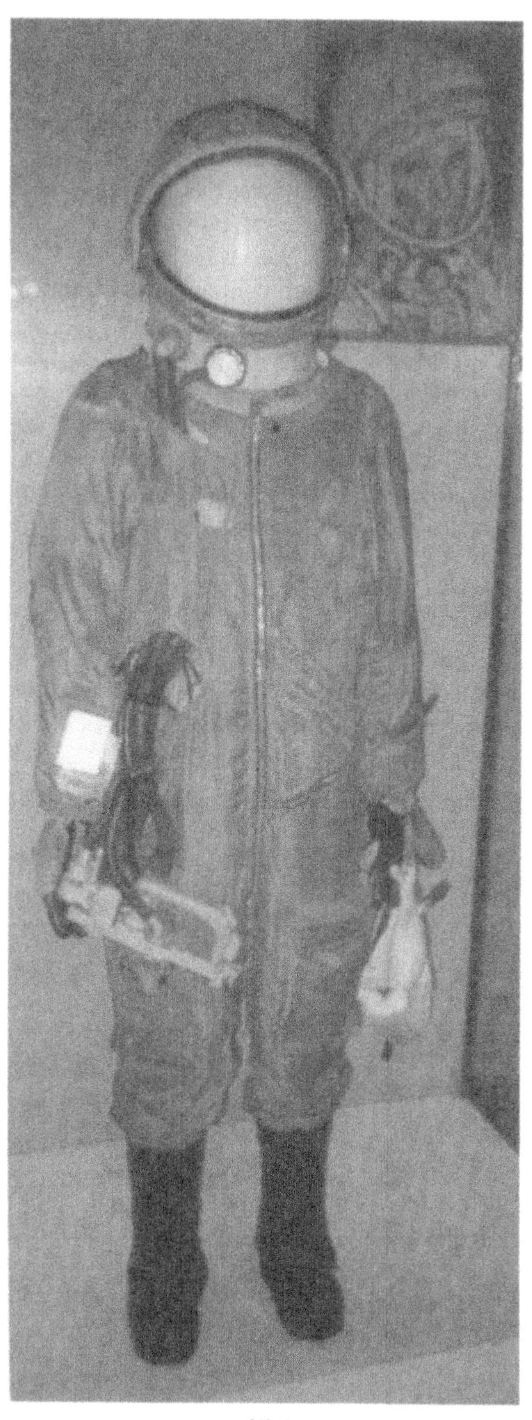

SK-1

SK-2

# VOSKHOD SUIT BERKUT

**Official name** BERKUT spacesuit.

**Description** The spacesuit was of the soft type, with a removable rigid helmet and an open-type ventilation system. The enclosure incorporated three layers: the outer restraint layer was made of Kapron and two pressure bladders, the main one was made of sheet rubber and the back-up (internal) bladder of porous rubber. A protective coverall with shield vacuum thermal insulation was put over the enclosure. The spacesuit operated in conjunction with a self-contained life support system (LSS) arranged in a removable backpack (KP-55 oxygen unit).

**Utilization (operations)** This pressure suit was designed to support an EVA by one of the two *Voskhod-2* crew members and rescue of both crew members in case of emergency depressurization of the spacecraft cabin. The BERKUT pressure suits were used by P. Belyaev and A. Leonov, the *Voskhod-2* crew members. On 18 March 1965, A. Leonov made his famous space walk.

**Development and operation dates** Development and tests: 1964–1965. Operations on board *Voskhod*: 1965.

**Technical characteristics** The spacesuit in conjunction with the on-board ventilation and oxygen supply system provided the crew member with:

- safe stay in a depressurized cabin for 4 hr;
- duration of a self-contained operation mode with the suit fed from the backpack LSS of 45 min;
- spacesuit operating pressure of 400 hPa (main mode) and 270 hPa (back-up mode);
- ventilating gas flow rate to the BERKUT suit from the on-board system of $150\,l\,min^{-1}$;
- spacesuit mass of 20 kg;
- backpack mass of 21.5 kg.

**Quantity of manufactured space suits**
Test and training models: 9.
Flight models: 4.

# SOYUZ 4/5 EVA SUIT YASTREB

**Official name** YASTREB spacesuit.

**Description** The pressure spacesuit is of the soft type, with a removable rigid helmet. The enclosure incorporated three layers: the outer restraint layer was made of Kapron and a two-layer pressure bladder, the main one was made of sheet rubber and the back-up (internal) bladder of porous rubber. A protective coverall with shield vacuum thermal insulation was put over the enclosure. The suits were manufactured in several standard sizes and featured additional adjustment of shoulder breadth and height. The pressure suit operated in conjunction with a self-contained life support system (SCLSS) of the regenerative type, arranged in the RVR-1P backpack.

**Utilization (operations)**
The YASTREB pressure suits were used by A. Yelisyev and E. Khrunov for transfer from *Soyuz-5* to *Soyuz-4* through free space on 17 January 1969.

**Development and operation dates** Development and tests: 1965–1967. Operations: 1969.

**Technical characteristics** The suit provided the cosmonauts with:

- duration of a self-contained operation mode with suit fed from the RVR-1P backpack for at least 2.5 hr;
- spacesuit operating pressure of 400 hPa (main mode) and 270 hPa (back-up mode);
- ventilating gas flow rate of 210 standard $l\,min^{-1}$;
- spacesuit mass of 20 kg;
- RVR-1P backpack mass of 31.5 kg.

**Quantity of manufactured spacesuits**
Test and training models: 18.
Flight models: 6.

# EVA SUIT PROJECT SKV

**Official name** SKV EVA spacesuit.
The acronym SKV [СКВ] stands for the "Extravehicular Activity Spacesuit" in Russian. The suit was designed for EVA from the OTSST orbiting heavy satellite/station.

**Description** The SKV spacesuit was of the semi-rigid type. The suit upper torso (i.e., the part above the groin), together with the helmet and backpack, were made of a composite material (glass-fibre-reinforced plastic). An entry hatch was located on the back. The backpack served as a hatch cover, where all the LSS components were located. The backpack hinged about the vertical axis. The hinges were located on the backpack/upper torso interface, at the left. At the right the latches that secured the backpack in the closed position were located. The latches were operated with one handle. The interface was sealed using the sealing hose located in the interface frame on the suit body. The suit arms and leg enclosures were of the soft type. The soft enclosure had two layers: the outer restraint layer was made of Kapron and the internal layer (pressure bladder) of rubber. The LSS was of the regenerative (closed-loop) type, with an evaporation heat exchanger. The LSS controls were located in the built-in control panel on the suit chest.

**Utilization (operations)** The SKV spacesuit was designed for EVA in Earth orbits within the altitude range from 450 to 36,000 km, including Earth's radiation belts.

**Development and operation dates** Development and laboratory tests of the experimental sample: 1961–1965. Note that, later on, the Moon spacesuit KRECHET was developed on the SKV basis.

**Technical characteristics** The suit provided the cosmonaut with:

- operating positive pressure of 400 hPa;
- operation time (from on-board systems) of 8 hr and (in self-contained mode) of 4 hr;
- spacesuit mass of 85 kg;
- time of donning of 5 min.

**Quantity of manufactured space suits** A total of five suits were made, including:

- 3 mock-ups for determining suit form and dimensions, and for strength and pressure-tightness tests;
- 2 fully functional mock-ups for man-evaluated ground tests.

# SOFT LUNAR SUIT PROJECT ORIOL

**Official name** ORIOL lunar spacesuit.

**Description** The pressure spacesuit had a Baikal LSS of the regenerative type, arranged in a backpack, and was designed to support cosmonaut EVA on the Moon. The suit was of the soft type. The outer protective enclosure had shield vacuum thermal insulation and incorporated the restraint layer made of Kapron and Orthofabric, a rubber bladder (main and back-up), a water-cooled garment, a ventilation suit and a removable rigid helmet. In the final phase of the suit's development programme, under consideration was a multi-purpose of the ORIOL spacesuit (to support both rescue in case of emergency and EVA).

**Utilization (operations)** Only experimental models were manufactured.

**Development and operation dates** Development and tests: 1966–1970.

**Technical characteristics** The suit provided the cosmonaut with:

- spacesuit operating positive pressure of 400 hPa (main mode) and 270 hPa (back-up mode);
- duration of SCLSS operation of 4 hr;
- mass of the spacesuit proper of ~20 kg;
- Baikal backpack mass of ~36 kg.

**Quantity of manufactured space suits**
Test models: 3.
There was also a mock-up of the Baikal backpack.

# SEMI-RIGID LUNAR SUIT KRECHET

**Official name** KRECHET and KRECHET-94 lunar spacesuits.
KRECHET was an experimental spacesuit and KRECHET-94 was a spacesuit for a Moon expedition as part of the N-1–L-3 programme.

**Description** The KRECHET spacesuit was of the semi-rigid type, with an integrated LSS. It was designed for EVA, including one on the Moon surface. The suit's HUT (i.e., the part above the waist) was integrated with the helmet and made of sheet aluminium alloy. The entry hatch was located on the HUT back. The backpack served as a hatch cover, where all the LSS components were located. The LSS was of the regenerative (closed-loop) type. The thermal control system incorporated a water-cooled garment and a heat exchanger/sublimator. The LSS controls were located on the suit chest. The suit arms and leg enclosure were of the soft type. The soft enclosure had three layers: the outer restraint layer was made of Kapron fabric and the two pressure bladders of rubber. The suit arms had shoulder and wrist bearings. A protective garment with shield vacuum thermal insulation was put over the enclosure. The antenna feeder device was set into the protective garment.

**Utilization (operations)** The KRECHET-94 spacesuit was designed for the N-1–L-3 programme, which foresaw a crew member EVA on the Moon surface.

**Development and operation dates** Development and tests: KRECHET 1966–1967 and KRECHET-94 1967–1972. The KRECHET-94 suit activities as part of the N-1–L-3 programme were suspended in 1972 and stopped in 1974. Note that the semi-rigid design was based on the experimental SKV pressure suit developed between 1962 and 1965.

**Technical characteristics** The suit provided the cosmonaut with:

- operating positive pressure (main mode) of 400 hPa and (additional mode) of 270 hPa;
- duration of self-contained operation for 10 hr;
- spacesuit mass, including LSS, of 106 kg.

**Quantity of manufactured spacesuits**
KRECHET spacesuit: 3 fully functional mock-ups for laboratory testing.
KRECHET-94 spacesuit: 22 for all types of testing and training (at the time of programme closure 9 more suits were under production).

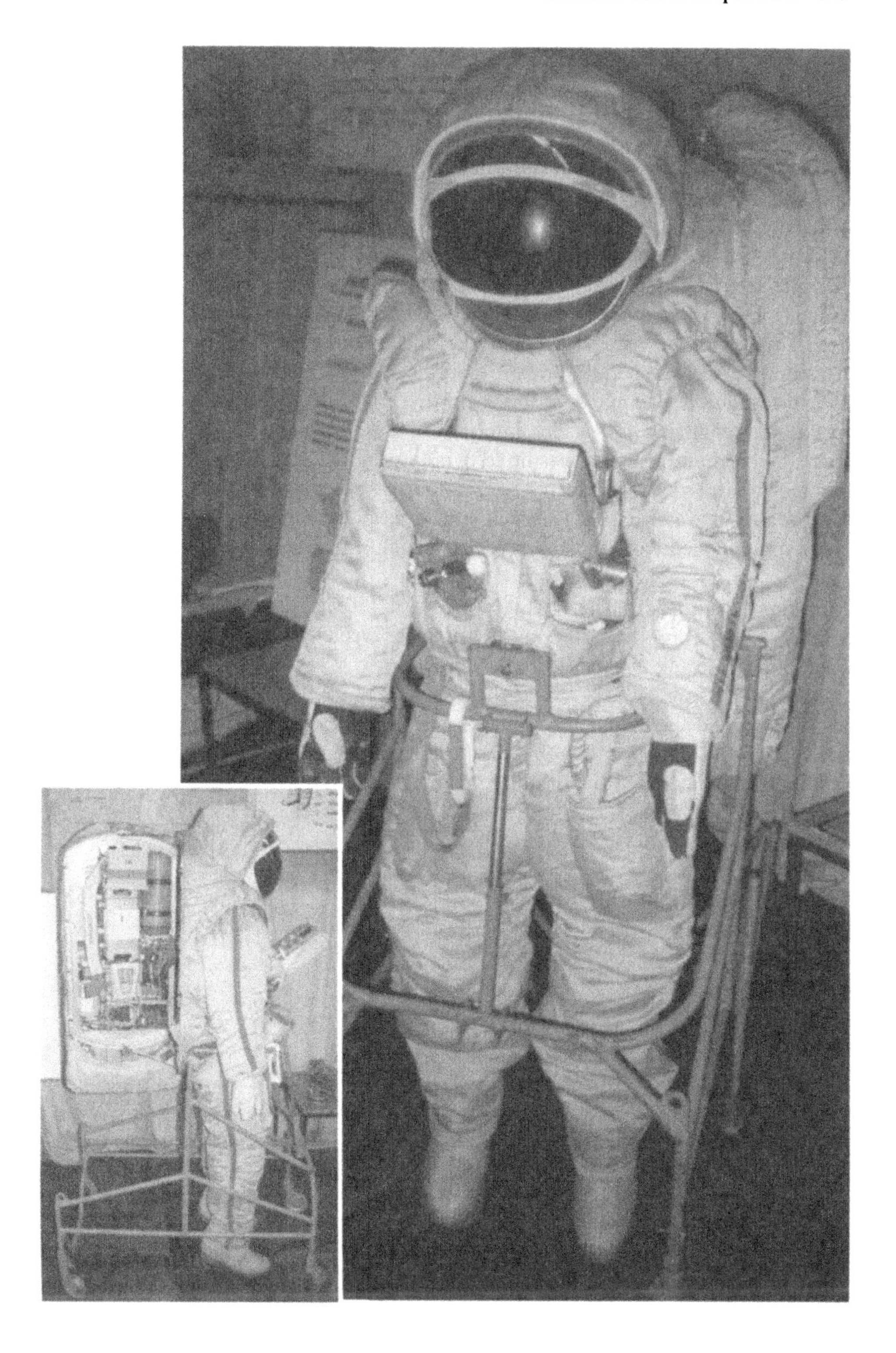

# LUNAR ORBITAL SUIT ORLAN

**Official name** ORLAN lunar orbital spacesuit.

**Description** The ORLAN spacesuit, like the KRECHET spacesuit, is of the semi-rigid type, with an integrated LSS. The enclosure incorporated a HUT, integrated with the helmet and made of aluminium alloy, and soft parts: suit arms and leg enclosures (lower torso assembly and suit legs). The soft enclosure had three layers: the outer restraint layer was made of Kapron and the two pressure bladders were made of rubber. A hatch located on the HUT back served as a door to the suit. The hatch cover was the backpack. The crew member opened and closed the hatch unassisted, using a special device. A protective garment with shield vacuum thermal insulation was put over the enclosure. The one-size suit could be used by any cosmonaut having a chest circumference of 96–108 cm and height of 164–178 cm. The suit's height adjustment was done by adjusting the soft parts of the enclosure. The LSS was of the regenerative (closed-loop) type. The thermal control system incorporated a water-cooled garment and a heat exchanger/sublimator. Electric power was supplied from the space vehicle via an umbilical.

**Utilization (operations)** The ORLAN pressure suit was designed to support crew member EVA from the orbiting Moon space vehicle as part of the N-1–L-3 programme. The suit was never used in flight.

**Development and operation dates** Development and tests: 1967–1971. The ORLAN suit activities were stopped in 1971 due to the commencement of work on its modification, as applicable to the orbiting station.

**Technical characteristics** See Appendix 2 on p. 319.

**Quantity of manufactured spacesuits**
Test and training models: 11.
Flight models: none manufactured.

# SALYUT SUIT ORLAN-D

**Official name** ORLAN-D spacesuit.

**Description** This spacesuit of the semi-rigid type was an ORLAN suit modification. It was designed for EVA from *Salyut-6* and *Salyut-7*. It was a reusable spacesuit. The LSS was of the regenerative (closed-loop) type, with exchangeable expendable components. Power supply, radio communication and telemetry were via a 20-m electric umbilical, which also served as a safety tether. The suit had a further safety tether with a snap-hook. It was a one-size suit, which could be used by any cosmonaut having a chest circumference of 96–108 cm and height of 164–180 cm.

**Utilization (operations)**
*Salyut-6*: 1977–1979 for 3 paired EVAs.
*Salyut-7*: 1982–1984 for 10 paired EVAs.
It was planned to use the suit on the *Almaz* orbiting station.

**Development and operation dates** Development and tests: 1969–1977. Nominal operations: 1977–1984.

**Technical characteristics** See Appendix 2 on p. 319.

**Quantity of manufactured spacesuits**
Test and training models: 27.
Flight models: 7.

# SALYUT/MIR SUIT ORLAN-DM

**Official name** ORLAN-DM spacesuit.

**Description** This space suit of the semi-rigid type was an ORLAN-D pressure suit modification. It was designed for EVA from *Salyut-7* and *Mir*. The ORLAN-DM was a transitional model to the fully self-contained ORLAN-DMA spacesuit. In the suit, the LSS components were modified and rearranged, a combined LSS control panel was developed and introduced, an emergency oxygen hose and a protective cask were introduced. The pressure bladder's reliability was improved, owing to usage of new materials, etc. Power supply, radio communication and telemetry were via a 20-m electric umbilical. The suit could be used by any cosmonaut having a chest circumference of 96–108 cm and height of 164–180 cm.

**Utilization (operations)**
*Salyut-7*: 1985–1986 for 3 paired EVAs.
*Mir*: 1986–1988 for 5 paired EVAs.

**Development and operation dates**
Development and tests: 1983–1985.
Nominal operations: 1985–1988.

**Technical characteristics** See Appendix 2 on p. 319.

**Quantity of manufactured spacesuits**
Test and training models: 5.
Flight models: 5.

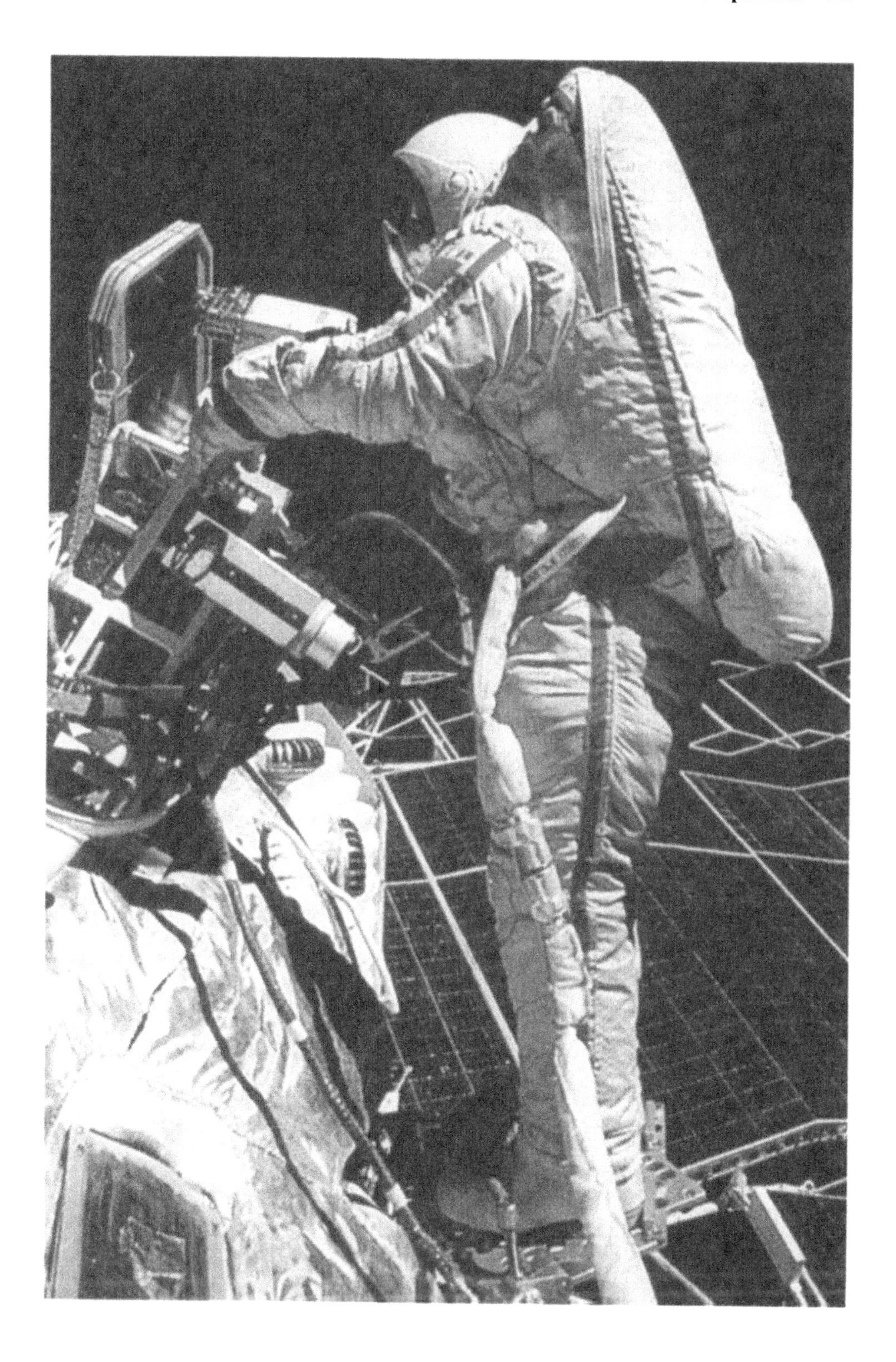

# MIR SUIT ORLAN-DMA

**Official name** ORLAN-DMA spacesuit.

**Description** This pressure suit of the semi-rigid type was an ORLAN-DM suit modification. Unlike the ORLAN-D and ORLAN-DM models, it could be used without the electric umbilical, because it was provided with a removable unit that incorporated an electrical power source (battery), radio and telemetry system, and an antenna-feeder device. The suit had an improved design and used new material for its soft parts. The HUT was modified: its inner volume was increased, a waist flange was introduced to secure the lower (soft) enclosure and the possibility of changing it on board the station was provided. The pressure gloves were updated and redundant sealing cuffs were introduced. A second safety tether was introduced. The capacity of the contamination and $CO_2$ control cartridge (CCC) located in the backpack was increased. Spacesuit/UPMK (index 21KS) interfaces were introduced. Power supply, radio communication and telemetry were available for self-contained mode (from the backpack) and via the 25-m electrical umbilical from the station. The radio antenna was set into the upper protective garment. The suit's height-adjustable enclosure could fit cosmonauts within the height range of 164–185 cm. The suit could be used by any cosmonaut having a chest circumference of 96–110 cm.

**Utilization (operations)**
*Mir*: 1988–1997 for 56 paired EVAs.
It was planned to be used on the *Almaz* orbiting station and the *Buran* spaceplane.

**Development and operation dates** Development and tests: 1985–1988. Nominal operations: 1988–1997.

**Technical characteristics** See Appendix 2 on p. 319.

**Quantity of manufactured spacesuits**
Test and training models: 16.
Flight models: 12.

АФГАНИСТАНА
НА ОРБИТЕ

# MIR/ISS SUIT ORLAN-M

**Official name** ORLAN-M spacesuit.

**Description** This spacesuit of the semi-rigid type was an ORLAN-DMA spacesuit modification. The ORLAN-M design took into account the experience of ORLAN-DMA operations on *Mir* and the additional requirements imposed by operations on the *ISS*. The suit underwent the following modifications:

- the suit's dimensions were enlarged in the waist area and the entry hatch was moved upward;
- an additional helmet-top window and protective glass for the main window were introduced;
- a calf bearing and the third pressure bearing (elbow) on the suit arm were introduced;
- one of the safety tethers was given variable length;
- CCC capacity was increased.

Power supply, radio communication and telemetry were available for self-contained mode (from the backpack) and via the 25-m electrical umbilical from the station. Owing to the above, the service characteristics (mobility, donning/doffing, field of view, etc.) were improved. The anthropometric ranges of chest circumference and height were improved (96–112 cm and 164–190 cm, respectively). The suit was provided with attachment points for SAFER.

**Utilisation (operations)**
*Mir*: 1997–2000 for 18 paired EVAs.
*ISS*: 2001–31 December 2002 for 9 paired EVAs.

**Development and operation dates** Development and tests: 1995–1997. Nominal operations: 1997–up to the present.

**Technical characteristics** See Appendix 2 on p. 319.

**Quantity of manufactured spacesuits (as at 31 December 2002)**
Test and training models: 17.
Flight models: 7.

*"Орлан-М" - четвертая модификация полужесткого скафандра для внекорабельной деятельности.*

*"Orlan-M", the fourth modification of the semi-rigid type EVA space suit.*

# MIR-2/HERMES SUIT EVA SUIT 2000

**Official name** ESA/RKA EVA SUIT 2000.

**Description** Semi-rigid EVA spacesuit with rear hatch entry; backpack door with integrated LSS; nominal suit pressure of 420 hPa (>95% $O_2$) for 7-hr operation; four years (30 sorties) in orbit with in-orbit maintenance; fail-safe concept (0.9995); unassisted don/doff; one-size torso for European anthropometric population range; in-orbit individual sizing of suit; 30-min pre-breath; overall dimensions to fit hatch openings of ⊖ 800 mm.

**Utilization (operations)** Planned for use by the European spaceplane *Hermes* for missions to *Mir-2* and on board *Mir-2*.

**Development and operation dates** Development started in 1993 and was terminated at the end of 1994. At the end of the project the first model of the suit enclosure (the suit demonstrator) had been successfully tested. Operation was planned to start in 2000–2004.

**Technical characteristics** The suit provided the cosmonaut with:

- suit enclosure comprising an HUT with integrated helmet/visor and backpack rear door; redundant bladder; scye, arm, wrist and ankle bearings; shoulder, elbow, hip and knee soft joints
- life support of the closed-loop regenerative type with in-orbit recharge of oxygen, cooling water and battery; replaceable LiOH canister; alternative metal-oxide regenerative system; metal-oxide batteries for $CO_2$ removal; body cooling by liquid-cooled garment;
- mass of <125 kg.

**Quantity of manufactured spacesuits**
Tests: 2.
Training: 0.
Flight: 0.

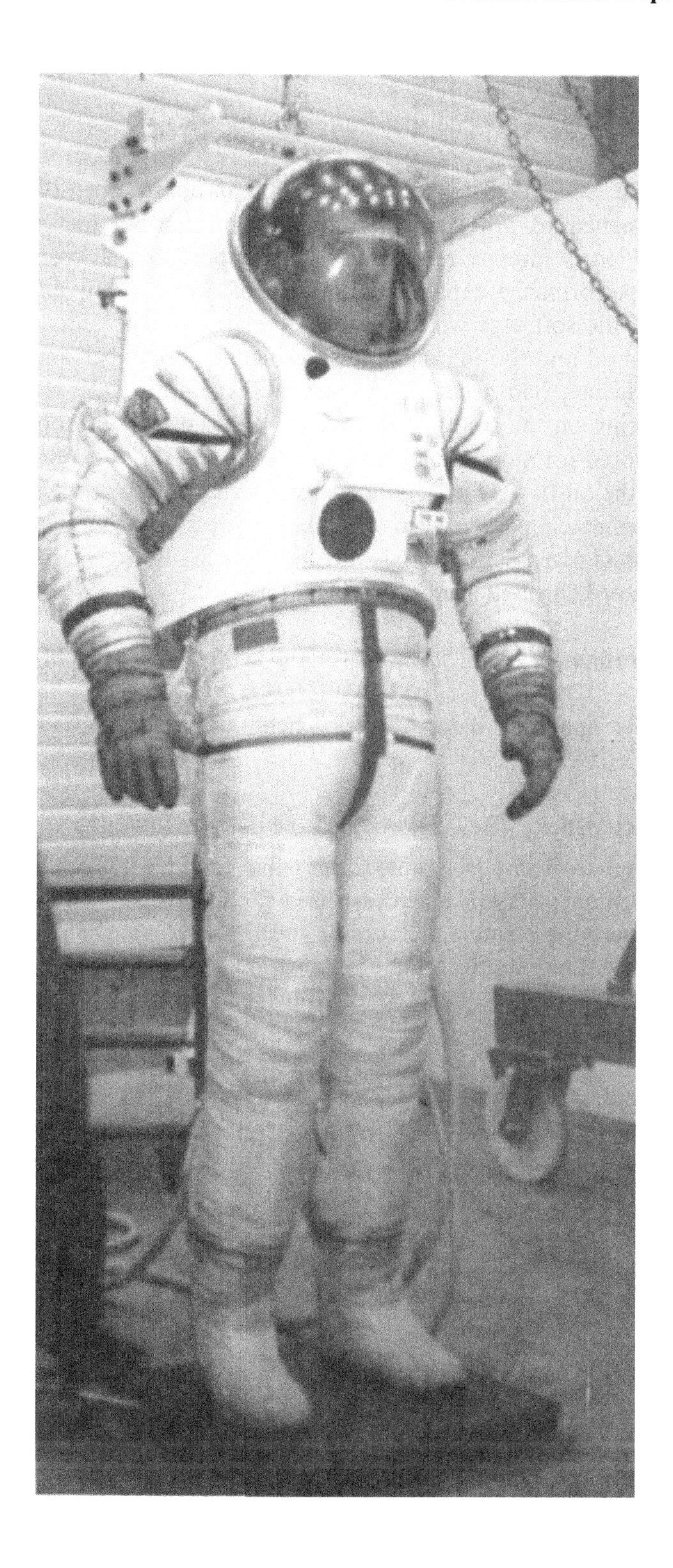

# SOYUZ RESCUE SUIT SOKOL-K

**Official name** SOKOL-K rescue spacesuit.

**Description** The SOKOL-K spacesuit with its on-board open-type ventilation system was designed to provide a cosmonaut with normal hygienic conditions during a flight in a pressurized cabin and support for a cosmonaut's vital functions and performance capabilities in case of *Soyuz* cabin depressurization. The suit was of the soft type with a two-layer enclosure (the outer restraint layer was made of fabric and the internal pressure bladder of rubber and a rubberized material). The helmet had a soft nape part and a sliding visor. The suit's front opening had lacing on the restraint layer and was used for suit-donning. The suit was custom-tailored for a cosmonaut to be seated in the shock-absorbing seat. In a nominal flight, the on-board fan ventilated the internal cavity of the suit with cabin air. In case of emergency (i.e., when the cabin depressurized or the cabin air was polluted), the pressurized suit was fed with the gas mixture (oxygen 40%, nitrogen 60%), which passed through a pressure regulator.

**Utilization (operations)** From *Soyuz-12* (in 1973) through *Soyuz-40* (in 1981).

**Development and operation dates** Development and tests: 1971–1973. Nominal operations: 1973–1981.

**Technical characteristics** The suit provided the cosmonaut with:

- time of suited crew stay in a pressurized cabin for up to 30 hr;
- time of suited crew stay in a depressurized cabin for up to 2 hr;
- spacesuit operating positive pressure of 400 hPa;
- ventilating air flow rate from the on-board fan of $\geq$150 standard $l\,min^{-1}$;
- gas mixture flow rate of 20 standard $l\,min^{-1}$;
- space suit mass of $\sim$10 kg.

**Quantity of manufactured spacesuits**
Test and training models: 66.
Flight models: 89.

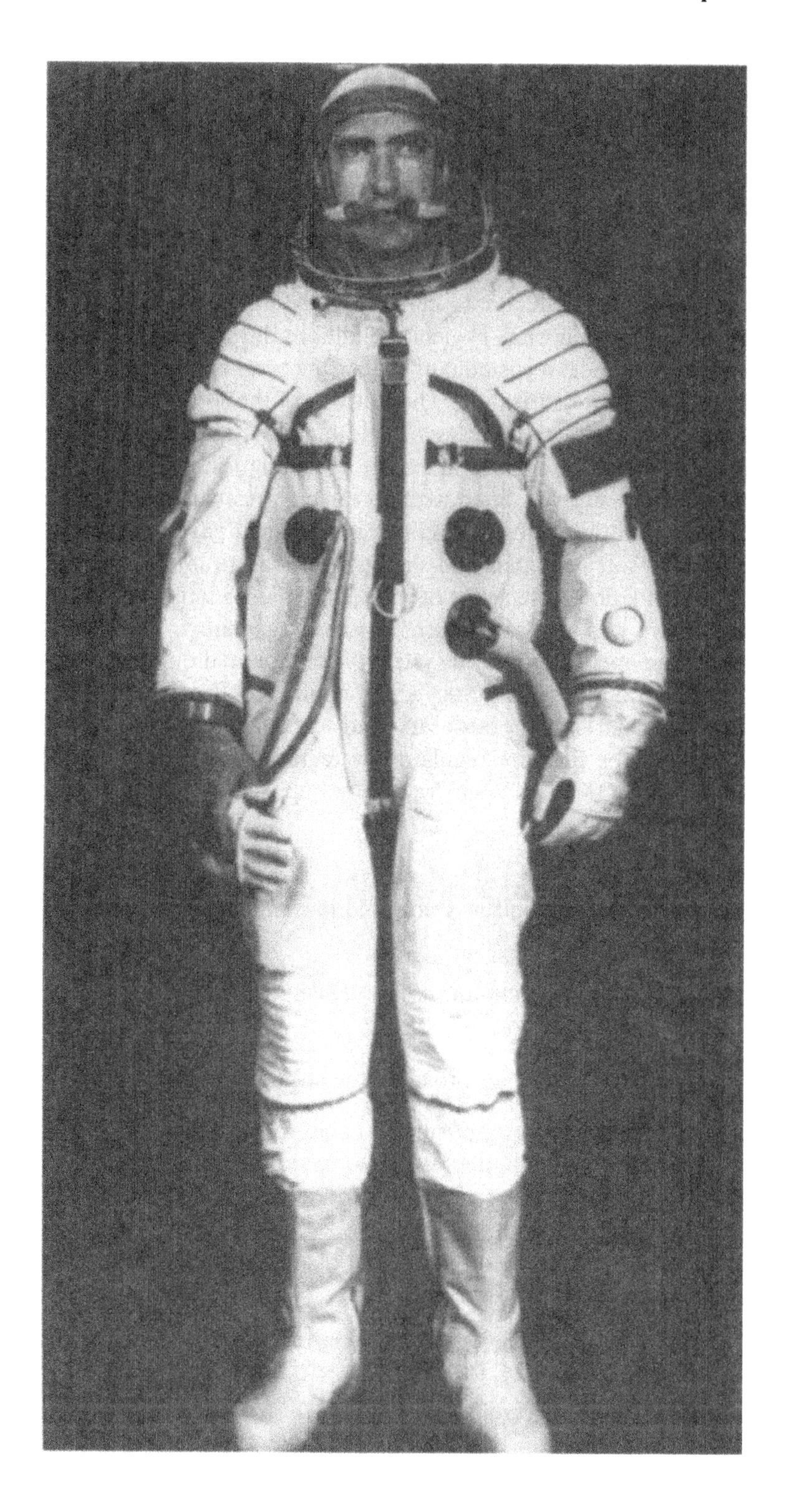

# SOYUZ RESCUE SUIT SOKOL-KV

**Official name** SOKOL-KV rescue spacesuit.

**Description** The SOKOL-KV spacesuit was an intermediate model between the SOKOL-K spacesuit and the SOKOL-KV-2 spacesuit. Like these suits it was designed for operation in transportation space vehicles of the *Soyuz-T* and *Almaz* programmes. The SOKOL-KV was connected to an on-board LSS and could provide a cosmonaut with normal hygienic conditions during a flight in a pressurized cabin; it could also support a cosmonaut's vital functions and performance capabilities in case of cabin depressurization. The suit was of the soft type with a two-layer enclosure (the outer restraint layer was made of fabric and the internal pressure bladder of rubber and a rubberized material). A distinguishing feature: in order to facilitate donning, the suit was fitted with an elastic, pressurized interface with two zippers, which divided the enclosure into two independent parts (shirt and trousers). The helmet with its soft nape part and sliding visor was an integral part of the suit enclosure. The ventilation system was of the open type. To improve heat removal, a water-cooled system was integrated during spacesuit assembly. The elastic tubes of both the ventilation and water-cooled systems were mounted on the KVO-11 undergarment. Lead-ins of the water lines, which connected the suit to the on-board system, were arranged on the trousers and the pneumatic lead-ins (air and oxygen), as well as the pressure regulator, were located on the shirt. Unlike the SOKOL-K suit, the SOKOL-KV suit was to be supplied with pure oxygen in case of cabin depressurization.

**Utilization (operations)** The suit was not used in nominal operations.

**Development dates** Development and tests: 1974–1979.

**Technical characteristics** The suit provided the cosmonaut with:

- time of suited crew stay in a pressurized cabin for up to 30 hr;
- time of suited crew stay in a depressurized cabin for up to 2 hr;
- spacesuit operating positive pressure of 400 hPa (main mode) and 270 hPa (back-up mode);
- ventilating air flow from the on-board fan of $\geq$150 standard $l\,min^{-1}$;
- oxygen flow rate of 20 standard $l\,min^{-1}$;
- spacesuit mass of $\sim$12 kg.

**Quantity of manufactured spacesuits**
Test and training models: 6.
In 1979 one of the SOKOL-KV suits was modified into a SOKOL-KV-2 mock-up.

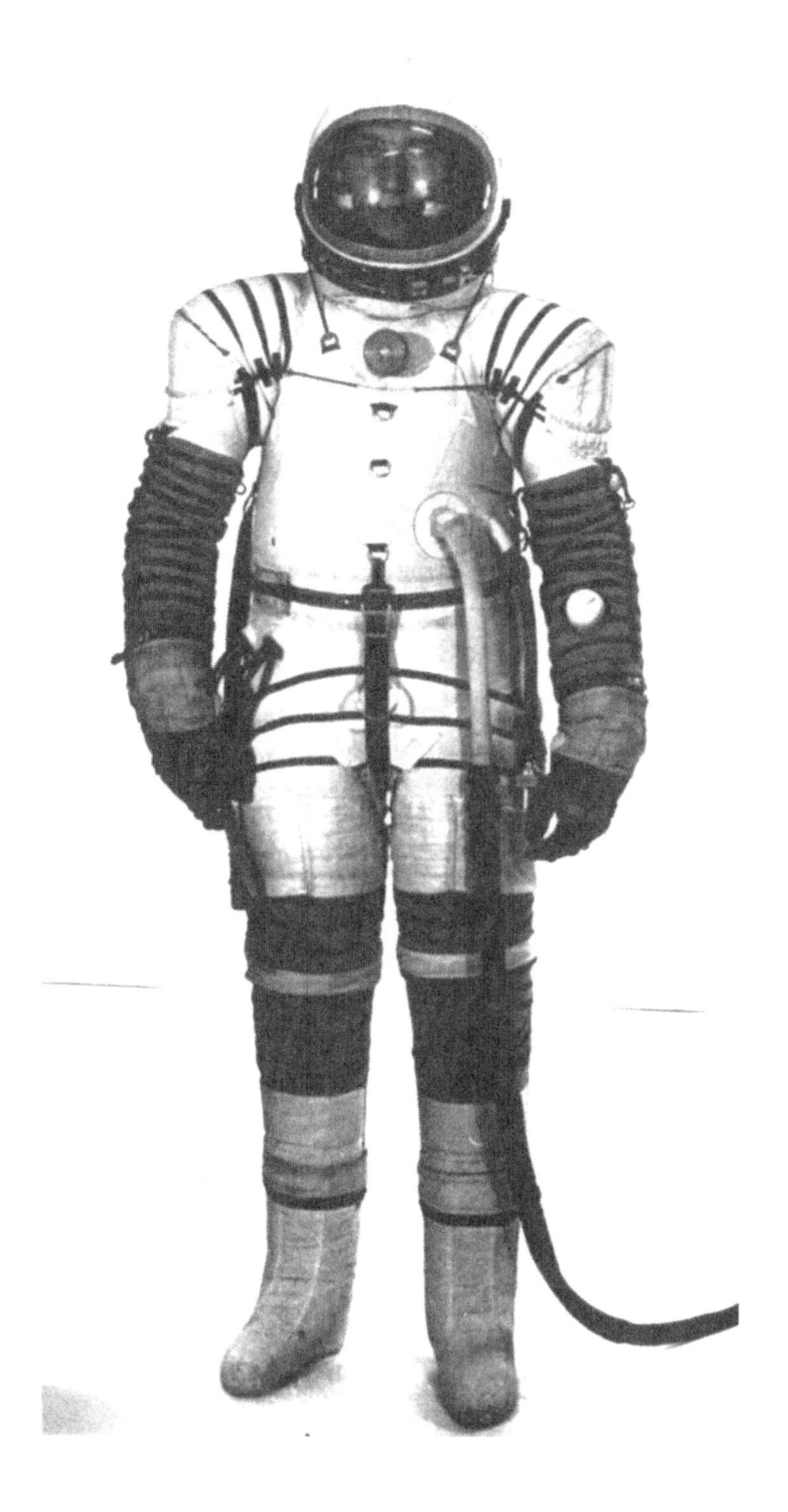

# SOYUZ-T/-TM/TMA RESCUE SUIT SOKOL-KV-2

**Official name** SOKOL-KV-2 rescue spacesuit.

**Description** The SOKOL-KV-2 spacesuit is connected to an on-board open-type ventilation system and is designed to provide a cosmonaut with normal hygienic conditions during a flight in a pressurized cabin and to support a cosmonaut's vital functions in case of cabin depressurization. The suit is of the soft type with a two-layer enclosure (the outer restraint layer is made of fabric and the internal pressure bladder of rubber and a rubberized material). The helmet has a soft nape part and a sliding visor and is an integral part of the suit enclosure. Differences between SOKOL-K and SOKOL-KV-2: the lacing on the front opening is replaced by two zippers; helmet and, thus, visor dimensions are increased; the pressure regulator is integrated with an in-leakage valve and is arranged in the centre of the body under the helmet (it is located at the side on the SOKOL-K suit). In case of cockpit depressurization, pure oxygen is supplied to the suit. The pressure regulator provides two suit pressure modes.

**Utilization (operations)** The suit was first used by cosmonauts Yu. Malyshev and V. Aksionov in 1980 aboard the *Soyuz-T-2* space vehicle. Suits have been used for all *Soyuz-T*, *Soyuz-TM* and *Soyuz-TMA* vehicles. The suits are delivered to the *ISS* aboard the Space Shuttle for each crew member and are intended to support a possible emergency descent aboard the *Soyuz-TM* or *Soyuz-TMA* vehicles.

**Development and operation dates** Development and tests: 1973–1979. Nominal operations: 1980–up to the present.

**Technical characteristics** The suit provides the cosmonaut with:

- time of suited crew stay in a pressurized cabin for up to 30 hr;
- time of suited crew stay in a depressurized cabin for up to 2 hr;
- spacesuit operating positive pressure of 400 hPa (main mode) and 270 hPa (back-up mode);
- ventilating air flow from the on-board fan of $\geq$150 standard $l\,min^{-1}$;
- oxygen flow rate of 20 standard $l\,min^{-1}$;
- spacesuit mass of $\sim$10 kg.

**Quantity of manufactured spacesuits (as at 31 December 2002)**
Test and training models: 63.
Flight models: 220.

# BURAN RESCUE SUIT STRIZH

**Official name** STRIZH rescue spacesuit.

**Description** The STRIZH spacesuit was designed for crew members on board the *Buran* reusable space vehicle. The suit was of the soft type with a built-in helmet and a two-layer enclosure. The outer restraint layer was made of high-strength, flame-retardant fabric. Besides a conventional restraint layer, the enclosure was fitted with a suspension/restraint system to retain the crew member in the ejection seat and to support a parachute descent. The internal pressure bladder was made of a rubberized Kapron. The enclosure featured a front opening, which was closed by two zippers. The helmet had a sliding double visor. An inflatable collar was attached to the enclosure. The suit operated in conjunction with a personal on-board life support system (PLSS). The PLSS operated simultaneously for two suited crew members. In a nominal flight the suit was ventilated with cabin air, but in an emergency the PLSS and the suit operated in the closed-loop mode.

**Utilization (operations)** The first unmanned mission of the *Buran* vehicle in 1988 used two nominal sets of the STRIZH spacesuit (with dummies and PLSS).

**Development and operation dates** Development and tests: 1981–1991.

**Technical characteristics** The suit provided the cosmonaut with:

- spacesuit operating positive pressure of 400 hPa (main mode) and 270 hPa (back-up mode);
- suited mode duration for a pressurized cabin with ventilating airflow rate up to $300\,\mathrm{l\,min^{-1}}$ (for two suits) for up to 24 hr;
- suited mode duration in an emergency with the PLSS operating in the closed-loop mode for up to 12 hr;
- crew member escape by ejection in the K-36RB seat from altitudes up to 30 km and for speeds up to $M = 3.0$;
- spacesuit mass of 18 kg.

**Quantity of manufactured spacesuits**
Test and training models: 27.
Flight models: 4.

# Bibliography

Abramov, I.P. (1970) Some results of self-contained life support system operation during the Soyuz-4 and Soyuz-5 flight. *Space Biology and Medicine*, Vol. 4, pp. 75–78. Meditsina, Moscow [in Russian].

Abramov, I.P., Severin, G.I., Stoklitsky, A.Yu. and Sharipov, R.Sh. (1984) *Space Suits and Systems for EVA*. Mashinostroeinie, Moscow [in Russian].

Abramov, I.P., Stoklitsky, A.Yu., Barer, A.S. and Filipenkov, S.N. (1994) Essential aspects of space suit operating pressure trade-off. *24th International Conference on Environmental Systems and 5th European Symposium on Space Environmental Control Systems, Friedrichshafen, Germany, 20–23 June 1994*, SAE Paper No. 941330.

Abramov, I.P. (1995) The experience in operation and improving the Orlan-type space suit. *Acta Astronautica*, **36**, 1–12.

Abramov, I.P. (1995) History of EVA space suit development. *10th Moscow International Symposium on the Aviation and Space History, Russia, 20–27 June 1995*, Paper No. X-MC-D-227 [in Russian].

Abramov, I.P. (1997) How the EVA space suits were created. *Zemlya i Vselennaya, Nauka, Russian Academy of Sciences*, **2** [in Russian].

Abramov, I.P., Glazov, G.M., Svertshek, V.I. and Stoklitsky, A.Yu. (1997) Ensuring of long operation life of the orbiting station EVA space suit. *Acta Astronautica*, **41**, 379–389.

Abramov, I.P., Albats, E.A., Glazov, G.M. and Elbakyan, A.Ts. (1998) Some issues of modification and development testing of the ORLAN-M space suit designed for the ISS. *49th International Astronautical Congress, Melbourne, Australia, 28 September–2 October 1998*, Paper IAA-98-IAA.10.1.02.

Abramov, I.P., Moiseyev, N. and Stoklitsky, A.Yu. (2001a) Concept of space suit enclosure for planetary exploration. *31st International Conference on Environmental Systems, Orlando, FL, 9–12 July 2001*, Paper No. SAE 2001-01-2168.

Abramov, I.P., Severin, G.I., Stoklitsky, A.Yu. and Skoog, A.I. (2001b) From Gagarin's space suit to orbital based space suits. *International Space Forum—2001, Moscow, 11–13 April 2001* [in Russian].

Abramov, I.P., Moiseyev, N. and Stoklitsky, A.Yu. (2002) Concept of mechanical interfaces for planetary space suit to airlock and rover. *32nd International Conference on Environmental Systems, San Antonio, TX, 15–18 July 2002*, Paper No. SAE 2002-01-2313.

Abramov, I.P., Pozdnyakov, S.S., Severin, G.I. and Stoklitsky, A.Yu. (2001) Main problems of the Russian ORLAN-M space suit utilization for EVAs on the ISS. *Acta Astronautica*, **48**, 265–273.

Abramov, I.P. (2002) Development and experience of operating the orbit-based extravehicular space suit. *Scientific and Technical Journal "Polyot" (Mashinostroeinie)*, **1**, 26–32.

Abramov, I.P., Glazov, G.M. and Svertshek, V.I. (2002) Long-term operation of "ORLAN" space suits in the "Mir" orbiting station: Experience obtained and its application. *Acta Astronautica*, **51**, 133–143.

Abramov, I.P. and Svertshek, V.I. (2002) Space suits and life support systems (short historical essay). *Aerospace Herald (All-Russia Aerospace Magazine)*, **5**, 38–44 [in Russian].

Abramov, I.P., Albats, E.A. and Glazov, G.M. (2002) The first results of the Russian EVA space suits operation in the International Space Station. *53rd International Astronautical Congress, Houston, TX, 10–19 October 2002*, Paper No. IAC-02-IAA.10.1.04.

Alekseyev, S.M. and Umansky, S.P. (1973) *High Altitude and Space Protection Suits*. Mashinostroeinie, Moscow [in Russian].

Alekseyev, S.M. (1987) *Space Suits. Yesterday, Today, Tomorrow*, Cosmonautics and Astronomy Series, Vol. 2. Znanie, Moscow.

Barer, A.S. and Filipenkov, S.N. (1994) Suited crewmember productivity. *Acta Astronautica*, **32**, 51–57.

Barer, A.S., Filipenkov, S.N., Katuntsev, V., Vogt, L. and Wenzel, J. (1995) The feasibility of Doppler monitoring during EVA. *Acta Astronautica*, **36**, 81–83.

Borisenko, I.G. (1984) *In Outer Space*. Mashinostroeinie, Moscow [in Russian].

Chertok, B.Ye. (1996) *Rockets and People*, Vol. 1—1994, Vol. 2—1996, Vol. 3—1997, Vol. 4—1999. Mashinostroeinie, Moscow [in Russian].

ESA Manned Space Programme Board (1994) *Status of the EVA Development*, ESA/PB-MS (94)49, 19 October 1994.

Hodgson, E., Etter, D., Abramov, I.P., Moiseyev, N. and Stoklitsky, A.Yu. (2000) Effects of enhanced pressure suit ankle mobility on locomotion on uneven terrain. *30th International Conference on Environmental Systems, Toulouse, France, 10–15 July 2000*, Paper No. SAE 2000-01-2481.

Ivanov, D.I. and Khromushkin, A.I. (1968) *Man Life Support Systems for High-altitude and Space Flights*. Mashinostroeinie, Moscow [in Russian].

Kamanin, N.P. (2001) *Skrytyi Kosmos* (Space Diaries): Vol. 1—1995, Vol. 2—1997 published by TOO Infortext, Moscow; Vol. 3—1999, Vol. 4—2001 published by Novosti Kosmonavtiki, Moscow [in Russian].

Keldysh, M.V. (1980) *Tvorcheskoye naslediye akademika S.P. Koroleva: Izbrannye trudy i documentry* (Heritage of Academician Korolev: Selected works and documents). Nauka, Moscow [in Russian].

Khromushkin, A.I. (1949) *Pressure Suits and Oxygen Life Support Equipment for High Altitude Flights*. Oborongiz, Moscow [in Russian].

Malkin, V.B. (1975) Barometric pressure and gas composition. In: *Foundations of Space Biology and Medicine* (Vol. II, Book 1, p. 35). Nauka, Moscow and NASA, Washington.

Moeller, P., Loewens, R., Abramov, I.P. and Albats, E.A. (1995) EVA Suit 2000, a joint European/Russian space suit design. *Acta Astronautica*, **35**, 53–63.

*Novosti Cosmonavtiki* (2002) **12**(10), August, p. 237 [in Russian].

Oberth, H. (1929) *Wege zur Raumschiffahrt*. Verlag Oldenbourg, Munich/Berlin.

Payne, L. (1991) *Lighter than Air*. Orion Books, New York.

Ross, A.I., Kosmo, J.I., Moiseyev, N., Stoklitsky, A.Yu., Barry, S. and Hodgson, E. (2002) Comparative space suit boot test. *32nd International Conference on Environmental Systems, San Antonio, TX, 15–18 July 2002*, Paper No. SAE 2002-01-2315.

Semeonov, Yu.P. (ed.) (1996) *Rocket and Space Corporation ENERGIA Named after S.P. Korolev, 1946–1996*. Rocket and Space Corporation Energia, Korolev, Russia [in Russian].

Severin, G.I., Svertshek, V.I., Abramov, I.P. and Stoklitsky, A.Yu. (1978) Salyut-6 extravehicular semi-rigid space suit. *29th International Astronautical Congress, Dubrovnik, Yugoslavia, 1–8 October 1978*, Paper No. 78-A-60.

Severin, G.I., Abramov, I.P., Barer, A.S. and Svertshek, V.I. (1984) Space suits. Ten periods of extravehicular activity from the Salyut-7 space station. *35th International Astronautical Congress, Lausanne, Switzerland, 7–13 October 1984*.

Severin, G.I., Abramov, I.P., and Svertshek, V.I. (1988) Crewman rescue equipment in manned space missions: Aspects of application. *38th International Astronautical Congress, 10–17 October 1987*, Paper No. IAA-87-576. [In: *Space Safety and Rescue 1986–87*, AAS Science and Technology Series 70. Univelt, Inc., San Diego, CA.]

Severin, G.I., Abramov, I.P. and Svertshek, V.I. (1988) EVA space suits: safety problems. *39th International Astronautical Congress, Bangalore, India*, Paper No. IAF-88-515.

Severin, G.I., Svertshek, V.I., Abramov, I.P. and Frolov, V.A. (1990) Autonomous EVA support complex designed for usage during space station assembly and maintenance; methods to increase the complex effectiveness. *41st International Astronautical Congress, Dresden, Germany, 6–12 October 1990*, Paper No. IAF-90-075.

Severin, G.I., Abramov, I.P., Svertshek, V.I. and Stoklitsky, A.Yu. (1991) *Problems of Space Suit Elaboration*, Gagarin Scientific Lectures in Astronautics and Aviation 1990, 1991, pp. 12–28. Nauka, Moscow [in Russian].

Severin, G.I., Abramov, I.P., MacBarron, J.W. and Vitsett, P. (1994) Personal life support systems for cosmonauts. EVA provision. In: *Space Biology and Medicine*, Vol. II, *Habitability of Space Vehicles*, pp. 275–329, Joint Russian–American edition. Nauka, Moscow.

Severin, G.I., Abramov, I.P. and Svertshek, V.I. (1995) Main phases of the EVA space suit development. *46th International Astronautical Congress, Oslo, Norway, 2–6 October 1995*, Paper IAA-95-IAA10.1.01.

Severin, G.I., Abramov, I.P., Svertshek, V.I. and Stoklitsky, A.Yu. (1995) Some results on modification of the EVA suit for Mir orbiting station. *25th International Conference on Environmental Systems, San Diego, CA, 10–13 July 1995*, Paper No SAE 951550.

Severin, G.I., Abramov, I.P., Svertshek, V.I., and Stoklitsky, A.Yu. (1996) Improvement of the extravehicular activity suit for the Mir orbiting station program. *Acta Astronautica*, **39**, 471–476.

Severin, G.I., Abramov, I.P., Doodnik, M.N. and Svertshek, V.I. (1999) History of creation of the Russian space suits, escape and life support means for space vehicle and space station crews. *50th International Astronautical Congress, Amsterdam, The Netherlands, 4–8 October 1999*, Paper IAA-99-IAA.2.1.07.

Skoog, A.I., Berthier, S. and Ollivier, Y. (1991) The European space suit, a design of productivity and crew safety. *Acta Astronautica*, **23**, 207–216.

Skoog, A.I. (1994) The EVA space suit development in Europe. *Acta Astronautica*, **32**, 25–38.

Skoog, A.I. and Abramov, I.P. (1995) EVA 2000: A European/Russian space suit concept. *Acta Astronautica*, **36**, 35–51.

Skoog, A.I., McBarron II, J.W. and Severin, G.I. (1995) Extravehicular activity space suit interoperability. *Acta Astronautica*, **37**, 115–129.

Skoog, A.I., Abramov, I.P., Stoklitsky, A.Yu. and Doodnik, M.N. (2002) The Soviet–Russian space suits. A historical overview of the 1960's. *Acta Astronautica*, **51**, 113–131.
Tsiolkovsky, K.E. (1958) *Vne Zemli*. Akademija Nauk SSSR, Moscow. [In *Beyond the Planet Earth*. Pergamon Press, Oxford, UK, 1960 (transl. by K. Syers).]
Umansky, S.P. (1970) *Man in Space*. Voyenizdat, Moscow [in Russian].
USSR Academy of Sciences (1966) Medical and biological problems of space flight. In: V.I. Yazdovsky (ed.), *Space Biology and Medicine*. Nauka, Moscow [in Russian].
Vogt, L., Wenzel, J., Skoog, A.I., Luck, S. and Svensson, B. (1991) European EVA decompression sickness risks. *Acta Astronautica*, **23**, 195–205.

# Index

*All individual spacesuit entries appear under "Suits"*

Printing: Mercedes-Druck, Berlin
Binding: Stein+Lehmann, Berlin

Printed in the United States
by Baker & Taylor Publisher Services